MICROCHEMICAL
ENGINEERING
IN PRACTICE

MICROCHEMICAL ENGINEERING IN PRACTICE

Edited by

THOMAS R. DIETRICH

A JOHN WILEY & SONS, INC., PUBLICATION

Published by John Wiley & Sons, Inc., Hoboken, New Jersey
Published simultaneously in Canada

For general information on our other products and services or for technical support, please contact our
Customer Care Department within the United States at (800) 762-2974, outside the United States at
(317) 572-3993 or fax (317) 572-4002.

Wiley also publishes its books in variety of electronic formats. Some content that appears in print may
not be available in electronic formats. For more information about Wiley products, visit our web site
at www.wiley.com.

Library of Congress Cataloging-in-Publication Data:

Dietrich, Thomas R., 1963-
Microchemical engineering in practice / Thomas R. Dietrich.
      p. cm.
Includes index.
ISBN 978-0-470-23956-8 (cloth)
1. Microreactors. 2. Microchemistry. I. Title.
  TP159.M53D54 2009
  660′.2832--dc22

                                                                            2008027979

Printed in the United States of America

10   9   8   7   6   5   4   3   2   1

CONTENTS

PREFACE

CHEMICAL ENGINEERING IN A WORLD OF SEVERE ECOLOGICAL AND ECONOMICAL PROBLEMS

Humanity is facing severe global problems, for example, global warming or decreasing natural resources. A typical example of this situation is the consumption of the world's oil resources.

It took nature several million years to "produce" the oil, which we burn in a few decades. The U.S. Department of Energy reported that 93% of the U.S. energy consumption was generated from non-renewable resources (oil, coal, gas, nuclear). The other 7% were renewable energies such as wind, solar, biomass, or hydroenergy (Source: EIA, Renewable Energy Consumption and Electricity Preliminary 2007 Statistics. Table 1: US Energy Consumption by Energy Source, 2003–2007 (May 2008)). The worlds energy consumption was 2005 at 1.4×10^{14} kW/h (Source: Energy Information Administration (EIA). International Energy Annual 2005 (June–October 2007), web site: www.eia.doe.gov/lea). It is expected to increase in 2030 to 2×10^{14} kW/h (Source: EIA, World Energy Projections Plus (2008)). It has been increasing more rapidly during the last few years due to growth in developing countries such as China and India.

In 2007 the world's oil consumption was approximately 4 billion tons. An ESSO study estimates the world's usable oil reserves at 181 billion tons. (ExxonMobil GmbH, Brochure "Oeldorado 2008", http://www.esso.de/ueber_uns/info_service/publikationen/downloads/files/oeldorado08_de.pdf) Even without any increase in oil consumption in the future, we will run out of oil in 45 years..

To make the situation even more critical, oil is the main source for most chemical products. Pharmaceuticals, polymers, and many of our daily consumer products are made from oil. Statistics published by the German Association of the Chemical Industry (VCI: "Chemiewirtschaft in Zahlen", 48.edition, 2006, p.98, tab.55) shows that the worldwide use of chemicals in the last 10 years doubled from 1.2 trillion € in 1995 to 2.05 trillion € in 2005. The chemical industry depends on oil and coal as natural resources for chemical products; therefore, it is not acceptable that most of these resources are simply "burned" to produce energy for our cars and households.

Due to these issues and to the environmental problems connected with the use of fossil resources, costs are increasing dramatically. As shown in the same statistical report (p.31, tab.12, the price of oil and other industrial resources increased from 2002 to 2005 by 120%.

THE NEW WAY: SUSTAINABILITY

The solution to these issues is described by one word: sustainability.

The report of the Brundtland Commission of the United Nations in 1987 about "Our Common Future" defined "sustainability" as a "development that meets the needs of the present without compromising the ability of future generations to meet their own needs". (UN Documents: "Report of the World Commission on Environment and Development: Our Common Future". chapter I.3., paragr. 27, http://www.un-documents.net/wced-ocf.htm). Many other definitions for sustainable behavior have been published since then.

At the moment we use our natural resources, e.g. for energy production or in the chemical industry, without replacing them. Most of these resources are limited. If we work and use these resources like we do it today, they will disappear from our planet in a short period of time. We have to find processes, which do not use up resources. We have to find and work with resources, which are renewable. Only then it will be possible to protect the environment and preserve nature, so that our children and grandchildren will still have a planet where it is worth living (see also Chapter 1).

However, it is not enough only to use alternative materials and processes; it is also necessary to develop and use more efficient chemical processes to save energy and resources. In many cases conventional chemical production is not efficient enough. New technologies are needed for "Process Intensification" (see Chapter 14). One of these is microreaction technology which is described in detail in the following chapters of this book.

Microreaction technology has been studied now for more than 15 years. Starting mainly in Germany and lead by institutes, for example, the Institute of Micro-technology in Mainz and the Karlsruhe Research Center, several companies have been formed producing microfluidic modules as well as whole microreaction systems.

The main idea behind this new approach is that within microstructures all chemical and physical parameters of a reaction [e.g., mixing (see Chapter 4) and heat exchange (see Chapter 5)] can be controlled much faster and better. Due to the flexible manufacturing processes of microfluidic structures (see Chapters 2 and 3), the reactors can be designed to fit to the required chemistry—in contrast to the conventional procedure, where the chemistry is "pressed" into a given plant. Separation procedures can be added if necessary (see Chapter 6).

It could be shown in many cases that this leads to higher selectivity and yield: the consumption of resources and solvents can be reduced; there are less byproducts and less waste; the use of energy is much more effective; and safety issues are reduced, due to the small internal volume of the microreactors.

This does not work against the interests of the chemical industry: processes with more efficiency, more yield, and more selectivity are also more cost effective. Costs for natural resources, waste deposition, and safety decrease with this technology. Moreover, a faster time-to-market gives the industry a huge additional benefit. The production can be done "on-site", where the products are needed. Large-scale production plants and long transportation to the point-of-use can be avoided.

All this leads to much more efficient processes and makes the new chemistry "green", helping to protect our environment. Governments all over the world have already realized the opportunities of this new technology for sustainability, safety, and environmental protection. Funding has been provided to develop the technology, especially from the German, the European and the Japanese Government. Developing countries such as China, India, or Mexico have also started developing technology.

NEW PARADIGMS IN CHEMICAL ENGINEERING

There are major differences between using a conventional reaction plant and a micro-reaction plant: using a microreactor to run a reaction under optimized conditions, the reaction parameters have to be known precisely, especially the kinetics and the thermodynamics of the reactions involved. Instead of controlling the reaction by time as in a usual batch process, they have to be controlled by the geometry of the microreactor. Therefore, one has to go through the following steps when starting a new project using microchemical engineering:

- Physical and chemical data of the reactions have to be collected.
- Simulations can help to calculate missing parameters. Software tools are available from different suppliers (see Chapter 7).
- Choosing the right technology using, for example, the following criteria:
  - Can particles be avoided or made small enough for microreactors?
  - Is the production volume in a suitable range for microreactors?
  - Are there issues with the yield or the selectivity in conventional plants?
  - Is the reaction time fast enough to profit from a microfluidic system?
  - Is a good temperature control of a microheat exchanger necessary?
  - Is a good mixing control of a micromixer necessary?
  - Is a good control of a dangerous reaction or dangerous components necessary?
- The right microfluidic modules (reactors, mixers, heat exchangers, separation modules) as well as the peripheral equipment, such as pumps (see Chapter 8)

and sensors (see Chapter 9) have to be chosen. This includes the choice of material as well as the dimensions of the microfluidics.

- To control and optimize the reaction scheme it is necessary to have suitable control software which is available from different suppliers (see Chapter 10).
- After the first experiments all parameters can be corrected and a laboratory plant with all the different modules can be built and optimized (see Chapter 11).
- For large-scale production a scaling-up procedure and appropriate manufacturing is necessary (see Chapter 13).

After the reaction has been developed, it is relatively easy to ramp-up to the production volume, either by parallelization of the microfluidic modules ("numbering-up") or by multiplying the microstructures in a larger housing ("equaling-up"). In both cases, a production set-up can be built and installed in a much shorter time than in a conventional ramp-up procedure. This again helps saving costs and shortening the time-to-market.

NEED OF EDUCATION

Even though the advantages of this technology are well established, not many production examples are known. The main reason for this is missing know-how in the chemical industry: there are no educated staff available. Universities usually do not teach microchemical engineering and often continuous processes are not part of the curriculum.

Chemists and chemical engineers do learn the fundamentals and principles in chemistry and chemical engineering, as they did in the past. It takes a long time for new technologies to enter into standard educational programs, especially if these new ideas involve expensive equipment. Universities usually do not have enough funding for investment in new equipment. Even though they should always be at the forefront of technological advances, it is difficult to get new ideas distributed within the academic community (see Chapter 12).

Knowing this, in 2007 the German government started a process to develop lectures and practical experiments for students which should help to introduce this new approach into chemistry departments. Grants have been given for equipment to be used at universities, mainly for education, but also for internal research projects. As always, the funding was sufficient only for a small number of projects at a small number of universities. But, with their experience, which will be published and can be used by any interested person, it should be possible to influence the curricula of other universities. German equipment manufacturers profit from this program and will be able to develop and produce tools for foreign research and education institutes.

Developing countries such as India, China, or Mexico have learned from the experience of Japan and Germany and started their own programs. Mexico is going to build and renew its chemical industry, for example. It has planned to invest in the newest and most effective technology. Lead by the Technológico de

Monterrey, courses on microchemical engineering will be introduced into bachelor and master studies. In parallel, cooperation with the chemical industry will bring these new ideas into practice.

It is very important to include these new technologies in the curricula of chemistry and chemical engineering schools. But it will take several years before this knowledge then reaches industry; therefore, it is necessary, to support chemists and engineers in the industry and make the new concepts available to them. This is done in different ways:

- National Chemical Engineering Associations, such as DECHEMA in Germany (www.MicroChemTec.de) or AIChE in the USA (Process Development, Areas 12e) or newly founded organizations such as the MCPT in Japan (www.mcpt. jp) are organizing workshops and seminars. They are collecting information and providing help to "beginners".
- Over the last 10 years the experts in this new field meet at the International Conference on Microreaction Technology (IMRET), discussing new developments and looking for partners to solve problems.
- There are already German, European and International Groups (ISO) working on standards in microchemical engineering. In particular, the connection of microfluidic chips from different suppliers shall be facilitated by standardized interfaces (see Chapter 15).

AIM OF THIS BOOK

Much research has been done in recent years on microchemical engineering. It is not possible to summarize all the results in one book. Therefore, the aim of this book is to give an overview of the advantages and the challenges of this new technology. It also aims to provide help to new users in getting started. Information is summarized to enable the reader to decide on the right reactor material as well as for the suitable dimensions. Examples in different fields are given, for example, for polymerizations (see Chapter 16), for photoreactions (see Chapter 17), for catalytic reactions (see Chapter 18), for enzymatic reactions (see Chapter 19), or for multiphase reactions (see Chapter 20). Cited literature will be helpful to gain deeper insight into certain subjects.

It is the hope of the authors that this book will help to accelerate the introduction of this new technology into the chemistry and chemical engineering departments of universities, into research and development institutes, and into the chemical industry—to the economical and ecological benefit of the world.

THOMAS R. DIETRICH

Mainz, March 2008

CONTRIBUTORS

JOËLLE AUBIN, Laboratoire de Génie Chimique UMR 5503 CNRS, 5, rue Paulin Talabot, BP-1301, 31106 Toulouse Cedex 1, France, E-mail: Joelle.Aubin@ ensiacet.fr

ALEXIS BAZZANELLA, Dechema e.V., Theodor-Heuss-Allee 25, 60486 Frankfurt, Germany, E-mail: Bazzanella@dechema.de

DIETER BOTHE, Lehrstuhl für Mathematik CCES, Center for Computational Engineering Science, Pauwelsstr. 19 52074 Aachen, E-mail: bothe@mathcces. rwth-aachen.de

ERIC A. DAYMO, Velocys, Inc., 7950 Corporate Boulevard, Plain City, OH 43064

JAN DZIUBAN, The Wroław University of Technology, ul. Janiszewiskiego 11-17, 50372 Wroław, Poland, E-mail: jan.dziuban@pwr.wroc.pl

JOHN EDWARD ANDREW SHAW, 45 Colne Avenue, West Drayton, Middlesex UB7 7AL, United Kingdom, E-mail: joshaw@tycoint.com

ANDREAS FREITAG (Glass), mikroglas chemtech GmbH, Galileo-Galilei-Str. 28, 55129 Mainz, Germany, E-mail: a.Freitag@mikroglas.com

J. G. E. (HAN) GARDENIERS, MESA+ Institute for Nanotechnology, University of Twente, P.O.Box 217, 7500 AE Enschede, The Netherlands, E-mail: j.g.e.gardeniers@utwente.nl

ASTERIOS GAVRIILIDIS, Department of Chemical Engineering, University College London, Torrington Place, London WC1E 7JE, United Kingdom, E-mail: a.gavriilidis@ucl.ac.uk

MARK GEORGE KIRBY, Heatric, Division of Meggit (UK) Ltd., 46 Holton Road, Holton Heath, Poole, Dorset, BH16 6LT, United Kingdom, E-mail: Mark.Kirby@HEATRIC.COM

FRANK N. HERBSTRITT, Ehrfeld Mikrotechnik BTS GmbH, Mikroforum Ring 1, D-55234 Wndelsheim, Germany, E-mail: frank.herbstritt@ehrfeld.com

TEIJIRO ICHIMURA, Department of Chemistry and Materials Science, Guraduate School of Science and Engineering, Tokyo Institute of Technology, W4-17, 2-12-1 Ohokayama, Meguro, Tokyo 152-8551, Japan, E-mail: tichimur@chem.titech.ac.jp

JEAN-MARC COMMENGE, GPM, LSGC-ENSIC, 1 rue Grandville, BP 20451, F-54000 NANCY, France, E-mail: commenge@ensic.inpl-nancy.fr

JEAN F. JENCK, ENKI Innovation, 3 chemin des Balmes, F-69110 Sainte-Foy, France, E-mail: jenck@enki2.com

ASIF KARIM, BASF AG, 67056 Ludwigshafen, Germany, E-mail: asif.karim@basf.com

DIRK KISCHNECK, Microinnova Engineering GmbH, Reininghausstrasse 13a, 8020 Graz, E-mail: dirk.kirschneck@microinnova.com

L. KIWI-MINSKER, Ecole Polytechnique Fédéral de Lausanne, SB-ISIC-GGRC, Station 6, CH-1015 Lausanne, Switzerland, E-mail: Lioubov.kiwi-minsker@epfl.ch

EUGENIA KUMACHEVA, Department of Chemistry, University of Toronto, 80 Saint George Street, Toronto, Ontario M5S 3H6, Canada; Institute of Biomaterials and Biomedical Engineering, Department of Materials Science and Engineering, University of Toronto, 4 Taddle Creek Road, Toronto, Ontario, M5S 3G9, Canada; Department of Chemical Engineering and Applied Chemistry, University of Toronto, 200 College Street, Toronto, Ontario M5S 3E5, Canada, E-mail: ekumache@alchemy.chem.utoronto.ca

HAB. LAURENT FALK, GPM, LSGC-ENSIC, 1 rue Grandville, BP 20451, F-54000 NANCY, France, E-mail: falk@ensic.inpl-nancy.fr

MARCEL A. LIAUW, ITMC, RWTH Aachen, Worringerweg 1, 52074 Aachen, Germany, E-mail: liauw@itmc.rwth-aachen.de

WOLFGANG LOTH, BASF AG, 67056 Ludwigshafen, Germany, E-mail: wolfgang.loth@basf.com

THOMAS MÜLLER-HEINZERLING, Siemens AG, Automation & Drives, Competence Center Chemical, Cement, Glass, G.-Braun-Str. 18, D-76187 Karlsruhe, Germany, E-mail: Thomas.Mueller-Heinzerling@siemens.com

MICHAEL MATLOSZ, GPM, LSGC-ENSIC, 1 rue Grandville, BP 20451, F-54000 NANCY, France, E-mail: matlosz@ensic.inpl-nancy.fr

YOSHIHISA MATSUSHITA, Department of Chemistry and Materials Science, Guraduate School of Science and Engineering, Tokyo Institute of Technology, W4-17, 2-12-1 Ohokayama, Meguro, Tokyo 152-8551, Japan

BERND NIDETZKY, Institute of Biotechnology and Biochemical Engineering, Graz University of Technology, Petersgasse 12, A-8010 Graz, Austria

BERND NIDETZKY, Institute of Biotechnology and Biochemical Engineering, Graz University of Technology, Petersgasse 12, A-8010 Graz, Austria, E-mail: bernd. nidetzky@tugraz.at

ZHIHONG NIE, Department of Chemistry, University of Toronto, 80 Saint George Street, Toronto, Ontario M5S 3H6, Canada

ALBERT RENKEN, Ecole Polytechnique Fédéral de Lausanne, SB-ISIC-GGRC, Station 6CH-1015 Lausanne, Switzerland, E-mail: Albert.Renken@epfl.ch

SVEND RUMBOLD, Heatric, Division of Meggit (UK) Ltd., 46 Holton Road, Holton Heath, Poole, Dorset, BH16 6LT, United Kingdom, E-mail: Svend. Rumbold@HEATRIC.COM

KOSAKU SAKEDA, Department of Chemistry and Materials Science, Guraduate School of Science and Engineering, Tokyo Institute of Technology, W4-17, 2-12-1 Ohokayama, Meguro, Tokyo 152-8551, Japan

NORBERT SCHWESINGER (Silicium), LS TEP FG Mikrostrukturierte mechatronische SystemeTU München, Arcisstraße 21,80333 München, Germany, E-mail: schwesinger@tum.de

TADASHI SUZUKI, Department of Chemistry and Materials Science, Guraduate School of Science and Engineering, Tokyo Institute of Technology, W4-17, 2-12-1 Ohokayama, Meguro, Tokyo 152-8551, Japan

MALENE S. THOMSEN, Institute of Biotechnology and Biochemical Engineering, Graz University of Technology, Petersgasse 12, A-8010 Graz, Austria

ANNA LEE Y. TONKOVICH, Velocys, Inc., 7950 Corporate Boulevard, Plain City, OH 43064, E-mail: tonkovich@velocys.com

DINA E. TREU, Germany, E-mail: treu@itmc.rwth-aachen.de

CATHERINE XUEREB, Laboratoire de Génie Chimique UMR 5503 CNRS, 5, rue Paulin Talabot, BP-1301, 31106 Toulouse Cedex 1, France, E-mail: Catherine.Xuereb@ ensiacet.fr

HONG ZHANG, Department of Chemistry, University of Toronto, 80 Saint George Street, Toronto, Ontario M5S 3H6, Canada

PART I

INTRODUCTION

CHAPTER 1

IMPACT OF MICROTECHNOLOGIES ON CHEMICAL PROCESSING*

JEAN F. JENCK

1.1 INNOVATION: AN ANSWER TO THE CHALLENGES OF SUSTAINABLE DEVELOPMENT

Sustainability was defined in 1987 by the World Commission on Environment and Development as "a development that meets the needs of the present without compromising the ability of the future generations to meet their needs." A classical approach is to say that development is sustainable so long as it takes care of the three P's of people, planet, and profit.

Has much been done in this direction, having a look at the current status of the chemical industry? Published under the umbrella of the European Chemical Federation, a tutorial review "Products and Processes for a Sustainable Chemical Industry" [1] shows that industrial sustainable chemistry is not an emerging trend, but already a reality through the application of "green" chemistry and engineering expertise.

On the other side, most of the basic pieces of equipment usually operated by the chemical industry are century-old designs; for instance, stirred vessels and impellers geometry do not seem to have changed since 1554 when G. Agricola pictured them in his book *Re Metallica* (Fig. 1.1).

The chemical industry progressed by building large plants, due to the economy of scale: Investment costs rise less than production capacity and rough estimates generally use the following formula ("gamma rule"):

$$(\text{cost})_A/(\text{cost})_B = [(\text{capacity})_A/(\text{capacity})_B]^{0.6 \text{ to } 0.7} \tag{1.1}$$

*Adapted from a lecture given by Jenck and coworkers at the Spring 2006 AIChE Meeting, Topical T1 Applications of Microreactor Engineering, Paper No. 23a, Orlando, Florida, April 24, 2006.

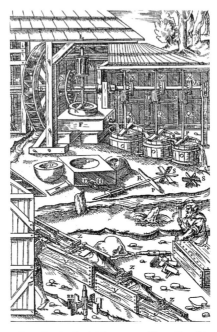

FIGURE 1.1 Process engineering in the 16th century (from Stankiewicz with permission [2]).

As depreciation figures tend to impact less and less standard manufacturing costs as production capacity increases, there has been a tendency to always make bigger manufacturing units.

However, "the classical world-scale plant is being phased-out," as stated by the president of corporate engineering at BASF, and it has been suggested that chemical engineering now follow the opposite direction, as a result of these trends:

- Need for a paradigm change in plant engineering
- Pressure of time-to-market
- Need for modular plant techniques
- "Microprocess engineering will have a role on plant philosophy more than on absolute size" [3]

A recent statement by the board of directors at Linde was in line with these trends [4].

1.2 PROCESS INTENSIFICATION: A NEW PARADIGM IN CHEMICAL ENGINEERING

Process intensification (PI), where the motivation is "doing more with less," is a design methodology aiming to minimize diffusion phenomena (mass and/or energy transfer). Its first goal is to build smaller, more compact, and cheaper production plants. PI started in the late 1970s when Colin Ramshaw developed the "Higee" technology at ICI. We can today bet on a future chemical plant based on modular elements to run a flexible miniplant [5–7]. The ongoing change introduced by process intensification may be depicted saying that the process, thus far, resulted from the optimization and balancing of four constraints:

Chemistry kinetics
Mass transfer
Heat transfer
Hydrodynamics

Through process intensification, transfer rates are maximized and the process is basically governed by chemical kinetics. It is no longer limited by diffusion. Fick law coefficients are maximized and global apparent kinetics closely approach intrinsic chemical kinetics [8].

Initially, the goal of the PI approach was to build "smarter" production plants relying on eco-efficient processes. Additional goals have materialized. On the one hand, running a chemical reaction in currently difficult—if not impossible conditions—will become possible. On the other hand, an emerging goal is to closely control the properties of products by mastering their production process. With process intensification, adapting the process to the chemical reaction becomes the leitmotiv with the following provisions:

Adapting the size of equipment to the reaction
Replacing large, expensive, inefficient equipment by smaller, more efficient, and cheaper equipment
Choosing the technology that best suits each step
Combining sometimes multiple operations in fewer pieces of equipment

To alleviate diffusion limitations, four principles of PI allow us to approach the intrinsic kinetics of phenomena:

1. Multifunctionality where unit operations are combined in a single piece of equipment
2. Alternative solvents (and even suppression of solvents) to increase the thermodynamic potential of reagents (activity and/or diffusivity)
3. Reduction of size by miniaturization, frequently using "microtechnologies" for equipment (microreactors, micromixers, microseparators), monitoring (microsensors), and control (microvalves)
4. Alternative energy fields: electromagnetic (microwaves, HF, photons), acoustic (ultrasounds), electric, and gravitational [9]. Beyond the process intensification strategy, some driving forces have been clearly identified:
   Process safety
   Continuous process replacing a batch process
   On-site on-demand production

1.2.1 Process Safety

The signs given by the chemical industry to the general public are that it does not always learn from its past:

The AZF explosion that killed 32 people and injured more than 2,000 in Toulouse, France, on September 21, 2001, was a reoccurrence of the BASF accident in Oppau, Germany, exactly 80 years earlier.

In Bhopal, India, on December 3, 1984, a cloud of 41 tons of MIC rose, killing thousands. A later report showed that an inventory of 10 kg of MIC would have been sufficient to run the plant [10].

Big inventories definitely are unsafe (Fig. 1.2). Smaller inventories and in-process volumes would have significantly lowered the magnitude of these accidents. Based on this point of view, everything should be done to lower in-process amounts of material and, consequently, hazardous material inventories. In so doing, the occurrence of large-scale accidents should become less likely. The in-process volume reduction will mainly result from a philosophical change, from a batch process to continuous process.

1.2.2 Continuous Process Replacing a Batch Process

Reaction can be more easily controlled in extreme conditions (low temperature, high pressure, etc.) when small amounts of material have to be instantaneously handled. Therefore, rather than running the reaction in a vessel of a few cubic meters, much better process control results by running the same reaction in a continuous process where much smaller amounts of hazardous materials and energy are instantaneously involved.

So doing brings about not only safer control of the process especially when the reaction media has to be kept in cryogenic conditions, but also a more specific

PLACE	DATE	CHEMICAL	ESTIMATED AMOUNT	CASUALTIES
Oppau/Ludwigshafen	September 21, 1921	ammonium sulfate, ammonium nitrate	4,500 t exploded	ca. 550 + 50 dead, 1,500 injured
Flixborough	June 1, 1974	cyclohexane	400 ton inventory, 40 ton escaped	28 dead, 36 + 53 injured
Beek	November 7, 1975	(mainly) propylene	>10,000 m³ inventory, 5.5 ton escaped	14 dead, 104 + 3 injured
Seveso	July 10, 1976	2,4,5- trichlorophenol, dioxin	7 ton inventory, 3 ton escaped	no direct casualties, ca. 37,000 people exposed
San Juan, Mexico City	November 19, 1984	LPG	>10,000 m³ inventory	5 + ca. 500 dead, 2 + 7,000 injured (mainly outside the plant)
Bhopal	December 3, 1984	methyl isocyanate	41 ton released	3,800 dead, 2,720 permanently disabled
Pasadena	October 23, 1989	ethylene, isobutane, hexene, hydrogen	33 ton escaped	23 dead, 130–300 injured
Toulouse	September 21, 2001	ammonium nitrate	200–300 ton	31 dead, 2,442 injured

FIGURE 1.2 Accidents in the chemical industry (from Stankiewicz with permission [11]).

FIGURE 1.3   Transition from batch to continuous mode.

reaction. A well-known and documented example is the Meck KgaA vitamin H process (Fig. 1.3). Resulting from all previous comments, as well as the philosophy of miniaturization, flexibility will bring about new facility concepts.

1.2.3   On-Site On-Demand Production

Making manufacturing units smaller, we can envisage delocalized productions, close to the customer, with much lower inventories (Fig. 1.4). The "on the road" man-ufacturing unit is no longer a dream. Online production devices have been designed in the following cases:

   Interox produces 1 ton per day peroxysulfuric acid (Caro acid) in a 20-cm^3 tubular reactor at 1-s residence time.

   Kvaerner provides modular phosgene $COCl_2$ generators, point-of-use and skid-mounted.

   Online on-demand generators have been designed for hydrogen cyanide, chlorine dioxide, ethylene oxide, etc.

FIGURE 1.4   Modular skid-mounted production unit (from Green (2005) with permission [12]).

Now that we have reviewed the driving forces of process intensification, one question comes to mind: What commercial venture on offer would support any process engineering intended to include process-intensified technology in the design?

1.2.4 Commercial Offer of Intensified Devices

Several types of equipment already exist, and their performance may be described in terms of their efficiency with regard to mass and heat transfers. Microreactors are special in their performances (Fig. 1.5). Intensified devices are already commercially

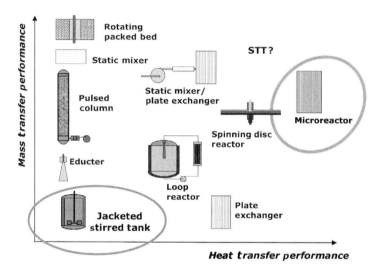

FIGURE 1.5   Transfer performances (from Fleet (2005) with permission [13]).

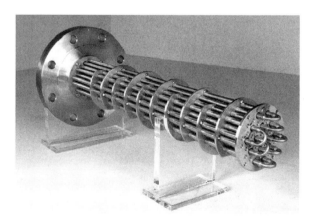

FIGURE 1.6   Compact heat exchanger (courtesy BHR Group).

available. A few examples appear in Figs. 1.6 through 1.9. Applications are illustrated for some of them in Fig. 1.10.

Monolith loop reactor (MLR) technology has experienced some developments in the case of supported catalysts, enzymes, or cells. An example of the industrial development of loop reactors is that advanced by Air Product which runs hydrogenations (Figs. 1.11 and 1.12).

Retrofitting a stirred tank reactor to replace slurry catalysts also increases productivity by a factor of 10 to 50. "Intensive" engineering may lead to the design of many external loop reactors.

Process intensification may be considered a new approach of process engineering based on the fact that mass and heat transfers are no longer limitations and that actual

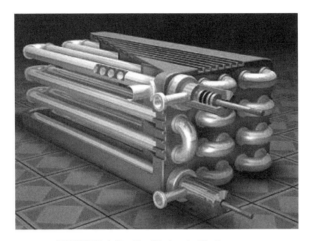

FIGURE 1.7   Oscillating baffled reactor.

FIGURE 1.8   Spinning disk.

FIGURE 1.9 Twin shaft mixer/kneader (LIST AG).

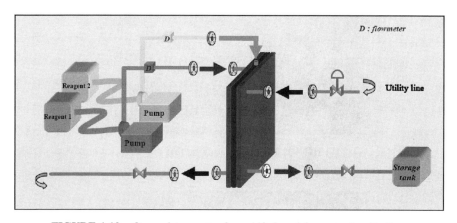

FIGURE 1.10 Open plate reactor from Alfa-Laval for aromatic nitration.

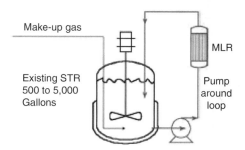

FIGURE 1.11 Monolith loop reactor.

FIGURE 1.12 Monolithic catalyst.

kinetics are now close to intrinsic kinetics. Having stressed the fact that commercial technologies are already commercially available, we now consider the specific contribution of microprocessing to process intensification.

1.3 MICROPROCESSING FOR PROCESS IDENTIFICATION

1.3.1 Supply of Microstructured Components

Today, microtechnologies are commercially available, thanks to world-class manufacturers and engineering firms such as IMM, Velocys, BTS Ehrfeld, Microinnova, Siemens, CPC, mikroglas, FZK, Heatric, Dai Nippon Screen, IMT, etc. This list is by no means exhaustive; a few examples are given in Fig. 1.13.

Microchannels with characteristic dimensions between 0.1 and 1 mm enable compact operations by reducing transport distances compared to conventional technologies where tube diameters fall in the range 10 to 100 mm.

Yole Development (http://www.yole.fr) has identified 21 microtechnology suppliers worldwide. Their locations are distributed among the major economic zones according to the chart shown in Fig. 1.14.

The field is constantly evolving not only in terms of materials and technologies, but also in terms of players: New companies start their activity; others stop it and reposition, raising their business targets.

1.3.2 System Integration: Selected Examples

Modularity becomes possible and opens new routes for unit engineering, as companies offer commercial systems where microelements are assembled. Here are four examples, in increasing order of integration:

Example 1: Bayer Technology Services Modules are clamped on a plate to assemble a miniplant (Fig. 1.15). With a base price of $\sim$500 € per module, a complete miniplant costs about 50–100 k€.

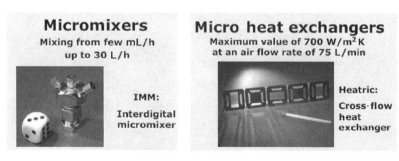

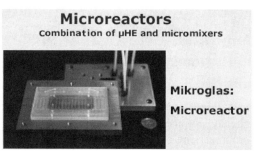

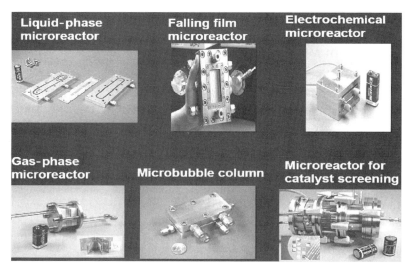

FIGURE 1.13   Microdevices.

Example 2: Dainippon Screen   See Fig. 1.16.

Example 3: Hitachi Ltd   By mounting several microreactors in parallel, a pilot plant has been developed to produce up to 72 tons of chemicals per year (Fig. 1.17).

Example 4: Siemens Automation and Drives   SiProcess is a process system for chemical syntheses, based on microprocess technology through a combination of modularity and automation (Fig. 1.18). The modules are compact and easily

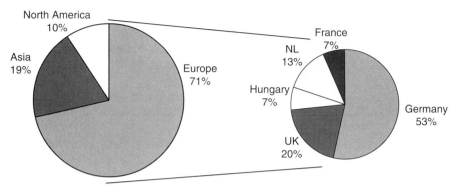

FIGURE 1.14   Microtechnology suppliers (from Yole Development with permission [14]).

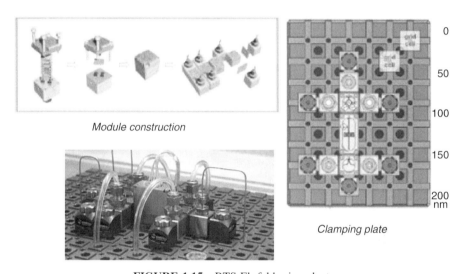

FIGURE 1.15   BTS Ehrfeld microplant.

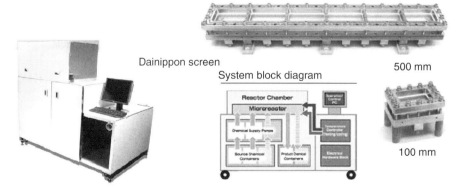

FIGURE 1.16   Dainippon microplant [15].

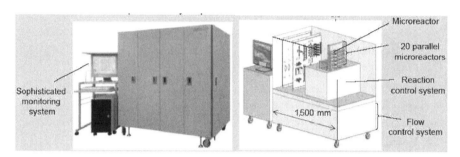

FIGURE 1.17 Hitachi microplant [16].

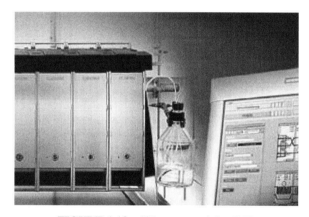

FIGURE 1.18 SiProcess modules [17].

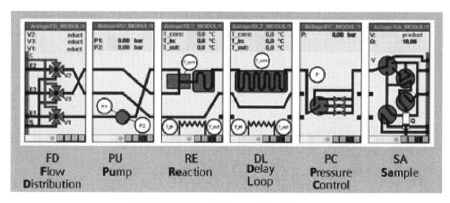

FIGURE 1.19 SiProcess flow diagram.

exchanged, and the system is designed so that end users can also insert their own components. The electronics of each module are connected to a higher-level system of automation. The system can be configured according to customer requirements using the modules shown in Fig. 1.19:

FD distributes the raw materials, as well as a solvent and a cleaning agent.

PU meters the raw materials.

RE chemical reaction (mixing and heating) occurs in a microreactor.

DL is a delay loop to complete the reaction.

PC controls pressure in the system.

SA is a sampling and quenching module.

1.4 INTENSIFIED FLUX OF INFORMATION (R&D) VS. INTENSIFIED FLUX OF MATERIAL (PRODUCTION)

The usual way to industrialize a new product or new process starts with results gathered at the laboratory scale. Critical parameters are worked out in the pilot plant. This set of information is then used to design and build the production unit. Microtechnologies are now changing the picture: The numbering-up principle is making industrialization move from the conventional scale-up to the scale-out (Fig. 1.20).

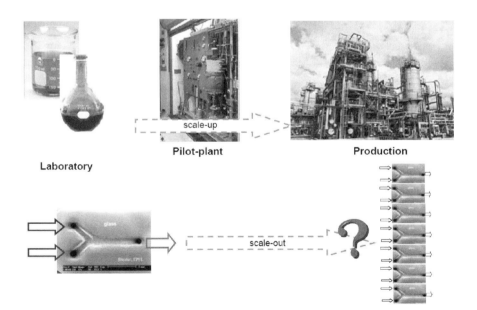

FIGURE 1.20   Scale-out (from Renken (2006) with permission [18]).

1.5 IMPLEMENTATION OF MICROTECHNOLOGIES IN CHEMICAL PROCESSING: A FEW SELECT EXAMPLES

1.5.1 Example 1

An interesting example of the pilot testing of microtechnologies has been unveiled by Degussa in association with Uhde. The objective of project Demis® was to run an epoxidation reaction to obtain propylene oxide from propylene and gaseous H_2O_2. See Figs. 1.21 and 1.22. Among the conclusions of this positive pilot testing, an important one was the verification of the numbering-up principle as a way to scale up the process.

1.5.2 Example 2: Fine Chemicals Plant in a "Shoe Box"

Forschungszentrum Karlsruhe (FZK) and DSM paricipated in a collaboration that led to a manufacturing "box" that is 65 cm high, weighs 290 kg, and has a 1,700 kg per hour throughput of liquid chemicals. "Micro" in its interior, as the device is made of micromixers and several ten thousands of microchannels, it can remove reaction heat of up to several 100 kW (Fig. 1.23).

It was announced at the 2006 ACHEMA in Frankfurt, Germany, that such a device was now in permanent operation at DSM Fine Chemicals in Linz, Austria. It was validated after a 10-week demonstration showing that 300 tons of high-value product could be manufactured with a better yield than with the conventional process and improved process safety conditions.

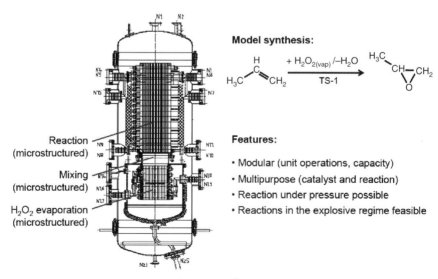

FIGURE 1.21   Demis® microprocessing.

FIGURE 1.22   Demis® reactor.

FIGURE 1.23   FZK microreactor [19].

1.5.3   Examples 3 and 4: Radical Polymerizations

Idemitsu Kosan claims that its polymerization pilot (with a size of 3.5×0.9 m) produces up to 10 tons per year (Fig. 1.24). It is still unclear which type of radical polymerization is handled in these microtechnologies. Figure 1.25 provides images of reaction units made up of microchannels. "Has the plant been scaled up at the industrial level?" remains an open question, but this pilot plant proves that micro-technologies can handle viscous flows, as already demonstrated previously by Siemens-Axiva in its Corapol process (Fig. 1.26).

FIGURE 1.24 Polymerization microplant in Japan [20].

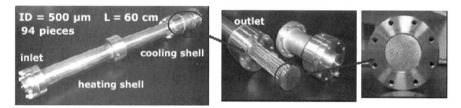

FIGURE 1.25 Microchannels for polymerization.

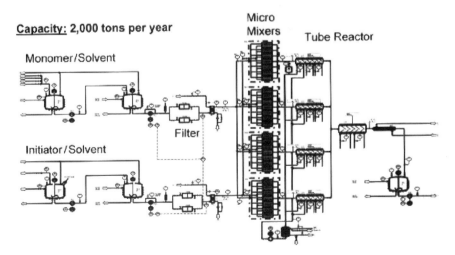

FIGURE 1.26 Polymerization in microreactor [21].

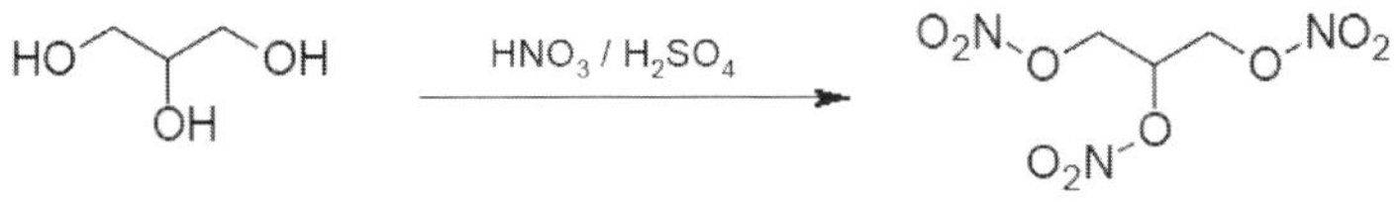

FIGURE 1.27 Trinitroglycerin synthesis.

1.5.4 Example 5: Nitroglycerin Microstructured Pilot Plant

In May 2005 Xian Chemicals started nitroglycerin (NG) production on a pilot plant level (15 kg/h NG, >100 L/h) in China (Fig. 1.27). Xian Chemicals invested ~5 M€ in a facility, developed by IMM in Mainz, Germany [22–23].

The finished material is used as medicine for acute cardiac infraction. Thanks to a microengineering philosophy in implementation:

- The product quality is at its highest grade.
- The plant operates safely and is fully automated.
- Environmental protection is ensured by advanced wastewater treatment and a closed cycle.

The Xian nitroglycerin microplant team appears in Fig. 1.28.

1.5.5 Example 6: Pharmaceutical Chemistry

Cellular process chemistry (CPC) is the basis of the Cytos® pilot system; it includes 10 (+1 spare) microreactors for a cost of ~1.2 M€. The microreactor capacity follows the user requirements on a contract basis.

FIGURE 1.28 Xian nitroglycerin microplant team [24].

FIGURE 1.29 CPC Cytos® microplant [25].

As an example, a commercial, large-scale, multiproduct plant near Leipzig, Germany, has been running since mid-2006, producing high added-value chemicals (niche applications for pharmaceuticals) with a range from 1 to 100 kg. Synthacon has started production of a multipurpose unit with 20 tons per year capacity.

Sigma-Aldrich has installed a standard Cytos® in Buchs, Switzerland. Many of Sigma-Aldrich's catalog products are produced under typical lab conditions in flasks up to 20 L. Out of the 2,000 compounds in this portfolio, about 800 could be produced in microreactors with little or no process modification. See Fig. 1.29.

1.6 CHALLENGE OF COST EFFICIENCY: BALANCE BETWEEN CAPEX AND OPEX

Based on the known examples of microtechnology implementations, the University of Eindhoven in the Netherlands has arrived at an interesting synthetic view of the field (Fig. 1.30). Microtechnologies have a rather high investment cost, at least as long, as there is no mass production of microdevices. How is the Capex cost offset by savings on operating costs, Opex?

1.6.1 Aniline by Hydrogenation of Nitrobenzene

Aniline is produced in a highly exothermic process. Following current practice, it is run in tubular fixed beds. This brings up several issues: poor performances, renewal

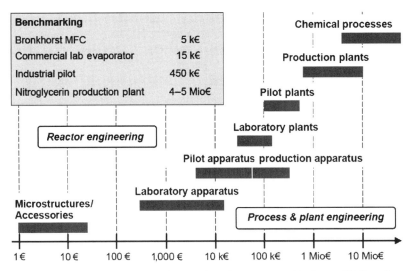

FIGURE 1.30   Process and plant engineering (courtesy Professor V. Hessel).

of catalyst that requires the operator to unload the old and reload the new, frequent catalyst regenerations. The microreactor technology (MRT) solution works:

Has an immobilized catalyst

Involves a lower hydrogen recycle rate

Avoids by-product formation because of better temperature control

Eliminates any previously necessary catalyst unloading and loading

Thus significantly lowers downtime

A cost analysis, conducted by CMD International in 2002 for a 50 kt per year unit, led to savings of about 200 kUS$ per year in favor of the MRT option. Would 5 US$ per t be a high enough reduction of the manufacturing costs to vindicate the MRT investment (and risk)?

1.6.2   Direct Hydrogen Peroxide

UOP arrived at a basic engineering quotation for a 160 kt per year plant operating with microstructures that showed operation in the explosive regime at low pressure

$O_2 : H_2$ ratio	3	1.50	1	6.79	1.89
Pressure (psia)	300	300	300	1.450	1.450
Conversion	0.9	0.9	0.9	0.25	0.25
Selectivity	0.85	0.8	0.65	0.8	0.8
Yield	0.765	0.72	0.585	0.2	0.2
TOTAL COST ($ MM)	68.27	61.54	64.00	140.07	98.85

FIGURE 1.31   Global cost of hydrogen peroxide plant [26].

Projected Production Cost (¢/lb)

Raw material less byproducts	4
Consumables	2
Utilities	3
Labor/Maintenance	1
Overhead	1
Capital recovery	6
Total	17

FIGURE 1.32 UOP hydrogen peroxide (copyright UOP LLC; all rights reserved).

is the least capital-intensive (Fig. 1.31). Explosive conditions are also those that guarantee significantly lower variable costs (Fig. 1.32).

1.6.3 Fine Chemicals

Lonza published the results of their detailed analysis of different type of reactions (Fig. 1.33). Type A are very fast and mixing-controlled; type B are rapid and kinetically controlled; type C are slow, but with a safety or quality issue. Of 86 reaction campaigns carried out at Lonza, 50% could benefit from a continuous process.

Concerning capital expenditures, microreactor costs are as high as or higher than traditional technology costs. However, this cost is compensated by high operating savings when the reaction is run continuously. As raw material costs contribute to 30 to 80% of the total manufacturing cost, higher product yield and quality may

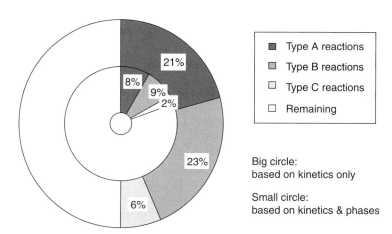

FIGURE 1.33 Typology of reactions [27].

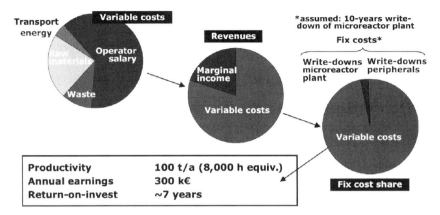

FIGURE 1.34   Operational costs.

have a significant impact. In addition, an automated process reduces QA/QC and labor costs.

1.6.4 Microreactor Process Cost Incentives

An assessment of microreactor process operation costs was made at the University of Iena, Germany based on the following:

Rough cost estimate grounded in a few similar estimates and a real case

Production scenarios in the field of fine chemicals, averaged to one plot

Only consideration of the reaction to the crude product with no purification [28]

The study's conclusions are clearly depicted in the following charts: operation costs as shown in Fig. 1.34 and capital expenditures as shown in Fig. 1.35.

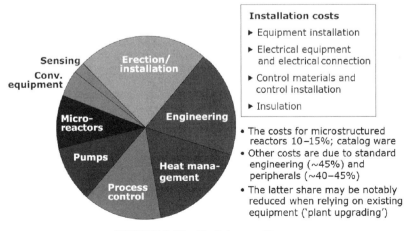

FIGURE 1.35   Capital expenditures.

Other available examples also stress the fact that run in inappropriate conditions, microtechnologies may be more expensive than regular technologies. However, micro-reactors open the way to such conditions that reaction selectivity may be significantly improved, making the operation cost drastically lower. Another defining example is the switch from batch to continuous operation to avoid cryogenic reaction conditions.

Great care must be taken to use the optimized performances of the microreactors in assessing their operating costs. Other than that, capital expenditures do not seem to be very different from regular process investment costs.

1.7 PERSPECTIVES

1.7.1 Opportunities for Microprocessing in Intensified Formulation

For fine chemicals and advanced materials, microprocessing implemented as part of product engineering will yield new properties because of the narrower distribution of molecular weight, particle diameter, etc. The challenge is to ensure a flow regime with as low an axial dispersion as possible. Some examples of these trends were presented at the 2006 AIChE Spring Meeting:

Organic nano particles (Paper No. 98a by Fuji)

Ultrafine powders (Paper No. 84a by EPFL and TechPowder, and Paper No. 98e by Microinnova)

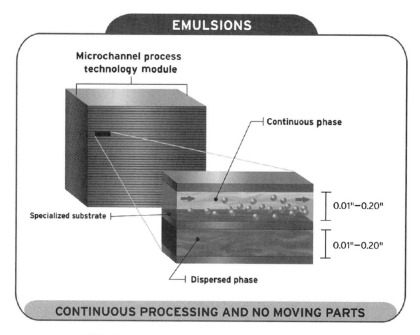

FIGURE 1.36 Velocys microchannel emulsification.

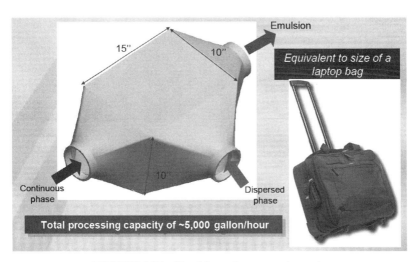

FIGURE 1.37 Emulsion microprocessing unit.

Block copolymers (Paper No. 62d by the University of Strasbourg)

Complex emulsions (Paper No. 140g by Unilever)

Those in the field of emulsion and L-L dispersion are particularly concerned about microtechnologies because of their intrinsic advantages over conventional techniques:

Higher energy efficiency, with therefore milder operating conditions

Narrower droplet size distribution

Controllability of droplet size

Cream synthesis plant for eight different chemicals and continuous use (top view)

Cream synthesis plant for eight different chemicals and continuous use (bottom view), not fully assembled

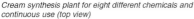

FIGURE 1.38 Microprocessing for cosmetics.

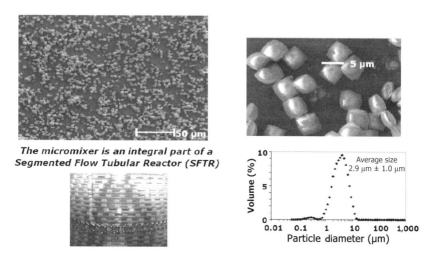

FIGURE 1.39   Segmented flow tubular reactor [29].

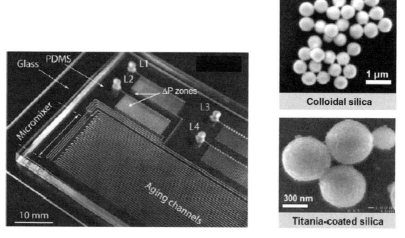

FIGURE 1.40   Microchannels for nanoparticles [30].

Velocys is developing a micromembrane emulsification technology (Figs. 1.36 and 1.37). IMM is using micromixers for cream synthesis (Fig. 1.38). In solid formulation, significantly narrower distributions of particle sizes can also be obtained. The efficiency of segmented flow microreactors is illustrated here: For copper oxalate in Fig. 1.39 and for, mono-disperse silica nanoparticles in Fig. 1.40.

1.7.2   Quantitative Assessment of Eco-Efficiency

Current studies focus on the much needed life-cycle assessment and positive impacts of microtechnology operations. A reaction operated in a microreactor process and

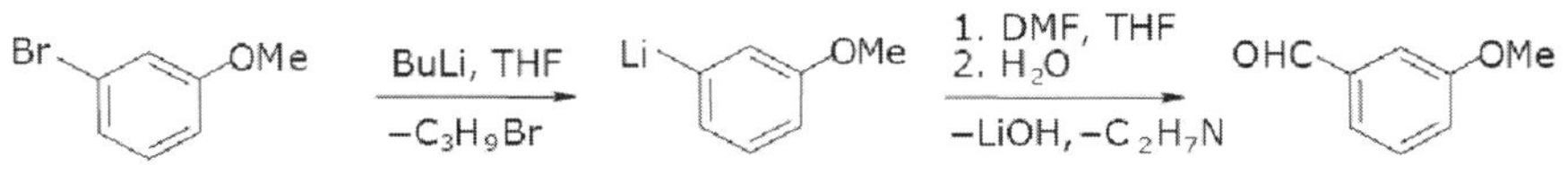

FIGURE 1.41 Synthesis of *m*-methoxy-benzaldehyde [31–32].

eco-efficient is shown in Fig. 1.41. The LCA clearly indicates significant ecological advantages achieved with the continuous synthesis in the microreactor (Cytos® Lab System) vs. the macro-scale discontinuous batch process (10-L double-wall reactor).

1.7.3 Market Segmentation of Microprocessing

Yole Development (http://www.yole.fr) see two different routes of development for microprocessing:

In Chemicals

New technology to develop and produce very high-quality molecules in fine chemicals

High-end differentiation of some chemical companies in a competitive environment, in which China continues its leading position in some market segments

Higher production yield and enhanced safety conditions

With MRT advancing it, the change from a batch to continuous flow process

In Pharmaceuticals (same drivers as for chemicals, but with some specificity)

New reaction conditions leading to new drugs

Increase in drug development pipeline profitability: More molecules will go through toxicological testing (Phases I to IV)

To conclude, we note these trends in the perspectives for microtechnologies:

- Technology validation is still to be assessed.
- The need for a better understanding of financial benefits continues to exist.

It is clear today that microtechnology is decisive for sustainable chemistry, as new routes are reconsidered. The approach is also emerging in the pharmaceutical industry, bringing benefit to the product development process. It also turns out to be useful for on-site applications (cosmetics, drugs, and testing). Whatever development route we consider, engineering methodologies and holistic approaches will be required.

BIBLIOGRAPHY AND OTHER SOURCES

1. J. Jenck, F. Agterberg, and M. Droesher, *Green Chem.*, 6 (2004), 544–556.
2. Image courtesy A. Stankiewicz, TU Delft, July 2003.

3. Dr. S. Deibel, *CHE Manager*, 2 (2006).

4. Dr. A. Belloni, *Process*, 13(4) (2006), 64.

5. C. Ramshaw, "Process Intensification and Green Chemistry," *Green Chem.*, 1 (1999), G15–G17.

6. A. I. Stankiewicz and J. A. Moulijn, "Process Intensification: Transforming Chemical Engineering," *Chem. Eng. Prog.*, 1 (2000), 22–34.

7. R. Jachuck, "Process Intensification for Responsive Processing," *Trans. IChemE*, 80, Part A (April 2002).

8. R. Bakker, *Reengineering the Chemical Processing Plant*, Marcel Dekker, New York, 2003.

9. R. Jashuck and J. F. Jenck, AIChE Process Development Symposium, Palm Springs, CA, June 12, 2006.

10. D. Hendershot, CEP, 2000.

11. Image courtesy A. Stankiewicz, TU Delft, July 2003.

12. Image courtesy A. Green, BHR Group, Sept. 2005.

13. Image courtesy S. Fleet, BRITEST, March 2005.

14. Image courtesy Yole Development, Lyon, France.

15. Photos in Chemical & Engineering News, Dec. 18, 2006, p. 38.

16. 71st Annual Meeting SCE, Tokyo, Japan, March 28–29, 2006. http://www.hqrd.hitachi.co.jp/global/news_pdf_e/merl060327nrde_microreactor.pdf.

17. http://www.siemens.com/siprocess.

18. Image courtesy A. Renken, EPF, Lausanne, Switzerland, Feb. 10, 2006.

19. http://www.fzk.de/idcplg?IdcService=FZK&node=1298&documentID_050873.

20. *Proc. IMRET*, 8, Atlanta, GA, April 2005.

21. T. Bayer, D. Pysall, and O. Wachsen, *Proc. IMRET*, 3, 2000.

22. http://www.imm-mainz.de/upload/dateien/PR%2020050405e.pdf?PHPSESSID=e8b7ef919907a581959f42cc890a8511.

23. *Chemie Ingenieur Technik*, 5 (May 2005), 77.

24. *Chemical & Engineering News* (May 2005), cover story.

25. http://www.cpc-net.com/cytosls.shtml.

26. P. Pennemann, V. Hessel, and H. Löwe, *Chem. Eng. S.*, 59 (2004), 4789–4794.

27. D. M. Roberge, L. Ducry, N. Breler, P. Cretton, and B. Zimmerman, *Chem. Eng. Tech.*, 28(3) (2005), 318–323.

28. U. Krtschil, V. Hessel, D. Kralisch, G. Kreisel, M. Küpper, and R. Schenk, "Cost Analysis of a Commercial Manufacturing Process of a Fine Chemical Using MicroProcess Engineering," *CHIMIA*, 60(9) (2006).

29. http://ltp.epfl.ch/page17388.html.

30. S. A. Khan, et al., *Langmuir*, 20 (2004), 8604–8611.

31. D. Kralisch and K. Kreisel, *Chem. Ing. Tech.*, 77(6) (2005), 62–69.

32. D. Kralisch and G. Kreisel, AIChE Spring Meeting 2006, Paper No. 23g.

PART II

MICROFLUIDIC METHODS

CHAPTER 2

MICROREACTORS CONSTRUCTED FROM METALLIC MATERIALS

FRANK N. HERBSTRITT

2.1 METALS AS MATERIALS OF CONSTRUCTION FOR MICROREACTORS

Aside from frequently scientific lab applications, where glass, semiconductor materials or plastics often play a domineering role (e.g., "Lab-on-a-Chip"), metal is probably the most important and most widely used category of materials of construction for components used in the field of microprocess engineering. This is not just attributable to designers' general familiarity with these materials. Unlike any other category of materials, metals—of course, also including a large variety of metallic alloys—combine a number of properties that are necessary in the construction of (micro) mechanical equipment. They are as follows:

Best All-round Workability   No other category of materials can be processed in as many ways as metals: Cutting techniques, electrical discharge machining, reducing and separating laser processes as well as form etching techniques allow precise shaping with great geometric latitude over a broad range down to the micrometer level. Stamping, casting, and forming processes make it possible to economically produce even complex building elements in large numbers. With galvanic, gas-phase, or vacuum-based deposition techniques, manifold functional coatings can be produced and, in combination with lithographic structuring techniques (LIGA), highly precise microstructures may be generated. Finally, a broad array of welding, soldering, and diffusion joining techniques make it possible to create high-strength inter-metallic bonds whose thermal and chemical stability often approach that of the base material.

High Mechanical Strength   In many cases, glass and ceramic materials outperform metals with regard to tensile and compression strength. However, particularly where

Microchemical Engineering in Practice. Edited by Thomas R. Dietrich
Copyright © 2009 John Wiley & Sons, Inc.

safety-relevant applications are concerned but also in terms of general use, the yield point or breaking elongation of a material is also of crucial importance for its applicability. Metals and especially metal alloys are highly ductile and elastic. They are therefore as equally well suited for the construction of heavy-duty pressure vessels as for the construction of filigree-type microstructure elements that must on a regular basis withstand robust handling (e.g., during cleaning operations) in laboratories, pilot plants, or production units.

High Thermal Stability   With regard to stability, most metallic materials relevant for process engineering applications (e.g., stainless steel, nickel-based alloys, or titanium) can without major restrictions be used under conditions exposing them to cryogenic temperatures up to about 400 to 700°C. Heat-resistant and high-temperature steels are available for applications that expose them to temperatures up to approx. 800 to 1,100°C, as far as corrosion stress is strictly limited. In those instances where more severe corrosion attack must be expected, refractory metals (particularly zirconium, tantalum, tungsten, and molybdenum) and their alloys cover a temperature range that may—with certain reservations regarding stability—clearly exceed 1,500°C. Unlike many brittle-rigid materials, such as most types of glass and many ceramics, nearly all metallic materials are highly resistant to temperature change. However, it must be considered, particularly when the sizing of safety-relevant design elements is concerned, that many metals are less ductile when exposed to low temperatures and have diminishing yield points as well as increasing corrosion sensitivity when exposed to high temperatures.

Good Chemical Resistance   Aside from a few precious metals, whose prices alone preclude in most cases their use in the manufacture of microstructure elements, none of the metallic materials are as chemical-resistant as fluorine polymers, most types of glass, and many ceramics. However, metallic materials of construction as a whole cover virtually the complete range of relevant applications in the chemical industry. Indeed, careful material selection is particularly important in this area.

2.2   MATERIAL SELECTION

With regard to the requirements a material of construction must meet, microprocess engineering is at first glance not basically different from classic process engineering. Function-relevant criteria, such as corrosion resistance, mechanical strength, and the temperature range in which a material can be used, must be considered to the same extent as the more economically important criteria of price, availability, and workability. Depending on the intended application, additional criteria, for example, heat conductivity, electrical or magnetic properties, wetability or bio-compatibility, may need to be examined as well. No single material fulfills all these criteria to a completely ideal extent (Fig. 2.1). For example, corrosion resistance to a certain medium may only be attainable with a particularly expensive or poorly workable material or certain component geometries may only be achievable by certain manufacturing methods to which, in turn, only a few materials lend themselves. But during the course of the history of chemical process technology, a broad and versatile assortment

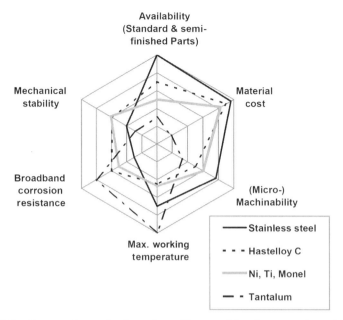

FIGURE 2.1   Comparative evaluation of metallic materials of construction with regard to criteria important for the construction of microreactors.

of metallic materials has evolved, with each of them covering a specific range of applications and many of them preferably used also for the construction of micro-processing elements.

Selecting one material from this assortment will therefore in most cases require a compromise—in the instance of the construction of conventional equipment as well as for the construction of microreactors. However, individual criteria will sometimes be evaluated differently for the construction of a microreactor than, for example, the construction of an agitated tank with a volume of several cubic meters. The price and strength of a material will in most cases play a lesser role if it is to be used to build a microreactor as compared to the construction of an agitated tank, because already in principle considerably less material is needed for the microreactor and any additional material necessary to attain some increased pressure resistance will only be of minor further impact on the device's cost. On the other hand, considerably stricter standards regarding corrosion resistance and workability of a material must in many cases be applied in the field of microprocess engineering than in the field of classical unit construction.

Although the material erosion of several 0.01 mm per year is acceptable on a conventional pressure vessel or pipeline for most applications, this rate of erosion would make many microstructure elements unusable within a few weeks or months. Moreover, in microreactor design, the use of corrosion-protective coatings, as it is practiced in classical unit construction, is normally not possible because, to assure durable protection, the thickness of these coatings often reaches the order of magnitude of the characteristic dimensions of microstructures. Even more so than in classical unit design, the corrosion resistance of a material in microprocess engineering is therefore in many cases the highest-ranking KO criterion for its use. The other criteria listed may have to be subordinated. This applies especially to the price of a material, albeit within certain limits. The extremely widely resistant precious metals, for example, gold or platinum and their alloys (which will resist considerably higher mechanical stress), are normally too expensive even for the construction of microstructure components.

As compared to these metals, which are actually only attacked by very few substances, the corrosion resistance of practically all technically and economically relevant materials for equipment in the area of process technology is based on the formation of a dense, stable corrosion layer (passive layer, generally oxidic), which protects the material under it from further chemical destruction. Therefore, corrosive attack on these materials is in most cases accompanied by damage to this passive layer, which may, for example, be caused by oxidizing or reducing effects, attack by acids or bases, or the formation of complex compounds that are soluble in the attacking medium. Metallic materials of construction that are protected by passivation therefore always have a more or less limited range of resistance and are sometimes preferably attacked by a whole category of media (pH value, oxidation potential, presence of certain ions).

Table 2.1 gives an overview of a number of metallic materials of construction that are of particular interest for microreactor design [1]. They are briefly introduced in the following paragraphs.

Austenitic stainless steels (e.g., AISI 316Ti) cover a solid basic spectrum of process engineering applications, particularly in the foodstuffs and pharmaceutical fields as well as in chemical applications with redox-neutral to slightly oxidizing and slightly acidic to basic organic and aqueous solutions. They are not resistant to many concentrated acids, reducing media, and halides, particularly chlorides, which cause increasing hole corrosion [2]. These steels are preferred for the manufacture of (especially modular) microreactor systems for laboratories and experimental plants with broader application spectra because of their low prices and—compared to other corrosion-resistant materials—good workability by all precision engineering and microprocess engineering methods (except LIGA) as well as by a large number of welding techniques.

In addition to nickel, *Hastelloy® Alloy C-276* (registered trademark of Haynes International, Inc.) contains chrome and molybdenum and small amounts of tungsten and iron. It belongs to the broad category of nickel-based materials with higher—in some cases, mutually complementary—resistance to a wider spectrum of media, particularly chloride-containing and slightly reducing substances, in comparison to stainless steel. Alloy C-276 covers practically the entire resistance spectrum of A4 stainless steels and significantly expands it to include halogens (except fluorine and, conditionally, chlorine), halogenides, some mineral acids, and reducing aqueous media. Like most nickel-based materials of construction, Alloy C-276 has a strong tendency to work-harden (see hardness values in Table 2.1). Although, on the one hand, this benefits robustness, particularly of thin-walled construction components,

TABLE 2.1 Comparison of the Physical Properties of Metals of Particular Importance for Microreactor Construction

Material	S/S Steel A4	Hastelloy C-276	Monel Alloy 400	Nickel 200	Titanium	Tantalum
Material no. (DIN)	e.g., 1.4571	2.4819	2.4360	2.4066	3.7025–3.7065	—
Material no. (U.S.)	AISI 316 Ti UNS S31635	UNS N10276	UNS N04400	UNS N02200	UNS R50250–R56400	UNS R05200–R05400
Density (kg/m^3)	7.9	8.9	8.8	8.9	4.5	16.6
Melting point/range	1,420–1,450	1,325–1,370	1,300–1,350	1,435–1,445	~1,670	3,000
Heat conductivity at 20°C (W/m^2 K)	15	10.6	26	70.5	20.1–22.6	54.4
Mean heat exchange coefficient, 20–200°C (10^{-6} K^{-1})	17.5	12.1	15.5	13.9	8.7–9.4	6.5
Yield limit at 20°C $R_{p0,2}$ (N/mm^2)	≥200	280–310	180–650	150–275	≥180–≥390	140
Yield limit at 200°C $R_{p0,2}$ (N/mm^2)	165	225–240	≥135	≥65	≥120	80
Tensile strength at 20°C, R_m (N/mm^2)	500–700	700–730	450–760	340–450	290–740	225–280
Young's modulus at 20°C E (kN/mm^2)	200	208	182	205	105	172
Hardness HV/HB, annealed	150–190	80–200	<150	80–110	60	90
Hardness HV/HB, cold-molded	250–300	≤240	210	165–230	120–200	200
Recommended temperature range for application	To 700°C	To 500°C	To 550°C	To 600°C	To 450°C	To ~900–1,100°C

it makes processing, especially by machining, markedly more difficult. On the other hand, microstructuring by electrical discharge machining or laser cutting of this material is possible with practically equal precision as stainless steel, albeit at somewhat greater expense. Aside from the price of the material itself, the processing costs usually make these components about 1.5 to 3 times more expensive than those made of stainless steel. Microstructuring by form etching, however, is not possible, at least not by methods that are generally commercially available.

Monel® *Alloy 400* (registered trademark of Special Metals Corp.) has a somewhat broader chemical resistance than pure nickel, but is not by far as chemical-resistant as Hastelloy C-276. However, compared to nickel, it is somewhat less expensive, can be used throughout a somewhat larger temperature range, and, due to work-hardening, can reach clearly higher strength values (especially 0.2% yield point; see Table 2.1). Compared to stainless steels, it is considerably more resistant to chloride-containing media and has better heat conductivity. Among other applications, this makes the material interesting for the construction of heat exchangers. Its workability is similar to that of Hastelloy C, with the exception that microstructuring by form etching is possible with Monel.

Nickel, commercially available at 99.2% purity as Alloy 200 or, with reduced carbon content, as Alloy 201, has a still narrower overall chemical resistance spectrum than stainless steel. However, unlike stainless steel, it is resistant to chlorine and hydrogen chloride (including wet hydrogen chloride; Alloy 201 is resistant to dry gases at temperatures up to 550°C), chlorides, fluorides, and etching alkalis (in concentrated solutions up to the melting point). Moreover, the heat conductivity of nickel as a pure metal is higher than that of the alloys at still relatively good strength values. The workability of nickel is similar to that of Monel. With regard to the LIGA process, nickel and some of its alloys are especially important in the field of microprocess engineering because of their processability by galvanic deposition techniques.

Titanium and its alloys have a similarly broad chemical resistance spectrum as Hastelloy C-276. However, the resistance properties of the two materials are in certain ways complementary. Titanium materials, in particular, are resistant to nitric acid and various soda solutions at practically all concentrations that would lead to corrosion attack on Hastelloy C, whereas the nickel-based alloy is the superior material when it comes to resistance to halogens. The mechanical strength of titanium materials is to a considerably higher degree dependent on the concentration of commonly present contaminants (e.g., Fe, O, N, C, or H) in the metal than that of the alloyed materials discussed here. This is the reason for differentiating among four grades even of the unalloyed material. In addition, numerous titanium alloys are technically relevant, of which especially those with an addition of a few tenths of a percent of palladium are particularly important for chemical applications because of their improved corrosion resistance. Although problematic reshaping properties as well as the elaborate and expensive welding requirements of titanium (welding is possible only in an inert atmosphere) impose narrow limits on its use in classical equipment construction, its use in microprocess engineering has so far been more limited by its relatively high price and poor availability as well as the difficult machining characteristics of the material.

2.3 MICRO- AND PRECISION ENGINEERING METHODS

Although the assortment of materials used in the construction of metallic micro-reactors is to a large extent the same as that used in classic unit construction—possibly with more emphasis on higher quality and more corrosion-resistant materials—due to the much smaller dimensions combined with the correspondingly higher precision requirements, the manufacturing processes used in the two fields will often be significantly different. Whereas larger-volume equipment is often constructed of cut, formed, and joined (sheet metal) parts, the manufacture of components for microprocessing equipment is in the majority of cases based on techniques of volume removal or the use of foil material of in most cases very precisely defined thicknesses. Machining, electrical discharge erosion, laser ablation, and laser cutting as well as form etching are among the most frequently used manufacturing methods in this area.

2.3.1 Machining Methods

A large portion of the manufacture of housing components for microreactors is classic precision engineering (Fig. 2.2). However, the production of a variety of microstructure components, particularly those made of metal, is also at least partially based on machining methods. With the aid of modern CNC machines, form tolerances and positional tolerances in a range from much less than 1 mm to many one-hundredths of a millimeter and the smallest structure dimensions, down to around 0.05 mm, can easily be attained. Special machines and tools can be used to reach yet smaller dimensions at higher precision. Machining techniques with defined cutting edges, such as

FIGURE 2.2 The housing components of many microreactor building elements such as this mixer are produced by machining. The microstructure components in the foreground are produced by laser cutting or LIGA technique (see Figs. 2.6 and 2.8) (courtesy of Ehrfeld Mikrotechnik BTS).

milling, turning, and drilling, are preferably used to shape the parts, whereas methods using undefined edges, such as grinding/sanding, lapping, and polishing, are normally used for surface treatments, serving to impart defined roughness or smoothness properties to the component surfaces.

Basically, nearly all metallic materials of construction can be machined. However, different strength and heat conductivity properties as well as the more or less strong tendency of many metals to work-harden or stick to or bond with tools make it necessary to carefully match processing parameters, construction materials, and tool geometries to the processing task at hand. This frequently results in large price differences if different materials are used to produce the same geometry. In particular, fine structures (e.g., smaller screw threads), which can be produced without problems on some materials, cannot be produced at all by machining on others. However, in general, the limits of the basically great geometric liberties allowed by machining techniques are on the one hand defined by the shape and size of the particular tool, which imposes a lower limit on lateral expansion and inside radii of grooves and bores. On the other hand, they are defined by the relatively high processing forces that impose upper limits on the aspect ratio (height to breadth) of free-standing structural elements and require stable clamping of the parts during processing.

It is also important to remember that machining is usually accompanied by high thermal input that may liberate mechanical tensions initially present in the material and could result in distortion especially where filigree-type geometries are concerned. Regarding their cost-effectiveness, machining techniques are well suited for relatively small quantities of more complex components with corresponding precision and geometry requirements as well as for medium quantities of components for which automated manufacture is rational (e.g., CNC-turned or milled parts requiring few mounting changes). However, structuring larger surfaces with filigree-type channels, as they are, for example, required for microplate heat exchangers or comparable reactor concepts, by machining is very difficult and expensive and is only justified if the requirements regarding geometry, precision, or surface quality cannot be met by less expensive methods (e.g., form etching) [3, 4].

Electrical discharge erosion, another precision engineering tool, is—like machining—considered a method of reduction. With this technique, the material removal results from electrical spark discharges between the workpiece and a work electrode in a liquid electrolyte that also serves as a cooling and flushing agent. This process is basically usable for all materials having some minimum electrical conductivity (from about 0.1 S/cm), among them of course particularly metals. Work voltages from several volts up to a few hundred volts, gap widths between electrode and workpiece surface in the micrometer range, and depths of reduction per discharge in the one-tenth of a micrometer area allow correspondingly precise processing [5, 6].

Depending on shape and guidance of the work electrode, one differentiates between wire erosion and die-sinking erosion. In wire erosion, a fine wire is guided through the workpiece, similar to the blade of a band saw or jigsaw, permitting multidirectional cutting. Saw kerf widths down to 30 to 50 μm are possible across depths that may reach the kerf width up to 300-fold (Fig. 2.3). Although producing closed internal contours necessitates making a starting hole and threading the cutting wire through the workpiece, wire erosion has particular advantages—for

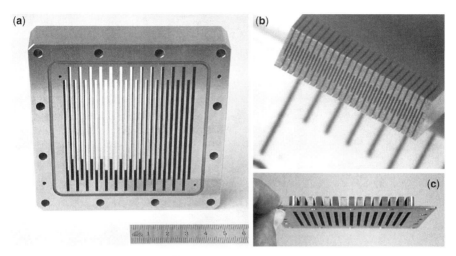

FIGURE 2.3 Due to the low effect of force on the workpiece, wire erosion allows the production of very thin-walled structures with high aspect ratios (a and c). Use of the finest cutting wires makes defined kerf widths of 50 to 70 μm possible (b) (courtesy of Ehrfeld Mikrotechnik BTS).

example, compared to laser cutting—where large aspect ratios (e.g., simultaneous cutting of stacked parts) or cutting of burr-free edges not requiring secondary processing are desired.

With die-sinking erosion, a shaped work electrode is slowly sunk into the workpiece, causing successive material erosion and leaving a negative imprint of its profile in the workpiece. Because skillful adjustment of the process parameters makes it possible to concentrate the material erosion mainly on the workpiece ($\Delta_{V\_Electrode}/\Delta_{V\_Workpiece}$: several $0.01-1\%$), the work electrode can in most cases be reused several times. The comparatively cost-intensive erosion methods are of particular interest in cases where machining methods meet their limitations, for example, generating bores with noncircular cross sections, cavities with very narrow inside radii (Fig. 2.4), long, deep grooves, apertures with sidewalls of small material thickness, or fine internal structures (e.g., screw threads) in difficult-to-machine metals.

2.3.2 Laser-Based Methods

Lasers are practically predestined for microprocessing of materials thanks to their ability to introduce very high-energy densities precisely positioned into very small volumes of material. In the production of microstructures for microprocessing applications, laser cutting and laser boring are particularly important. Pulsed Nd:YAG lasers are employed for most of these applications. Form erosion techniques (ablation), which allow structural dimensioning beyond the lower limits of machining and electrical discharge erosion methods, are, however, only used in special cases of microprocessing because they are relatively expensive. Whereas form erosion and most boring techniques are essentially based on the targeted, pulselike evaporation

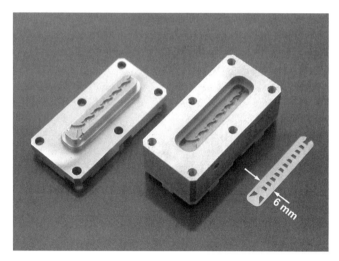

FIGURE 2.4 The faceted mixing channel in this static micromixer was produced by die sink erosion. This technique allows acute-angled geometries with angle radii of less than 10 μm to be generated (courtesy of Ehrfeld Mikrotechnik BTS).

of locally very exactly defined amounts of material, laser cutting normally involves only a melting process, possibly with partial oxidation of the molten material, and simultaneous removal of the melt by a rapid gas stream (Fig. 2.5). With ablation, targeted movement of the laser focus on the surface of the workpiece or mask projection techniques allow relieflike structures to be generated, while laser cutting always generates complete breaks through the material—in most cases, foils of some 0.01 to 10 mm in thickness. With both techniques, the generation of structures to dimensions of about the size of the laser focus is possible, depending on type of material and cutting depth, starting at approximately 10 μm, at aspect ratios of about 5 to 15 (for bore holes even higher). The depth resolution of form ablation techniques corresponds

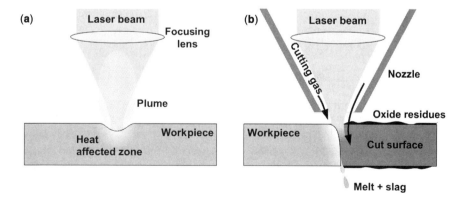

FIGURE 2.5 Schematic description of laser ablation (a) and laser cutting (b).

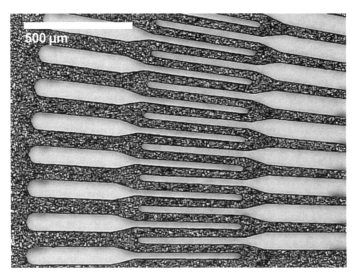

FIGURE 2.6   This microstructure was produced by laser cutting of a 0.3-mm-thick stainless steel foil (see slices in the foreground of Fig. 2.2). The overlapping ends of the club-shaped channels are about 25 μm wide (courtesy of Ehrfeld Mikrotechnik BTS).

approximately to the ablation depth per pulse, which is normally around a few tenths of a micrometer.

Laser boring and cutting methods are unbeatable particularly when a very large number of openings, possibly of complex shape and correspondingly small dimensions, and aspect ratios >1 must be generated in metallic foils (Fig. 2.6). In any case, precision and speed are materials-related. For example, it may be possible to form some structures of stainless steel quite easily, whereas it might be more difficult to make them of Hastelloy and almost impossible if tantalum is used [6–8].

2.3.3   Photolithography-Based Processes

In photolithography, location-selective light exposure and subsequent development of a light-sensitive polymer material (resist) are used to generate a mask on the surface of the workpiece. This mask is then used as a template for one or more additional structuring steps. Of course, microprocess engineering also makes use of the essential advantages of photolithography: its potential to achieve very high structural precision, even on larger surfaces, at relatively low costs, as well as the characteristics that make it useful as a replicative production method.

However, with this method, it is normally necessary to transfer the structure of the resist mask onto a sufficiently stable, heat- and corrosion-resistant material. This can basically be done by the addition or subtraction of material. To build up sufficiently thick metal layers of at least several one-hundredth up to several tenths of a millimeter thickness, galvanic deposition (e.g., the LIGA process, from the German

Lithographie-Galvanoformung-Abformung, meaning "lithography-electroforming-casting") has proven useful, while the corresponding subtractive structural transfer is in most cases attained by erosive wet chemical or electrochemical processes. The LIGA process (Fig. 2.7) is practically unsurpassed with regard to precision and achievable aspect ratios. But, if structure heights greater than a tenth of a millimeter are to be generated, the process requires relatively difficult and costly masking and exposure techniques (synchrotron radiation). Also, its use is by its nature limited to metals or alloys that can be electrodeposited (particularly nickel, NiFe, NiCo, or NiW as well as gold, silver or copper; see Fig. 2.8).

The wet chemical (almost exclusively isotropic) etching processes available for many metallic materials (except, e.g., Hastelloy C) are limited with regard to aspect ratio (in most cases, <0.6), depth precision (approximately $\pm 10-20\%$), and surface roughness of the generated structures. Correspondingly, form etching techniques are preferably used in replicative structuring of flat, possibly laterally expanded channel systems of shallow depths (a few one-hundredths to a few tenths of millimeters) and moderate requirements regarding structure precision, for example, the channel plates of plate heat exchangers (Fig. 2.9). On the other hand, LIGA technology is used particularly when very small structures (upwards of approximately 5 μm), high dimensional accuracy (measuring tolerances of approximately ± 0.2 to 1 μm, depending on structural height), and large aspect ratios (up to 50 and in special

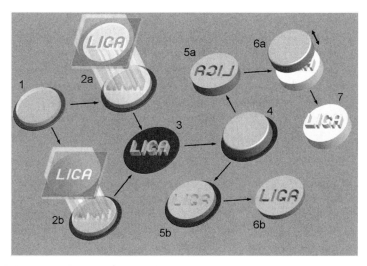

FIGURE 2.7 Process steps of the LIGA process: A wafer is coated with a resist layer (1). After hardening, the resist is exposed through a mask with high-energy radiation (2a and 2b). After development, the unexposed ("positive" resist, 2a) or the exposed ("negative" resist, 2b) area of the resist rest on the wafer (3). By electroforming a metallic mold is grown around the resist stencil (4). After removal of the resist and, if necessary, some further processing steps, the metal part can be used either as a mold for a casting process (5a–7) or as a functional metallic micropart itself (5b–6b).

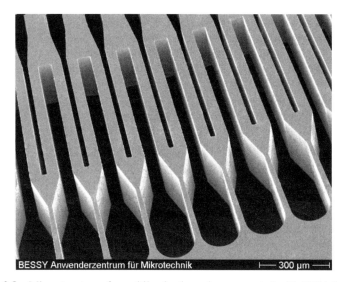

FIGURE 2.8 Microstructure of a multilamination mixer generated with LIGA (see Figs. 2.2 and 2.6). The NiFe part has been generated by the process 1–6b sketched in Fig. 2.7 (courtesy of BESSY-AZM).

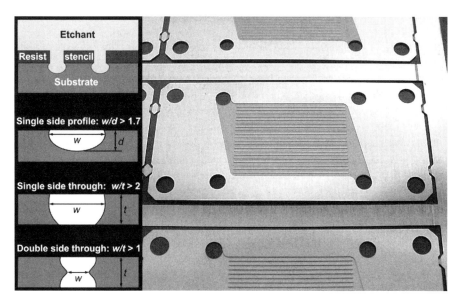

FIGURE 2.9 A micro heat exchanger channel plate produced by form etching of a 0.15-mm-thick stainless steel foil. The channels are 80 μm deep and 500 μm wide. On the left, the etching process and some channel profiles, which can be obtained by this method, are sketched (courtesy of Ehrfeld Mikrotechnik BTS).

cases even above 100) under moderate corrosion load are important. Micromixers for selected areas of application are an example [6, 7].

A series of replicative deforming and forming techniques—among them, for example, stamping, investment casting, and powder injection molding—have the potential to produce microstructure components for microreactors as far as material spectra, geometric possibilities, and achievable precision are concerned. However, high tool costs and comparatively expensive process optimization make these processes interesting only if a few hundred up to several tens of thousands of parts are to be produced. This has so far only been achieved for a few individual components of microprocess building elements. These processes are therefore not widely used at this time, but they will certainly be increasingly employed in the years to come.

2.4   JOINING AND MOUNTING TECHNIQUES

When selecting a joining process for the assembly of the components of a micro-reactor, it is first necessary to decide whether the parts should be joined nonperma-nently or permanently. A compromise will have to be made in any situation: If a permanent connection, for example, by welding, soldering, or diffusion bonding, is chosen in order to save the costs of sometimes expensive sealing and connection elements, access to the microstructures or the component for cleaning or inspection purposes will normally not be possible later on. Comparatively, nonpermanent join-ing, for example, with screws, will increase the design costs and frequently also the manufacturing costs. Every separable joint in contact with liquids will require a sealing element, but maintaining these components will usually be much easier. For example, microreactors designed for flexible use in laboratories, experimental plants, or for other special applications where certain requirements regarding cleaning (e.g., GMP requirements) have to be met, designs that can be completely dismantled are normally the better choice.

The assembly tolerances necessary in such designs require the use of secondary frictional connection gaskets—preferably gaskets with O-rings made of elastomers—that, however, to a certain extent limits their application possibilities. One of the factors limiting the use of gasketed/sealed connections is the space requirement. For example, gasketing of stacked very thin microstructure foils, as they are used in very compact micro heat exchangers or plate reactors, cannot be handled with this type of sealing concept. Another factor is that the temperature spectrum for the use of relevant elastomer materials in chemical applications is limited to a range of about -50°C to approximately 250°C or, in exceptional cases, to just above 300°C. PTFE gaskets can be used at lower temperatures and only metallic sealing elements are usable at higher temperatures. Both of these types of gaskets must be changed out each time the component is disassembled, and especially metallic sealing elements also require much higher restraining forces than elastomer seals.

If permanent component bonding is chosen, various welding, soldering, and diffusion joining methods are available, depending on what material a component is made of. Welded bonds are normally the strongest and most corrosion-resistant.

When expertly performed and accompanied by careful preparation and after-treatment, the welds often match the properties of the base material to a large degree. Basically, welded connections conforming to standards are recommended for bonding tasks on reactors that must meet special safety and strength requirements (e.g., EU Pressure Equipment Directive or special plant standards set by the user). It is important as early as the design stage to make the immediate areas between components properly suitable for welding and to also consider the possible effects of the welding process on the geometry of the parts, for example, through partial melting or distortion. For welding of filigree-type or thin-walled parts, Laser or electron beam welding processes are available. With these processes, the heat input can be concentrated on a smaller material volume than is normally the case with gas or arc welding processes.

Soldered bonds are practical, among other applications, when the mechanical and chemical stability of the bond is less important, but the heat input must be limited or higher position tolerances of the parts to be bonded are required than can be achieved by welding. One interesting example is oven soldering. With this technique, the parts to be soldered are fixed in the intended position with a soldering medium applied to the joint areas (e.g., as a paste, powder, or foil), and then the component is placed in an oven, if necessary under a gas blanket or vacuum, where the soldering medium is melted. After the soldering medium has distributed over the areas to be joined, the component is cooled down again. Unlike welding, soldering also lends itself to

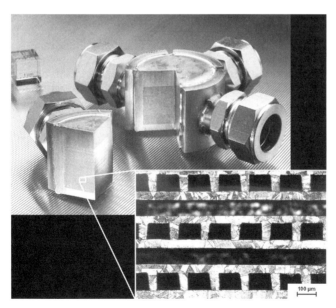

FIGURE 2.10 Cross-flow microplate heat exchanger, made by diffusion welding of micro-machined channel stainless steel foils. Devices like this withstand pressures of several hundred bars and exhibit heat exchange coefficients of some $10,000\,\mathrm{W/Km^2}$ (courtesy of Forschungszentrum Karlsruhe).

joining different materials with each other. However, each soldering medium only works for certain groups of materials.

In diffusion joining, the parts to be joined must have exactly form-fitting—in most cases, flat—joining surfaces. They are pressed directly to each other at high temperatures under high pressure (10–100 bars). Here too, differentiation is made between diffusion welding and diffusion soldering, depending on whether or not an auxiliary medium (solder, with a melting point lower than that of the basic material) is used. In diffusion soldering, the soldering layer—in most cases, only a few micrometers thick—is normally applied by electrodeposition, whereby multiple layers can also be built up. The joining temperature in diffusion soldering is somewhat higher than the melting temperature of the solder, whereas in diffusion welding is just a little lower (approx. 90%) than the melting temperature of the base material. Both methods, by their nature, require a high degree of planeness as well as high form-locking and surface quality and cleanliness of the areas to be joined. Of course, these requirements are highest in diffusion welding because there is no soldering medium to even out any flaws. Diffusion-joining methods require relatively extensive equipment and very good knowledge of the process. They are preferably used in the production of very compact microplate heat exchangers and reactors of similar design, where a large number of microstructure foils are welded into a compact block (Fig. 2.10). Microreactors produced this way are very highly pressure- and temperature-resistant and allow the highest surface-to-volume ratios of the channel systems built into them.

BIBLIOGRAPHY AND OTHER SOURCES

1. Data compiled from Material Data Sheets of UGINE & ALZ, Krupp Edelstahlprofile GmbH, ThyssenKrupp VDM GmbH, Deutsche Titan GmbH, Jäckel & Co. Edelstahl–Metalltechnik GmbH, Bibus Metals AG, LN Industries, Goodfellow GmbH, and Rembar Co., Inc.

2. Siebert, O. W., and Stoecker, J. G. (1997). Materials of construction. In: R. H. Perry, D. W. Green, (eds.), *Perry's Chemical Engineer's Handbook*, 7th ed., Sect. 28, pp. 1–64. New York: McGraw-Hill.

3. König, W., and Klocke, F. (1999). *Fertigungsverfahren 1—Drehen, Fräsen, Bohren*, 6. Aufl. Berlin: Springer-Verlag.

4. König, W., and Klocke, F. (1996). *Fertigungsverfahren 2—Schleifen, Honen, Läppen*, 3. Aufl. Düsseldorf, Germany: VDI-Verlag.

5. Dürr, H., Pilz, R., Herrbach, S., and Seliga, E. (2003). Trennen. In B. Awiszus, J. Bast, H. Dürr, and K.-J. Matthes (eds.), *Grundlagen der Fertigungstechnik*, pp. 123–210. München, Germany: Fachbuchverlag Leipzig in Carl Hanser Verlag.

6. Brück, R., Rizvi, N., and Schmidt, A. (2001). *Angewandte Mikrotechnik*. München, Germany: Carl Hanser Verlag.

7. Ehrfeld, W. (ed.) (2002). *Handbuch Mikrotechnik*. München, Germany: Carl Hanser Verlag.

8. Bäuerle, D. (2000). *Laser Processing and Chemistry*, 3rd ed. Berlin: Springer-Verlag.

CHAPTER 3

MICROREACTORS CONSTRUCTED FROM INSULATING MATERIALS AND SEMICONDUCTORS

NORBERT SCHWESINGER and ANDREAS FREITAG

3.1 SILICON MICROREACTORS

3.1.1 Introduction

Silicon (Si) is one of the best investigated materials. As a doped semiconductor, single crystalline Si is the basic material in microelectronic technology. Along with its excellent electrical properties, it also has outstanding mechanical and thermal properties. Although microelectronic circuits occupy only areas near the surface of the silicon, the use of other properties is connected with complete volume usage. This restriction had to be overcome by suitable three-dimensional structuring technologies. It was only a question of time before mechanical products of crystalline silicon would be launched in the market. First, pressure sensors with thin Si membranes as mechanical elements conquered the market very successfully. This was the starting point of silicon micromachining or micro-electro-mechanical-systems (MEMS). Several other successful developments, such as acceleration sensors, Gyros, thermo-piles, ink jet print heads, etc., have followed. Consequently, the development of microreactors focused on Si as a basic material. Several key advantages of Si make it the favored material in chemical process technology:

1. Si can be structured by means of well-known and approved technologies.
2. Si can be covered with different materials using thin film technologies.
3. Si is chemically stable in most solutions.
4. Si is thermally stable up to 1,200°C.
5. Si has excellent thermal conductivity.

Microchemical Engineering in Practice. Edited by Thomas R. Dietrich
Copyright © 2009 John Wiley & Sons, Inc.

6. Si allows batch processing.
7. Si allows the integration of electronic circuits.
8. Si allows bonding techniques with Si or glass without any additional glue materials.

This section will give a short description of the properties of Si, typical structuring technologies, useful deposition technologies, information on the design of Si microreactors, and typical assembly procedures.

3.1.2 Properties of Silicon

Silicon crystallizes in a diamond lattice. It exhibits brittle behavior but no plastic deformations and no mechanical hysteresis. It can be deformed only elastically at room temperature [1]. The yield strength of silicon is comparable with that of stainless steel. Not only these mechanical properties make Si of interest for microreactors. Some of its properties are listed in Table 3.1. Obviously, silicon has excellent chemical properties as well and is therefore a favored material for microreactors. Silicon is stable in pure acids but is etched in some mixtures. Halogens like fluorine and chlorine attack the material. In the same way, silicon is etched in several alkaline solutions, especially at temperatures above room temperature.

3.1.3 Basic Conditions for the Microstructuring of Silicon

The microstructuring of silicon generally requires masking of the Si wafers. These masks protect the areas that should not be etched from contact with the etchants. Furthermore, they contain microwindows, which allow the etchants to dissolve the basic material and to produce microstructures. Masks can be made of materials that resist chemical dissolving of the etchant for as long as etching takes place. Silicon dioxide SiO_2 and silicon nitride Si_3N_4 are favored masking materials. Both materials

TABLE 3.1 Selected Properties of Silicon

Property		Value
Density		$2.33 \, g/cm^3$
Melting point		$1,417°C$
Yield strength		$7 \, Gpa$
Thermal conductivity		$1.46 \, W/cm \, K$
	Solution	
	Alcohols	Very good
	Ketones	Very good
	Fats/oils	Very good
Chemical Stability	Aromatic hydrocarbons	Very good
	Halogens	Poor . . . good
	Acids	Very good
	Alkaline solutions	Poor . . . good

can be deposited on the surface of Si wafers by thin film technologies. Silicon wafers are available on the market in three different surface orientations: <001>-Si, <111>-Si, and <110>-Si. The numbers characterize the crystalline plane that forms the surface of the wafer. <111>-oriented Si wafers etch in alkaline solutions with a very slow etch rate. Better results are expected with <001>- and <110>-oriented Si wafers. <110>-Si wafers are comparatively expensive as they are not used in microelectronics. The favored basic or so-called substrate material is therefore <001>-Si. Special requirements for doping and electrical resistivity do not exist.

3.1.4 Structuring of Silicon

One can distinguish two different methods of structuring silicon. Wet etching processes are historically older. Modern processes are based on dry etching techniques with high accuracy, precision, and reproducibility.

Wet Etching of Silicon   Wet etching of Si is a basic technology. Two different methods exist that differ in the chemicals used and in the resulting structures. Silicon can be etched in acidic solutions and in alkaline solutions.

Wet Etching in Acidic Solutions   Wet etchants such as mixtures of nitric acid HNO_3 (70%), hydrofluoric acid HF (49%), and water or acidic acid CH_3COOH, respectively, can etch silicon at room temperature. The ratio of the different acids in the mixture can vary within a wide range. Water is added up to about 20% in weight. Silicon can be etched in all directions with the same reaction speed. Etch rates of 0.2 μm/min to 1,000 μm/min can be observed, depending on the concentration of the two acids. Very smooth surfaces can be achieved at low etch rates with a high content of HNO_3. The increasing content of HF in connection with a reduction in HNO_3 leads to higher etch rates, sharp edges, and rough surfaces. The etching process is exothermical and autocatalytical, too. Thus, controlling the process gets more complicated the higher the etch rates are. Three-dimensional structures can be produced when the silicon is covered with a Si_3N_4 masking layer. Si_3N_4 is suitable only as a masking material. SiO_2 disappears after relatively short process times [2]. A typical result of this process is shown in Fig. 3.1. All wet etched structures are shaped similarly as depicted in Fig. 3.1. Deep etch rates and etch rates underneath the masking layer are comparable. The influence of the crystalline planes' orientation on the shape is not noticeable. The process leads to great tolerances and is only useful after anisotropic wet etching to round sharp edges.

Wet Etching in Alkaline Solutions   Silicon can be etched in several organic and inorganic solutions. KOH (25 to 40%) is the most frequently used etching solution. Instead of KOH, one can also use CsOH, NaOH, and NH_4OH as inorganic etchants or ethylendiamine, hydrazine, and TMAH as organic solutions. Etching is performed at temperatures between 60 and 110°C. The etch behavior differs completely from that in acidic solutions. Instead of rounded shapes as seen in the previous chapter, one can achieve very sharp shapes and edges as well. The etching is strongly

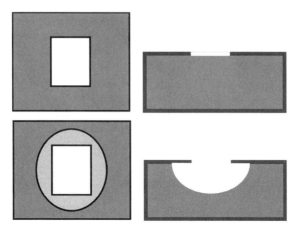

FIGURE 3.1   Wet chemical etching of Si in acidic solutions. Left: top view, right: cross section. 1. Row: window in masking layer; 2. Row: shape after any etching time.

dominated by the crystalline planes of the single crystal. Some planes (100 and 110) etch with high etch rates; other planes (111) will not be influenced by the solution. Thus, it is necessary to know the positions of the planes in a single-crystal Si wafer. Considering a cube, one can easily distinguish the different planes as shown in Fig. 3.2. Single crystalline <001>-oriented Si wafers possess a phase, called flat, that is oriented in the <110>-direction. Therefore, all (111)-planes intersect the (001)-surface parallel or perpendicular to the position of the flat. As the angle of intersection, one can find a value of 54.74°. In that case, all free-standing (001)-planes will be etched as a kind of deep etching. If the surface is covered with a masking material, etching is possible only inside the windows of the mask. Etching stops at all (111)-planes that intersect, as the last plane, the window edge. Deep etching continues as long as (001)-planes remain. Consequently, one achieves rectangularly shaped deep etched structures with tilted sidewalls. The angle of tilting is 54.74°. When etching is continued until the last (100)-plane has disappeared, the tilted sidewalls intersect each other and the etching stops [3].

As an illustration of this behavior, consider Fig. 3.3. Only the edge on the bottom of the opening is parallel to the flat direction. All other edges are tilted against this direction. Thus, the masking layer is underetched in several positions. In practical applications, it is convenient to design the openings in such a way that all edges

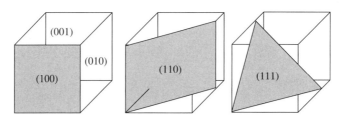

FIGURE 3.2   Position of planes in single crystalline solids.

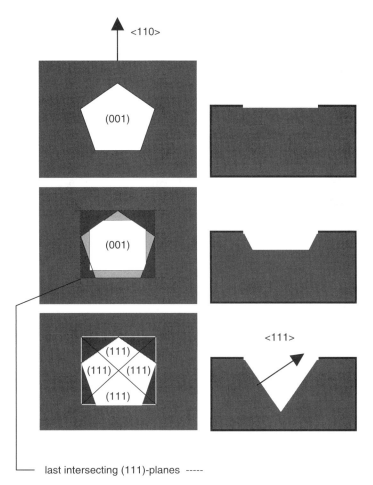

FIGURE 3.3 Anisotropic wet etching of Si. Left: top view; right: cross section. 1. Row: window in masking layer; 2. row: shape after a short etching time; 3. row: shape at the etch stop.

are parallel or perpendicular to the $<110>$-direction of the flat. This prevents the underetching of the mask. Unfortunately, the design of structures is limited to shapes perpendicular or parallel to the flat direction due to the strong dependency of the etch behavior on crystalline directions in Si. Therefore, this process is called the anisotropic wet etching of Si. When this the technology is used, it is possible to fabricate microstructures with very high precision. Limitations result from the lithography process, which is required to define the openings in the masking layer. With common lithography machines, one can count on a tolerance of about ± 1 μm. This is more than sufficient for microfluidic applications in microreactors. Channels produced with this process are shown in Fig. 3.4. Convex corners of the mask are underetched by fast etching planes (white arrows). For microfluidic applications in microreactors, this is a useful behavior. Omitting the underetching requires complex corner compensation methods [4–6].

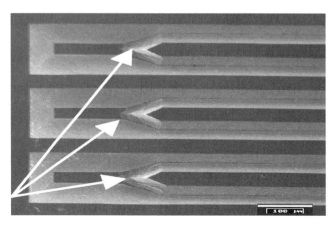

FIGURE 3.4 180° turns of two V-shaped channels.

Dry Etching of Silicon Anisotropic dry etching of silicon in plasmas has meanwhile become a well-understood technology. Currently, several dry etching methods can be identified. Nearly all of them are based on the reactive ion etching (RIE) process. Ions and free radicals are produced in the plasma of this vacuum process. The etching mechanisms are chemical etching using free reactive radicals and physical etching using the kinetic energy of ions accelerated and directed to the substrate to be etched. Etching is combined in all cases with a simultaneous or stimulated deposition process. Although sidewalls are protected because of deposited layers, the layers at the bottom of the structures are first sputtered by bombarding with ions. Second, free surface atoms are etched away in a chemical process. The intensity of these processes depends on the design of the etching machine, gas flow regime, gas composition, process pressure, process power, and the way the plasma is coupled into the reactor chamber. Therefore, one encounters names like high aspect ratio micromachining (HARM), deep reactive ion etching (DRIE), or advanced silicon etching (ASE). The gases for etching are CF_4, CHF_3, or SF_6 in combination with O_2 [7–9]. Main problems are caused by the masking layers. They are also sputtered during the process. Therefore, it is important to use layers that are stable during the entire process time. In some cases, useful materials are photoresist, SiO_2, and metals.

Similar to anisotropic wet etching processes, plasma dry etching allows the production of microstructured shapes. An important difference from wet etched structures is the possibility of perpendicular sidewalls and therefore structures with a high aspect ratio. This value is defined as the ratio of the structure width to the structure depth. It is possible to generate structures with an aspect ratio of 1 : 30 in silicon. There are no limitations regarding the arrangement of the openings in the masking layer. The main advantage of this process technology is the possibility of generating any type of structure in any direction, in contrast to anisotropic wet etched structures. Unfortunately, equipment and process chemicals are considerably more expensive than in wet etching processes. Tolerances of this process are limited by the

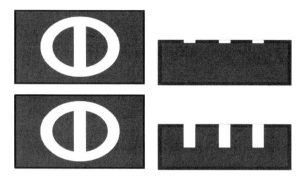

FIGURE 3.5 Anisotropic dry etching of Si (RIE). Left: top view, right: cross section. 1. Row: window in masking layer; 2. row: shape after any etching time.

photolithography only. Therefore, structures with high reproducibility and precision can be fabricated. An end result of the process is shown in Fig. 3.5.

3.1.5 Assembly of Microreactors

In general, microreactors contain microchannels or grooves. These channels can be shaped in different ways depending on the final application, that is, a reactor, mixer, or staying stretch. Fabrication of the channels/grooves is a two-step process. First, the microchannels are etched using the technology described above. If required, a selective deposit of different materials onto the surface of the channels is possible, using what is commonly known as CVD or PVD technologies. Second, a plate positioned above the structured substrate closes the channels. Main requirements are fulfilled for several applications with glass as a covering plate. Connecting silicon with glass can be achieved using the so-called anodic bonding (AB) technique [10, 11]. Special Pyrex glass wafers (Corning #7070) with a thermal expansion coefficient similar to silicon are used in this process. Glass is connected with the positive electrode of a voltage source (Fig. 3.6). Applying a voltage of about 100 to 1,000 V (depending on the thickness of the glass wafer) and a temperature of about 400°C, one will get a leakproof compound of both wafers after a process time of 20 to 30 min.

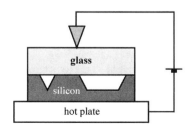

FIGURE 3.6 Closing of channels in Si using anodic bonding technique.

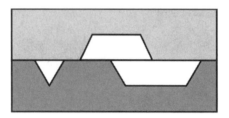

FIGURE 3.7   Two structured Si wafers bonded together.

For some applications, glass is not sufficient. It has a relatively low melting point and it cannot be structured precisely. In this case, silicon is used as a covering plate, too. If the covering plate acts as a substrate as well and if it is structured in the same way as the substrate on the contact side, a great variety of shapes can be achieved. Figure 3.7 illustrates how the two substrates are brought into contact. Applying this technique, one can staple several structured substrates over each other. Very complex microfluidic systems may be fabricated. Unfortunately, some limitations do exist. Connecting two or more substrates requires a silicon direct bonding (SDB) process. No glue or other adhesives are needed. First, two substrates are aligned and brought into contact. This sandwich must be tempered for 4 to 6 h at a temperature of about 800°C. As a result, one gets a leakproof compound of two substrates [12–14]. Repeating this process with a third wafer finally delivers a compound of three substrates. High internal stresses prevent further substrate bonding processes. Fluidic interconnections between the substrates can be realized by dry or wet etching processes from the side opposite the structured surface. All etching and assembly processes may be carried out at wafer level.

3.1.6   Design Rules

As described above, Si as a basic material for microreactors can be sufficiently structured with anisotropic dry and wet etching methods. Nevertheless, wet chemical

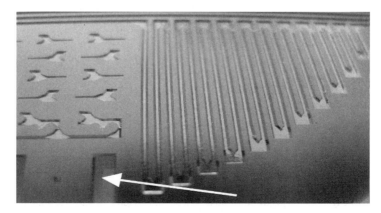

FIGURE 3.8   Different shapes in Si, produced in one anisotropic wet chemical etching step.

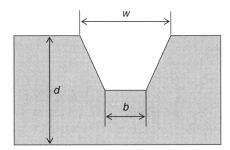

FIGURE 3.9 Cross section of etched Si structure.

anisotropic etching is the favored technology in most cases. Although the process is limited in terms of freedom of shapes, it allows the economical production of micro-reactor elements. All structures shown in Fig. 3.8, including holes etched through (white arrow), have been produced in just one etching step. This main advantage is linked to the following important design rules:

1. Depth of all structures is half the thickness of the wafer $d/2$.
2. Distance between two structures on the surface is a minimum of 200 μm on each side.
3. Masking layers contain opening windows on both sides of the wafer aligned to each other.
4. The width w of openings can be calculated using the following equation:

$$w = bf \frac{d}{\sqrt{8}}$$

5. Etching is performed simultaneously from both sides until half of the wafer thickness $(d/2)$ is reached.
6. Convex corners negatively influencing the flow behavior have to be compen-sated until half of the wafer thickness is reached $(d/2)$.

Applying this process technology, one can fabricate most of the desired microreactors very cost-effectively. See Fig. 3.9.

3.2 GLASS MICROREACTORS

3.2.1 Introduction

Glass offers many advantages compared to other materials when it is used in chemi-cal applications: It is chemically inert, hydrophilic, and optically transparent. There are three main glass types [15]:

- Borosilicate glass is very commonly used. It can resist strong acids, saline sol-utions, chlorine, bromine, iodine, and strong oxidizing and corrosive chemicals.

Even for a longer period of time and at temperatures above 100°C, it exceeds the chemical inertness of most metals and other materials. The glass is produced in a float process; it can be diced manually or with a laser. Borofloat glass is mass-produced, and therefore not expensive. However, the glass is not available in every desired gauge.

- Soda-lime glass is also a common type of industrially produced, inexpensive glass. It usually contains 65 to 75% silica, 12 to 18% soda, 5 to 15% lime, and small amounts of other materials to provide particular properties. Soda-lime glass is resistant neither to high temperatures nor sudden thermal changes, nor corrosive chemicals.
- Fused silica glass is fused in bars that are diced in slices. Fused silica is treated by lapping, rough polishing, and fine polishing. The glass can be used in applications with a wavelength from 193 nm. Borosilicate glass in comparison needs a minimum wavelength of 375 nm. Fused silica glass also has a high optical transmission of 99%. It is an expensive type of glass and hard to treat.

Different techniques, such as standard photolithography in combination with wet etching or mechanical methods like ultrasonic drilling or sandblasting, have been used to manufacture microfluidic devices. In this area, Foturan glass is a unique material because it is photosensitive. As a result of this property, it allows the user to manufacture very fine structures with a high aspect ratio.

3.2.2 Microfabrication of Glass

Isotropic Etching of Glass   Standard photolithography is a well-known process that derives from semiconductor technology. Usually, float glasses like Borofloat from Schott are used for this process. Such materials offer an already well-defined surface quality that is necessary for the subsequent process steps. First, one side of the glass wafer is coated with an adhesion layer. Typically, this layer is of titan or chromium and has a thickness of approx. 100 nm. On top of this layer a photo resist is spin-coated. The coating is then selectively irradiated with light (usually ultraviolet) through a stencil, or mask, that is designed to allow light to fall only on the desired places. The light causes a chemical change in the exposed region. Depending on the system, it is possible to wash away either the exposed or unexposed regions selectively, using the appropriate fluid, called the developer. When the exposed region is removed by the developer (i.e., exposure makes the photoresist more soluble), the process is called positive tone, and when the developer leaves the irradiated region behind (i.e., irradiation makes the photoresist less soluble), the process is called negative tone.

After this step, the adhesion layer is removed and the cleared glass surface can be etched. In most cases, hydrofluoric acid is used for this purpose. HF is commonly used for cleaning metals. Furthermore, different additives (such as HCl or HNO_3) are used to supply a homogeneous etch rate by removing surface layers that are formed when one uses pure HF solution. A 10% hydrofluoric acid concentration

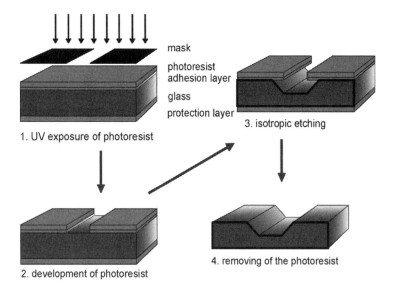

1. UV exposure of photoresist

mask
photoresist
adhesion layer
glass
protection layer

3. isotropic etching

2. development of photoresist

4. removing of the photoresist

FIGURE 3.10 Isotropic etching of glass.

gives an etching rate between 0.5 and 3.0 μm per minute. With this an etching depth between 10 and 300 μm is achievable. The maximum etching depth is mainly limited because not only the glass is etched but also the adhesion layer. This leads to a stripping of the photoresist [16, 17].

Wet etching of glass is an isotropical etching technique (Fig. 3.10). The material is etched in all directions with the same speed because it has an amorphous structure. This results in structures with a width more than twice their depth, and in addition the corners are rounded. The bottom of the structure stays smooth and optically transparent. Nevertheless, HF etching is an economical solution for fast, flexible, high-quality prototyping and volume production. Finally, an appropriate reagent removes (strips) the remaining photoresist, and the substrate is ready for the next process steps.

Dry Etching of Glass Dry etching is a process during which elements of the material are transferred into the gas phase. For this, the bindings between the atoms and molecules of the material have to be overcome. This is achieved by exposing the material to a bombardment of ions (usually a plasma) that dislodge portions of the material from the exposed surface. Unlike in wet etching, the dry etching process is typically nonisotropic. The plasma is created by using high-frequency electric fields to ionize a low-pressure gas phase. The pressures of the gaseous species are typically below 1 mbar. The decision on which gas to use mainly depends on the material that should be etched. For etching glass usually C2F6 is used.

Microstructuring of Photoetchable Glass

Using Blanket Exposure   Photoetchable glasses belong to the group of lithium-aluminum-silicate glasses. To make the material sensitive for UV light, it contains traces of noble metals. The exact composition varies from glass type to glass type. The broad range is given in Table 3.2.

The elements cerium and silver create a direct sensitivity in the glass against UV light. This effect makes it possible to carry out a direct exposure of the glass with UV light. During the exposure all parts that should not be structured are covered with a standard chromium mask which contains the required structures. In the heat treatment step that follows, all exposed parts crystallize, while the nonexposed parts stay in regular glass form. Afterward these crystallized parts can be removed using a wet etching bath that contains a solution of hydrofluoric acid. In it the exposed parts are etched approx. 20 to 30 times faster than the unexposed parts. By etching a substrate from both sides, it is possible to achieve holes with an aspect ratio between 3 to 10. The smallest structure size that can be achieved with this material falls in the range of 25 μm. It is mainly limited through a crystal size of approx. 3 to 5 μm. In addition, this crystal size defines the roughness of the etched walls, which is approx. 1 μm [16, 17]. See Fig. 3.11.

To convert the open structures into closed channels, the etched structures have to be closed afterward with a top and bottom layer. This is done in a thermal diffusion bonding step. To prepare the system, two or more glass layers with well-polished surfaces are adjusted on top of each other. In a subsequent step, this sandwich is put into an oven and under pressure the glass layer bonds together. To carry out this process successfully the glass layer has to have the same coefficient of thermal expansion (CTE).

Laser Patterning Process   The laser patterning process described here is also based on using photoetchable glasses. The main difference in the blanket exposure is the use of a pulsed UV laser system with modest energy levels for the exposure of the glass. The laser beam is focused into the glass, and in the focus point a photochemical reaction starts. By controlling the depth of the focus and also

TABLE 3.2   Composition of Photoetchable Glass

SiO_2	75–85%
Li_2O	7–11%
K_2O	3–6%
Al_2O_3	3–6%
Na_2O_3	1–2%
ZnO	0–2%
Sb_2O_3	0.2–0.4%
Ag_2O	0.05–0.15%
CeO_2	0.01–0.04%

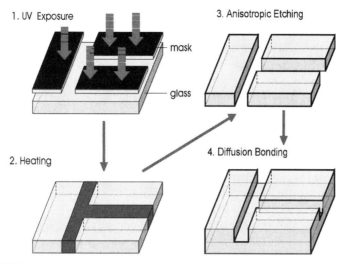

FIGURE 3.11   Principal processes during the structurization of photoetchable glass.

moving the glass substrate along a specified path, a true three-dimensional structure can be exposed into the glass [18–23]. All the following steps that are necessary to achieve the final structure are similar to the blanket exposure described above. See Fig. 3.12.

In addition to the process described above, different laser types (such as CO_2, Nd:YAG, or Excimer lasers) can also be used to structure regular glass materials by ablation. This process is well established to structure metals or polymers. The disadvantage is that high laser power is required to structure glass materials with a laser. This energy is mainly transformed into heat to melt and evaporate the material.

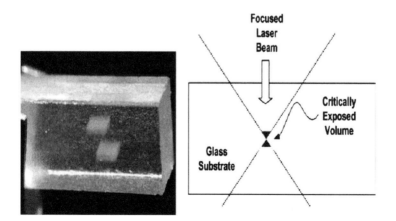

FIGURE 3.12   Laser exposure of photostructurable glass.

To increase the absorption of the laser light, special glasses have been developed for direct laser ablation.

Mechanical Structurization Methods

Ultrasonic Drilling   Ultrasonic drilling is a mechanical process by which a tool is negatively reproduced in the glass material. The tool swings with a high frequency in direct contact with the material surface. To support the ablation of surface material, an abrasive slurry is applied. With this technology, all kinds of materials (not only glass) can be structured mainly perpendicular to the surface. Due to the fabrication process, the production of three-dimensional structures is limited. The density and accuracy of structures depend on the fabrication technology of the tool. The tool is not thermally stressed and works therefore nearly free of distortion. But to achieve this high quality, the tool has to be serviced frequently. This results in high tool costs. For this reason, ultrasonic drilling is mainly used for the production of larger quantities.

Sandblasting   Sandblasting is a flexible, cost-effective, and accurate technique for producing submillimeter structures in glass and other materials. Because the use of a mask is required, position and channel size accuracy fall within 10 μm. There are no limits to the shape of the holes and trenches. In the first step, the design of the channels and holes is transferred onto the glass by means of photolithography: A photoresist film is laminated on the glass and the mask with the design is positioned above the film. The film is then illuminated with UV light through the mask. Illuminated areas will be removed from the glass during development. Afterward the wafer is powderblasted. During the exposure of the wafer to the powder, the areas not covered by film are removed by ablation, while the covered areas deflect the powder. The rate of ablation and depth can be controlled by controlling the time and particle speed. With this technology, it is possible to structure glass plates with a thickness up to 4 mm. The minimum structure size depends on the material thickness by a factor of 0.3 to 0.4 with spacing in the same range. Features that are not processed completely through the material have a wall angle of approx. 20°. For through holes, this angle can be reduced to 5°. The main advantage of this process is that it can be used for all glass types. Even ceramic or silicon can be structured. Due to the fact that the parts are processed at room temperature, no thermal stress is induced and the parts are free from distortion [24].

3.2.3   Techniques for Joining Structured Parts

There are different levels within a complete microreactor setup. It does not only contain the fluidic component itself but also the necessary equipment to integrate it into a superior environment. Taking this into account, one has to decide where to define the interfaces. For the connection of glass material with other materials, different processes will next be described.

Permanent Bonding

Diffusion Bonding   To form closed structures like channels to build a fluidic component, it is necessary to bond together different structured layers. To do so, mainly thermal diffusion bonding is used. It is a well-established method of joining materials that have identical thermal behavior with extremely well-polished and clean surfaces by employing pressure and heat for a period of several hours. During this time, atoms diffuse between the substrates, forming new chemical bonds between the surfaces. The advantage of this process is that there is no need for a glue or any other material. Therefore, it is a chemically very stable interconnection.

Soldering   Solder glasses are special glasses with a particularly low softening point. They are used to join glass to other glasses, ceramics, or metals without thermally damaging the materials to be joined. Soldering is carried out within a temperature range between 350 to 700°C. To achieve satisfactory soldering, the solder glass must flow well and sufficiently wet the parts to be joined. Flow and wetting are temperature- and time-controlled; the higher the temperature, the less time is required for sufficient flow, and vice versa. Thus, soldering at high temperatures may take only a few minutes, whereas at low temperatures reaching sufficient flow takes a very long time and usually can be achieved only under additional mechanical load.

As with all sealings involving glass, adapting the thermal expansions of the components to be joined with solder glass is a necessary prerequisite for stable, tight joints. As a rule, the coefficient of thermal expansion of the solder glasses should be smaller by 0.5 to 1.0 $\times$ 1.0 to $-6K$ than the expansion coefficients of the sealing partners. Solder glass can be supplied in the form of powder, granulate, and sintered compacts (e.g., rings, rods, etc.) and as a suspension [25].

Gluing   If the chemical stability of the connection between the materials is not a problem, it is viable to use a glue instead of the above-described thermal methods. For a direct glass–glass connection, usually UV curing glues are used. But gluing also works well for the assembly of glass with other materials like metals or plastic.

Mounting of Fluidic Components   The connection between the different parts within a complete microreaction setup should be manufactured from inert materials. For a temperature range between -20 and 120°C, Teflon tubes can be used. For this mikroglas works with standard HPLC fittings. These can be used for pressures up to approx. 20 bar. The microreactor made from glass therefore is put between two frame plates made from metal. After that the tubings are pressed by the fittings on top of the glass chip. See Fig. 3.13.

The main concern with this connection technology is that a number of connections have to be checked regularly because that the fittings are "working." To avoid this, other technologies have been investigated. One new approach is the integration of standard O-ring seals. These are placed between the glass chip and the frame. Therefore, the frame material has to be chemical inert also (PVDF). This seems to be very reliable, especially if a stack of a number of microreactors needs to be built (Fig. 3.14).

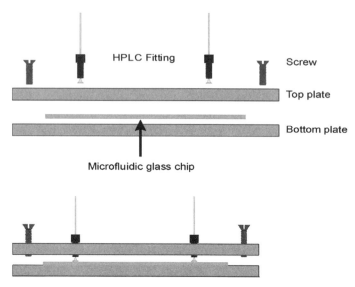

FIGURE 3.13 Connection technology of Teflon tubes to the microreactor.

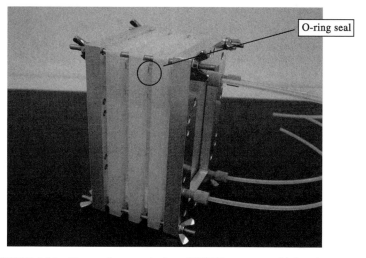

FIGURE 3.14 Frame plates made from PVDF between multiple microreactors.

3.2.4 Examples for Microfluidic Components

By using the above-described technologies, different microfluidic components such as static mixers, micro heat exchanger, and microreactors have been developed and manufactured. The advantage of glass material is its resistance against corrosive chemicals and optical transparency. It is also stable in a wide temperature range. Only the maximum applicable pressure is limited. It depends on the maximum open structure and can be approx. 40 bar for isotropic etched channel systems only made from two layers, down to approx. 3 bar for complex designs made from several structured layers and containing channels in the millimeter range.

BIBLIOGRAPHY AND OTHER SOURCES

1. Frühauf, J. *Werkstoffe der Mikrotechnik*. Leipzig: Fachbuchverlag, 2005.
2. Schwesinger, N. "The anisotropic etching behaviour of so-called isotropic etchants." *Microsystem Technologies* (1996): 481–486.
3. Elwenspoek, M., and Jansen, H. *Silicon Micromachining*. New York: Cambridge University Press, 1998.
4. Mayer, et al. "Fabrication of non-underetched convex corners in anisotropic etching of (100)-silicon in aqueous KOH with respect to novel micromechanics elements." *Journal of the Electrochemical Society*, 137, no. 12 (1990): 3947–3951.
5. Wu, X., and Ko, W. "Compensating corner undercutting in anisotropic etching of (100) silicon." *Sensors and Actuators*, 18 (1989): 207–215.
6. Puers, B., and Sansen, W. "Compensation structures for convex corner micromachining in silicon." *Sensors and Actuators*, A21–23 (1990): 1036–1041.
7. Nayve, R., et al. "High-resolution long-array thermal ink jet printhead fabricated by anisotropic wet etching and deep Si RIE." *Journal of Microelectromechanical Systems*, 13, no. 5 (2004): 814–821.
8. Ayón, A., Zhang, X., and Khanna, R. "Anisotropic silicon trenches 300–500 μm deep employing time multiplexed deep etching (TMDE)." *Sensors and Actuators*, A91 (2001): 381–385.
9. Pandhumsoporn, T., Feldbaum, M., et al. "High etch rate, anisotropic deep silicon plasma etching for the fabrication of microsensors." In *SPIE Proceedings*, Vol. 2879, 1996, pp. 94–102.
10. Jung, E. "Packaging options for MEMS devices." *MRS Bulletin*, January 2003, pp. 51–59.
11. Beißner, S., Sichler, P., and Büttgenbach, S. "Passivation layer for anodic bonding of silicon to glass." *Sensors and Materials*, 15, no. 4 (2003): 191–196.
12. Gong, S. "Fabrication of pressure sensors using silicon direct bonding." *Sensors and Materials*, 16, no. 3 (2004): 119–131.
13. Mjkia, N., et al. "Multi-stack silicon-direct wafer bonding for 3D MEMS manufacturing." *Sensors and Actuators*, A103 (2003): 194–201.
14. Gui, C., et al. "Selective wafer bonding by surface roughness control." *Journal of the Electrochemical Society*, 148, no. 4 (2001): G225–G228.
15. Pfaender, H. G. *Schott Glaslexikon*. Landsberg am Lech: mvg Verlag, 1997, pp. 28–30.
16. Dietrich, T. R. "Photostrukturierung von Glas." In *Handbuch Mikrotechnik* (W. Ehrfeld, ed.), pp. 407–429. Wien: Hanser Verlag, 2002.
17. Dietrich, T. R., Ehrfeld, W., Lacher, M., Krämer, M., and Speit, B. "Fabrication technologies for microsystems utilizing photoetchable glass." *Microelectronic Engineering*, 30 (1996): 497–504.

18. Karam, R. M., and Cassler, R. J., "A new 3D, direct-write, sub-micron microfabrication process that achieves true optical, mechatronic and packaging integration on glass-ceramic substrates." Sensors 2003 Conference, Rosemont, IL, June 2–5, 2003.

19. Livingston, F. E., Hansen, W. W., Huang, A., and Helvajian, H. "Effect of laser parameters on the exposure and selective etch rate in photostructurable glass." In *SPIE Proceedings*, San José, CA, USA, 21 January 2002, Vol. 4637, pp. 404–412.

20. Livingston, F. E., and Helvajian, H. "True 3D volumetric patterning of photostructurable glass using UV laser irradiation and variable exposure processing: Fabrication of meso-scale devices." In *SPIE Proceedings*, Osaka, Japan, 27 May 2002, Vol. 4830, pp. 189–195.

21. Hansen, W. W., Jansen, S. W., and Helvajian, H. "Direct-write UV laser microfabrication of 3D microstructures in lithium alumosilicate glass." In *SPIE Proceedings*, San José, CA, USA, 10 February 1997, Vol. 2991, pp. 104–112.

22. Fuqua, P. D., Taylor, D. P., Helvajian, H., Hansen, W. W., and Abraham, M. H. "UV direct-write approach for formation of embedded structures in photostructurable glass-ceramics." In *Materials Development for Direct-Write Technologies* (D. B. Chrisey, D. R. Gamota, H. Helvajian, and D. P. Taylor, eds.) pp. 79–86. *Materials Research Society Proceedings*, Warrendale, PA, USA, 2001, Vol. 624, pp. 79–86.

23. Helvajian, H. "3D microengineering via laser direct-write processing approaches." In *Direct-Write Technologies for Rapid Prototyping Applications* (A. Piqué and D. B. Chrisey, eds.) pp. 415–474. New York: Academic Press, 2002.

24. Corporate information. "Mikrosandstrahlen." Little Things Factory GmbH, October 2006.

25. Corporate information. "Schott technical glasses." Schott AG, December 2000.

CHAPTER 4

MICROMIXERS

JOËLLE AUBIN and CATHERINE XUEREB

4.1 INTRODUCTION

Today, micromixers play a significant role in both the chemical and chemical engineering fields. Initially, miniaturized mixers were integrated into small fluidic chips and employed in analytical applications (micrototal analysis systems, μTAS), such as high-throughput screening and quench flow analysis. Their advantage of providing the extremely fast mixing of very small fluid volumes had proven to be particularly valuable for numerous and/or high-value samples. Later, the performance characteristics of such mixers were also recognized to be advantageous in chemical processing applications. For this, micromixers, as chiplike mixer components, were developed for carrying out traditional mixing tasks such as blending, emulsification, and dispersion, but also for the fabrication of products with highly controlled physical properties and as chemical reactors with integrated heat exchangers. Industrial-scale processing with miniaturized mixers is now a reality, and to provide sufficient throughput, the micromixers used at the production scale include a network of microstructured channels in larger housing (correctly termed microstructured mixers).

Because of the vast application field of micromixers and microstructured mixers (homogenization, chemical reaction, dispersion, emulsification etc.), the efficiency of these devices is very important to the definition of process performance. Indeed, it will affect various process parameters including heat and mass transfer rates, process operating time, cost and safety, as well as product quality. For this reason, it is important to not only understand the mixing process occurring in micromixers but also characterize and evaluate their mixing performance.

Microchemical Engineering in Practice. Edited by Thomas R. Dietrich
Copyright © 2009 John Wiley & Sons, Inc.

In this chapter, our aim is to provide the reader with sufficient knowledge on micromixers for their application in chemical engineering processes by addressing aspects such as:

Mixing problems in miniaturized devices and the means possible for improving the mixing process

Technical characteristics of micromixers and commercially available devices

Evaluating the performance of the micromixer and characterizing the mixing quality

Multiphase mixing

4.2 MIXING PRINCIPLES AND FLUID CONTACTING

4.2.1 The Laminar Mixing Principle

As for all kinds of mixers, the objective of a micromixer is to obtain a uniform distribution of the components being mixed down to the smallest scale possible, and this as quickly as possible. Due to the small dimensions of the internal structures in micromixers, the flow is predominantly laminar in such devices. It is therefore the very slow mechanism of molecular diffusion that limits the mixing time, not the faster mechanism of convection, which dominates in turbulent flows. This transport by molecular diffusion obeys Fick's law, which can be rearranged to show that the mixing time t_{md} is proportional to the characteristic diffusional path l and the diffusion coefficient D:

$$t_{md} \propto \frac{l^2}{D} \tag{4.1}$$

From this equation, it can be seen that the mixing time is highly dependent on the characteristic length scale of the system, which is determined by the dimensions of the microchannels. At extremely small scales, where the channel dimensions of the micromixer are on the order of a few tens of microns, miscible liquid streams will mix by molecular diffusion in less than a second. However, when the dimensions of the channels are several hundreds of microns, mixing purely by molecular diffusion can take tens of seconds, which is obviously limiting when dealing with fast chemical reactions. Furthermore, when liquids are being mixed, the mixing process is even longer than for gases because the diffusion coefficient is around 1,000 times smaller than that of a gas.

Therefore, in order to effectively mix at this scale and in a reasonable time, the fluids must be manipulated such that the interfacial areas between the fluids is increased massively and the diffusional path is decreased, thereby enhancing the molecular diffusion mechanism to complete the mixing process. As a result, the design of the mixing device and the type of fluid contacting employed are the determining factors for micromixer performance.

All further considerations concerning mixing in microdevices are valid for single-phase systems, either liquid or gas. The specificity of gases in microchannels lies in the hypothesis of continuity of the fluid. This is linked to the fact that the distance between two molecules in the gas phase can be on the order of magnitude of the channel width (for a Knudsen number, Kn, greater than 0.1). However, this limit is reached with channels of widths less than 1 μm, which is rarely the case in devices used for industrial applications.

4.2.2 Types of Fluid Contacting

Numerous fluid contacting mechanisms exist for enhancing the mixing process in micromixers. Certain devices use the energy of the flow (e.g., pressure gradients resulting from pumping action or hydrostatic potential) and the geometry of the mixer in order to manipulate the fluids; these are called passive mixers. Active mixers, on the other hand, use an external source of energy to facilitate the mixing process, such as ultrasound, various types of vibrations, or pulsed flow rates, among others. Table 4.1 gives an example of different active and passive mixing mechanisms; however, others may exist. Whether an active or a passive mixing mechanism is used in the micromixer, the objective of each remains the same: to increase the interfacial area between the species and to renew the fluid interface, thereby promoting mixing by molecular diffusion.

In this section, only the most commonly employed fluid contacting mechanisms in chemical engineering applications will be presented in more detail. For more details on the different mixing mechanisms that exist, the reader may refer to [1, 2].

TABLE 4.1 Means for Inducing Active and Passive Mixing

Active Mixing Methods	Passive Mixing Methods
Electrokinetic instabilities	Y- or T-type contacting
Periodic flow perturbations	Multilamination
Acoustic vibrations	Hydrodynamic focusing
Magneto hydrodynamic forces	Flow splitting and recombination
Ultrasound	Structured channel geometries
Piezoelectric actuators	Channel packings or obstacles
Mechanical impellers	Impinging jets
Micropumps and valves	Micronozzle injection

Y- and T-Type Contacting Y- and T-mixers (Fig. 4.1) are extremely simple contactors and relatively efficient if correctly designed. If the dimensions of the microchannel are sufficiently small, mixing by diffusion can be very fast due to the minute size of the characteristic length scale of the system, as previously explained by Eq. (4.1). However, in order to mix quickly in larger-sized microchannels, the fluid velocity must be increased sufficiently such that unstable vortex structures are created, thus enhancing the mixing process [3, 4].

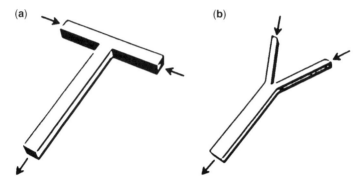

FIGURE 4.1   (a) T-mixer. (b) Y-mixer.

Multilamination   Multilamination is one of the most widely studied contacting mechanisms and, as a result, a number of micromixers of this type are now commercially available. The principle of this mixing mechanism is to contact multiple fluid streams of different species alternately (Fig. 4.2). This can be done in either a cocurrent or countercurrent mode. The mixing time in such contactors depends on the width of the fluid lamellae, which is controlled directly by the dimension of the microchannels. In mixer geometries whereby the initial width of fluid filaments is relatively important ($>100\,\mu m$), mixing can be further promoted by coupling the multilamination technique with hydrodynamic focusing.

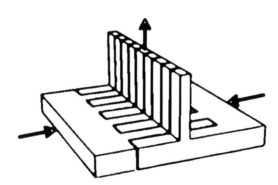

FIGURE 4.2   Countercurrent multilamination mechanism.

Hydrodynamic Focusing   Hydrodynamic focusing refers to a decrease in the diffusional path that is perpendicular to the direction of the flow. This mechanism is generally used as a complement to multilamination mechanisms. By reducing the width of the microchannel either abruptly or progressively, the fluid lamellae are forced through a narrower channel cross section. This decreases the characteristic distance for diffusion and therefore enhances the mixing process (Fig. 4.3).

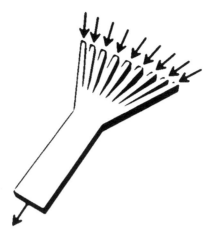

FIGURE 4.3 Decreasing the diffusional path via hydrodynamic focusing.

Splitting and Recombination of Fluid Streams The repetitive mechanism of flow division and recombination is based on the mixing principal of traditional static mixers. In this type of mixer, the fluids are repeatedly stretched, split, and then recombined in order to decrease the thickness of the fluid filaments and increase the interfacial area between the different fluid species. Very often, the flows generated by split and recombine mechanisms have chaotic characteristics, which enable fast and efficient laminar mixing.

Chaotic Mixing Chaotic flows can be generated by either passive or active mixing mechanisms and are characterized by the stretching and folding of fluid elements, which produces exponential growth of the fluid interface and thus a divergence from initial conditions. Passive chaotic micromixers are based on three-dimensionally structured channels that generate transverse flows and recirculation patterns, which lead to exponential increase of the fluid interface and therefore assist mixing. Figure 4.4 shows two examples of three-dimensional channel structures that create chaotic flow passively.

(a)   (b)

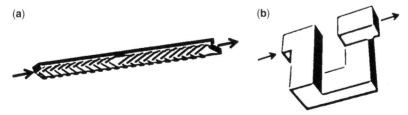

FIGURE 4.4 Three-dimensionally structured micromixer elements. (a) Staggered herring-bone micromixer. (b) Serpentine mixer.

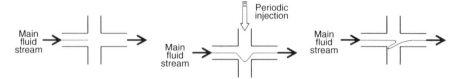

FIGURE 4.5   Periodic injection of a fluid stream into the main flow.

In the case of active mixers, chaotic flows can be generated by applying a periodic instability to the flow via an external source. Some examples are the periodic injection of a fluid stream into the main flow, which enables fluid filaments to be stretched and folded (Fig. 4.5), and use of nonuniform electric potentials on the walls of the microchannel, creating electro-osmotic flow.

4.3   TECHNICAL ASPECTS

4.3.1   Typical Dimensions of Micromixers and Microstructures

The dimensions of the channels are generally defined as a function of the desired performances of the mixer, especially if heat transfer must be achieved at the same time as mixing. In this case, the transfer area, and more specifically the ratio transfer area/volume of fluid, will be the limiting parameter and therefore needs to be taken into account to choose the correct dimensions. Evidently, the result of this analysis must lead to reasonable pressure loss and feasible flow rates.

Depending on the type of fluid contacting, different ranges of channel widths and channel depths may be encountered—with either a low or high ratio between these dimensions—within the physical limits that may be imposed by the material (e.g., the thickness of a silicon wafer). Generally, channel widths lie between 10 and 1,000 μm, and channel depths between 10 and 400 μm. However, the most common dimensions fall in the range of 100 to 500 μm and 50 to 300 μm. A thorough review is given by Nguyen and Wu [5].

In the cases where mixing is not entirely governed by the fluid contacting area, the length of the microchannels can be determined by the level of mixing required for the operation. In these cases (e.g., in passive multilamination mixers), it is the molecular diffusion that controls the final degree of segregation. Most apparatus provide satisfactory mixing within the first 20 to 30 mm of channel length.

4.3.2   Typical Flow Rates

The flow rates passing through the micromixer evidently depend on the pressure loss that is considered acceptable, the required residence time, and also the capacity of the system for heat and mass transfer.

Typically, for micromixers with channel widths in the range of 10 to 1,000 μm, one can expect flow rates of $mL \cdot h^{-1}$ to $1 \cdot h^{-1}$, which results in a large range of fluid velocity in the channels. It should be pointed out that in the case of chaotic advection micromixers, usual velocities fall in the range of 1 to $30 \, \text{mm} \cdot \text{s}^{-1}$, which leads to intermediate Reynolds numbers (5–50) and provides better mixing efficiency than at lower Reynolds numbers. In multilamination micromixers, fluid velocities are generally higher; Wong et al. [6] have demonstrated velocities up to $10 \, \text{m} \cdot \text{s}^{-1}$ in ceramic cross-shaped mixers.

4.3.3 Comparison of Different Materials (Fabrication Techniques, Type of Material vs. Application)

It should be noted that the geometries of microchannels are rarely cylindrical. This is due to the fact that the fabrication techniques are specific to these microdevices and are not at all the same as those encountered in macrodevices. Most often, microchannels have a rectangular or trapezoidal cross section. We will not detail here all of the fabrication techniques; however, for more specific information the reader should consult [7].

The most common materials used for micromixers are silicon, glass, and some metals. Silicon is well known because of its application in microelectronics, and the same traditional microfabrication technologies are also used for manufacturing microchannels: Masking followed by chemical etching or the plasma technique called deep reactive ion etching (DRIE), with the latter enabling almost vertical walls to be obtained, are both common methods. Moreover, silicon is valued for its good chemical compatibility, mechanical properties, and heat transfer capacity. In order to close the channel, a final bonding step is necessary. This can be done in a plasma chamber using silicon or another material (e.g., a plate of glass). Closing the channel with a transparent material has the advantage of allowing flow visualization via microscopy.

Micromachining techniques offer the possibility of making microchannels in pure metals, like the LIGA process (X-ray lithography). This is particularly interesting for the fabrication of industrial purpose micromixers, which may be used for harsh chemical reactions or operarting conditions.

When choosing the fabrication material of a micromixer, there are two important elements to take into account. The first one is linked to the nature of the material and its wettability with respect to the liquid(s) used. This can heavily modify the flows and thus the mixing capacity of the device if slipping occurs at the walls. In the case of reactions occuring at the walls, roughness can thus induce an effective surface far from the theoretical one calculated on the basis of a smooth one. The second element is linked to the geometrical uncertainties: A small imperfection can lead to a high increase in pressure drop and enhance mixing! But this also can bring about errors in predictions or wrong interpretations of the phenomena occurring in the microchannels.

4.3.4 Examples of Commercially Available Micromixers

With the integration of micromixers in industrial processes, a considerable number of devices are now commercially available through specialized manufacturers. As an

TABLE 4.2 Selected Commercially Available Micromixers

Mixer Type	Mixing Principal	Applications	Manufacturer
Cascade-type mixer	Split and recombine	Miscible mixing, generation of dispersions, flows with particles <50 μm, viscous fluids	Ehrfeld Mikrotechnik BTS, http://www.ehrfeld.com
Comb-type mixer	Multilamination	Miscible mixing, generation of dispersions	Ehrfeld Mikrotechnik BTS, http://www.ehrfeld.com
Microjet mixer	Impinging jets	Precipitation reactions, particle generation	Ehrfeld Mikrotechnik BTS, http://www.ehrfeld.com
Valve-assisted mixer	Injection of multiple substreams	Miscible mixing, precipitation reactions, emulsion generation	Ehrfeld Mikrotechnik BTS, http://www.ehrfeld.com
Slit plate mixer	Multilamination	Miscible mixing, generation of dispersions	Ehrfeld Mikrotechnik BTS, http://www.ehrfeld.com
Caterpillar mixer	Split and recombine	Miscible mixing, generation of dispersions, precipitation reactions, fine slurry processing	Institut für Mikrotechnik Mainz, http://www.imm-mainz.de
Star laminator	Multilamination	Miscible mixing, viscous fluids (<10 Pa · s)	Institut für Mikrotechnik Mainz, http://www.imm-mainz.de
Slit interdigital micromixer	Multilamination and geometrical focusing	Miscible mixing, generation of dispersions, precipitation reactions, chemical reactions, extraction processes	Institut für Mikrotechnik Mainz, http://www.imm-mainz.de
Superfocus interdigital micromixer	Multilamination and geometrical focusing	Miscible mixing	Institut für Mikrotechnik Mainz, http://www.imm-mainz.de
Impinging-jet micromixer	Impinging jets	Precipitation reactions	Institut für Mikrotechnik Mainz, http://www.imm-mainz.de
HT-micromixer (SR)	Split and recombine	Miscible mixing, emulsion generation	Little Things Factory GmbH, http://www.ltf-gmbh.de
HT-micromixer (ST)	Split and recombine	Miscible mixing, emulsion generation	Little Things Factory GmbH, http://www.ltf-gmbh.de
HT-micromixer (MM)	Multilamination	Miscible mixing	Little Things Factory GmbH, http://www.ltf-gmbh.de
Interdigital micromixer	Multilamination and hydrodynamic focusing	Miscible mixing, generation of dispersions	mikroglas chemtech GmbH, http://www.mikroglas.com
Cyclone micromixer	Multilamination	Miscible mixing, multiphase contacting	mikroglas chemtech GmbH, http://www.mikroglas.com

example, and perhaps a starting point for the user, Table 4.2 lists a range of selected micromixers that are commercially available, their manufacturer, as well as the mixing principal used and some of the possible mixing applications.

4.4 EVALUATING THE PERFORMANCE OF A MICROMIXER

As previously discussed (Section 4.2), a large number of mechanisms exist for enhancing mixing in microchannels, and therefore choosing the most appropriate micromixer for a particular application may not be a straightforward task. Initially, the choice may be based on the compatibility of the technical aspects of the mixer and the given application. However, the user will ultimately want to evaluate the performance of the chosen mixer and compare it with that of others. To do this, various performance characteristics (e.g., pressure drop, flow patterns, mixing efficiency, residence time distribution, and the presence of dead zones, etc.) of the device need to be determined. These performance characteristics can be evaluated either experimentally or virtually via the numerical simulation of the flow. It is clear that the experimental testing of a wide range of micromixers could be a timely and costly affair, and may require specific equipment that is most often found in research laboratories. For this reason, the numerical simulation (via computational fluid dynamics) of flow within micromixers is an attractive alternative, enabling different geometries to be quickly designed and tested in a virtual manner. This way, the user can select the most suitable micromixer geometries for the application and then evaluate their performance experimentally.

4.4.1 Pressure Drop

As in all chemical engineering applications, pressure drop is an important performance characteristic, which one aims to minimize. Considering the small dimensions of micromixers, one would expect relatively high-pressure losses since the pressure drop is proportional to the channel length and inversely proportional to the hydraulic diameter. In addition, for micromixers whereby the geometries are relatively complex compared with a simple straight channel, the pressure drop is expected to be even higher. However, when evaluating the real performance of a micromixer, one must compare the pressure drop relative to the mixing efficiency (e.g., mixing time or mixing length).

In recent years, there have been great breakthroughs in the development of experimental techniques available for pressure measurement in microsystems. Today, many different types of pressure sensors exist, enabling either local or global measurements in stationary or transient flows [8]. Determining the pressure drop across a micromixer is a simple procedure that requires the pressure difference between the outlet and inlet to be measured, as done in traditional mixing devices. Although specialty micropressure sensors exist, traditional equipment such as differential pressure sensors based on membrane deflection or piezo-resistive sensors is often sufficient, provided that it is well adapted to the pressure range being measured.

The pressure drop can also be determined very easily from CFD simulations. In fact, the pressure field is a direct result of solving the Navier–Stokes equations for these types of flow simulations.

4.4.2 Hydrodynamics

The hydrodynamics or flow patterns within a micromixer can be determined either numerically or experimentally. CFD simulations of the flow within the micromixer enable the velocity components in all three directions to be determined throughout the entire volume of the device. Subsequently, vector fields and streamlines can be constructed, which enable better understanding of the flow patterns created in the micromixer.

The increased interest in miniaturized devices over recent years has also led to the development of specific experimental techniques that allow the performance of these devices to be characterized. In particular, microparticle image velocimetry (micro-PIV) has been developed specifically for the determination of fluid flow fields within microchannels [9]. The principles of micro-PIV are based on those of classical PIV: Microtracer particles are introduced into the flow and illuminated by laser light; two successive images (separated by a very short time interval) of the particles are then taken; these two particle images are correlated in order to determine the distance traveled by the particles during the time interval and subsequently the instantaneous velocity field in the flow. However, in contrast to PIV, the entire volume of the micro-channel flow is illuminated in micro-PIV, and a microscope is used to focus on a nearly two-dimensional measuring plane, whereby the velocity field is determined. This type of measuring equipment is particularly interesting because it allows local analysis to be carried out without disturbing the flow at a high spatial resolution (~ 1 μm). Furthermore, using a microscope objective that has a small depth of field enables velocity fields at several heights within the channel to be measured. Advances in stereoscopic micro-PIV and micro-PIV with confocal scanning laser microscopes are being made, thus enabling 3-dimensional velocity fields to be measured in microdevices.

It should be pointed out that the micro-PIV technique is predominantly limited to laboratory studies because the optical nature of the measurement requires that the microdevices are made of transparent material (e.g., glass or polymer), which may not always be adapted to the aggressive conditions of some industrial applications.

4.4.3 Residence Time Distributions

The residence time distribution (RTD) is a performance characteristic that enables temporal inhomogeneities to be evaluated. It is particularly relevant to micromixers because the parabolic velocity profile of laminar flow gives rise to large temporal inhomogeneities, which translate into wide residence time distributions. Ideally, a mixer should be designed to give a high degree of plug flow, thus decreasing temporal inhomogeneities.

The RTD of a micromixer can be determined both experimentally and numerically, by injecting a known amount of tracer (real or virtual) at the mixer inlet and

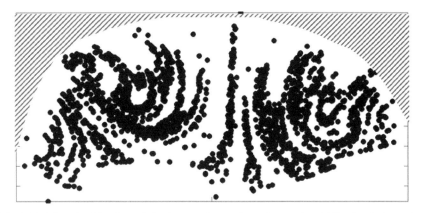

FIGURE 4.6 Identification of dead zones in a micromixer using CFD simulations. Dead zones are designated by the area shaded in grey.

then measuring the tracer concentration at the outlet over time. Experimentally, the concentration of the tracer at the outlet can be measured using traditional methods, such as conductivity probes or mass spectrometry [10], and with specifically developed sensors [11]. To determine a RTD using CFD, a Lagrangian approach can be used to track the trajectories of a known number of mass-less tracer particles from the mixer inlet. At the outlet, the number of particles exiting the mixer as a function of time is recorded to construct the RTD [12].

RTD and particle trajectories can also be used to identify the presence of dead volumes in the micromixer. By plotting particle trajectories on a two-dimensional plane perpendicular to the flow, the spatial distribution of the tracer enables zones that are poorly mixed to be identified. An example of this is given in Fig. 4.6: Badly mixed zones are represented by the area shaded in grey where no particle trajectories are apparent. The presence of a dead volume is confirmed by analysis of the RTD if the mean residence time of the tracer is less than the theoretical residence time (reactor volume/volumetric flow rate).

4.4.4 Mixing Quality

Mixing quality can be evaluated experimentally and numerically using a number of different approaches, giving both qualitative and quantitative information.

Visualizing and Evaluating the Homogeneity of a Mixture Visual representations of mixing allow qualitative analysis and are particularly useful for understanding the mixing phenomena in the device. Such data can then be further analyzed to quantify the mixing performance.

The most common and simplest way to visualize mixing experimentally is by observing and recording the flow in the micromixer using a microscope and video camera. Of course, this approach requires that the micromixer have at least one transparent wall so that the flow is visible. Typically, two types of visualization experiments exist:

- Dilution-type experiments, whereby a colored or fluorescent stream is contacted and diluted with a transparent stream.
- Reaction-type experiments, whereby the reaction between two miscible fluids leads to the formation of a colored product or color change (e.g., an acid–base reaction in the presence of an indicator).

In such experiments, the color change is followed along the micromixer or observed at the mixer outlet, enabling different mixing devices to be compared qualitatively.

Although these experiments are relatively simple, their disadvantage is that they provide a global evaluation of the mixing quality that is averaged over the depth of the micromixer. As a result, no local information on the structure mixing process can be obtained. Scanning confocal microscopy is one possible solution for obtaining local depth-wise information on the mixing pattern [13, 14]. It scans the flow throughout the depth of the device in a fast point-wise method, enabling reconstruction of the three-dimensional mixing pattern. However, this type of equipment is costly and may be limited to R&D-type analyses of micromixer devices.

Alternatively, CFD simulations of mixing in micromixers can be carried out. For visualizing the mixing phenomena, two approaches exist. The first mimics experimental techniques by simulating the diffusion of a passive tracer of known concentration via an Eulerian approach. The homogeneity of the tracer concentration is followed and analyzed along the micromixer and/or at the outlet. Although this kind of simulation is relatively simple to set up, it is not entirely recommended due to problems of numerical diffusion and the subsequent erroneous solutions that are encountered when simulating at such small scales. These inherent problems of numerical diffusion can be largely avoided by using a Lagrangian approach, whereby mass-less tracer particles are tracked throughout the fluid flow. In order to visualize the mixing phenomena in the micromixer, the particle locations at different positions within the mixer or at the mixer outlet are plotted on three-dimensional graphs. This type of representation allows the fluid contacting mechanism occurring in the mixer to be better understood and enables the mixing quality at various positions in the mixer to be followed. Figure 4.7 gives an example of this type of visualization method for a chaotic micromixer.

The visual methods described above can provide important information about the mixing mechanism within a micromixer. However, one often seeks to quantitatively compare different mixers by evaluating the homogeneity of mixing at the outlet. Typically, the homogeneity of a mixture is quantified via the statistical analysis of concentration samples from the mixture, which is based on Dankwerts's "intensity of segregation" concept [15]. The intensity of segregation approach is based on the variance σ^2 of the concentration at different regions in space with respect to the mean concentration. The most common measure of segregation in mixing applications is the coefficient of variance, CoV (see, e.g., [14, 16–21]). The CoV is the ratio of the standard deviation σ of the M concentration samples to the mean

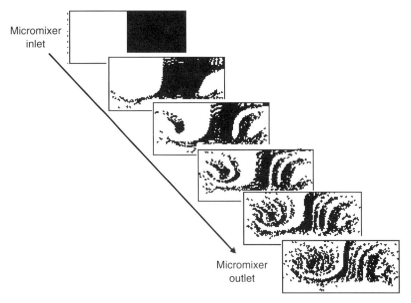

Micromixer inlet

Micromixer outlet

FIGURE 4.7 Visualizing mixing along a staggered herringbone mixer using mass-less tracer particles to track the flow.

concentration $\bar{c}$, as expressed by Eq. (4.2):

$$\text{CoV} = \frac{\sigma}{\bar{c}} = \frac{\sqrt{\dfrac{1}{M}\displaystyle\sum_{i=1}^{M}[c_i - \bar{c}]^2}}{\bar{c}} \tag{4.2}$$

In the case of mass-less tracer particles, a number-based CoV_n can be calculated by placing an $n \times m$ grid of equal-sized cells over the mixer cross section, as expressed by Eq. (4.3):

$$\text{CoV}_n = \frac{\sigma}{\overline{N}} = \frac{\sqrt{\dfrac{1}{M}\displaystyle\sum_{i=1}^{M}[N_i - \overline{N}]^2}}{\overline{N}} \tag{4.3}$$

where the number of samples $M = n \times m$, N_i is the number of particles in each sample, and $\overline{N}$ is the mean number of particles per sample.

Although the CoV is widely employed, it has some weaknesses: It is highly dependent on the size, shape, and number of sampling areas used and it filters out all information on the scale of segregation, which is illustrated by the classical checkerboard problem. These limitations of the CoV have led to the development of alternative methods for analyzing mixing when using tracer particles whereby the sampling unit is a point. These methods differ from the CoV analysis in that they are based on the distance between each point and a neighboring event and that they do not require the studied region to be divided into sampling areas. The method described in [12] provides information on the spatial mixing quality with respect to a certain scale of segregation. It measures the distance x_i from each point in a chosen lattice of p sample points to the nearest tracer particle event. The variance of the point-event distances around a distance x_R provides a means of evaluating the "mixedness" of a system with respect to a "well mixed" criterion. This predefined criterion is set by x_R and corresponds to the scale of segregation whereby two events are considered spatially close enough to be mixed. The point-event variance is given by

$$\sigma_{\mathrm{PE}}^2 = \frac{1}{p-1} \sum_{i=1}^{p} (x_i - x_R)^2 \tag{4.4}$$

If $x_R \neq 0$, any value of $x_i < x_R$ is considered "better mixed" than the desired scale of segregation. However, according to Eq. (4.4), this will give rise to unphysical increases in the variance. To avoid this, Aubin et al. [12] suggested the use of a threshold such that in the case where $x_i < x_R$, x_i is assigned a value equal to x_R. This means that σ_{PE}^2 will tend toward a value of zero as the spatial distribution of the population approaches uniformity in the limit of the defined scale of segregation.

Striation Thickness   In laminar flow mixing, the thickness of fluid striations or filaments is extremely important as it directly determines the mixing time, as given by Eq. (4.1), and therefore the efficiency of the micromixer. Striation thickness can be measured experimentally or determined numerically using CFD, either directly or indirectly.

Concentration profiling is one means to determine the thickness of a fluid striation experimentally. Typically, the online optical properties (photometric or fluorescence) of the flow are measured across the projected channel cross section [22]. This gives concentration profiles that are averaged over the depth of the micromixer. Measuring several concentration profiles at different distances along the mixer allows information on the rate of striation thinning to be obtained. Confocal scanning microscopy may also be used to determine striation thickness. The two-dimensional confocal micrographs may be used to measure the striation thickness along a line transect.

Using CFD simulations, the striation thickness can be determined directly using mass-less tracer particles or assessed indirectly via chaos analysis. The direct method consists of analyzing the distance between tracer particles on a two-dimensional particle plot (e.g., Fig. 4.7) [12]. Sampling is carried out along a line transect and the distance between two consecutive tracer particles Δz is measured. The interparticle distances are then tested with respect to a threshold value Δz_T, which corresponds to the scale of segregation whereby two consecutive particles are considered to be part of the same striation or not:

$\Delta z \leq \Delta z_T$: Consecutive particles belong to the same striation, $f(x) = 1$.
$\Delta z > \Delta z_T$: Consecutive particles belong to different striations, $f(x) = 0$.

Figure 4.8 gives an example of striation thickness determination using this method.

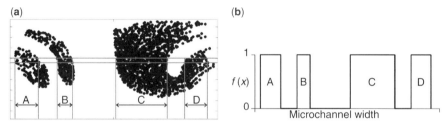

FIGURE 4.8 Determination of striation thickness along a line transect using CFD simulations. (a) Fluid striations on a cross section of a micromixer as depicted by mass-less tracer particles. (b) Calculation of the striation thicknesses with $f(x)$.

Alternatively, chaos analysis can be used to indirectly assess striation thickness (e.g., [17, 18, 23]). This involves evaluating the stretching of a small fluid element, which is inversely proportional to the striation thickness. Typically, stretching distributions and the rate of stretching or the Lyapunov exponent are determined, the latter being representative of the upper limit of the rate of striation thinning. Although such calculations are rich in information and detail, they are unfortunately very computationally intensive and have so far proven most appropriate for the fundamental study of laminar flows.

Micromixing Efficiency Micromixer performance may also be evaluated using competing parallel test reactions. The most frequently used method is the Villermaux/Dushmann reaction, whereby iodine is formed by an acid-catalyzed redox reaction between iodide and iodate [24], which has been adapted for micromixer applications [25, 26]. In this system of two competitive reactions, one is quasi-instantaneous, whereas the other is fast and highly dependent on the micromixing process. In the case of nonideal mixing, the slower reaction is promoted, forming iodine, which can be detected by UV-Vis spectrophotometry. This method mainly enables qualitative comparisons of micromixers, but does not directly give quantitative data, such as mixing times or mixing lengths. However, further analysis of such data can provide information on mixing time. An example of this is given in [27] whereby the micromixing data from [26] are expressed in terms of micromixing time t_m as a function of specific power dissipation, ε in $W \cdot m^{-3}$. The analysis shows that the experimental values reported by Panic et al. [26] for a number of different micromixers with low viscosity liquids can be roughly correlated to give the following relationship for mixing time:

$$t_m \cong 100 \left(\frac{\mu}{\varepsilon}\right)^{1/2} \tag{4.5}$$

4.5 MULTIPHASE MIXING

Multiphase mixing is very often required for carrying out various chemical or physical operations, with or without reaction. In this section, we focus on multiphase dispersions in which both the operating conditions and geometrical aspects of the micromixer are of great importance.

4.5.1   Liquid–Liquid Mixing

Liquid–liquid mixing applications in micromixers are the same as those in classical apparatus, but the great advantage of micromixers is the possibility of obtaining a very narrow droplet size distribution. This enables the fabrication of emulsions with highly controlled physical and rheological properties.

Interdigital micromixers are very interesting for the fabrication of such emulsions [28]. Some general trends have been pointed out by these authors concerning the resulting emulsion as a function of the operating conditions:

- In a given apparatus, the mean droplet diameter decreases almost linearly with the increase of the total volume flow, and the droplet size distribution becomes narrower. This is a consequence of the amount of energy dissipated into the system.
- The reduction of the channel width leads to a significant decrease in droplet size, as with the reduction of the feeding slit height [29].
- For the formation of emulsions with a narrow drop size distribution, it is highly desirable to operate with a high ratio between the two liquid flow rates.

Well-adapted operating conditions can lead to droplets about 1 μm in diameter [30], and to a large decrease in and even a suppression of the use of surfactants for emulsion stability.

4.5.2   Gas–Liquid Mixing

Promoting good contact between a gas and liquid in a microchannel is often linked to the mass transfer occurring at the interface. A high interfacial area between the two phases therefore has to be created and in a sufficiently stable way, such that there is enough time for the transport process or the reaction to occur. In simple microchannels, gas dispersions can be generated through the high-speed impact of a gas and liquid in a T-junction, or by feeding the main channel filled with liquid by a cross-junction filled with the gas. The use of a capillary immersed in the microchannel just before an orifice can also lead to the creation of monosized microbubbles [31] with diameters between 5 and 120 μm. Interdigital micromixers also give very regular bubble sizes of less than 100 μm in diameter [32].

Because of the high importance of surface phenomena in microdevices, the geometry of the microchannel also has some influence on the nature of the dispersion. In particular, the hydraulic diameter of the channel is not sufficient for predicting the characteristics of the dispersion. For example, a rectangular or round channel cross section with the same hydraulic diameter will not systematically lead to the same result: When the flow regimes are mapped, the boundaries between the different regimes are displaced as a function of the geometry of the channels. A thorough study of flow regime classification and mapping has been given in [33]. Four highly different regimes can be observed (Fig. 4.9):

- *Stratified Regime*   Occurs at very low values of liquid velocity, and the liquid and gas phases are entirely separate.
- *Intermittent Regime*   At intermediate values of the liquid velocity and low values of the gas flow rate, elongated gas bubble plugs are surrounded by a liquid film and separated by liquid slugs.

Stratified regime: Wavy flow pattern

Intermittent regime: Elongated bubble pattern

Intermittent regime: Slug flow pattern

Annular regime: Wavy-annular pattern

Annular regime: Annular pattern

Dispersed regime: Bubble flow pattern

Dispersed regime: Dispersed flow pattern

FIGURE 4.9   Description of gas-liquid flow regimes and patterns [33].

- *Annular Regime*   When the gas flow rate is increased, the intermittent regime becomes annular, with the gas continuously flowing in the center of the channel and the liquid flowing along the channel walls.
- *Dispersed Regime*   Obtained at high liquid velocities, the gas is dispersed into bubbles, which are much smaller than the channel diameter. As this regime is obtained at turbulent Reynolds numbers (based on the liquid phase), it may never be observed in very small channels.

The addition of surfactants or additives to thicken the liquid may increase the stabilization of the gas dispersion. Hessel et al. [34] have measured specific interfacial areas in the range of 9,000 to 50,000 m^2/m^3, which is about 100 times greater than what can be expected in a traditional bubble column.

4.5.3   Systems with Solids

In industrial applications involving solids, two main cases can be identified: operations in which solid particles have to be introduced into the micromixer and operations in which solid particles are generated inside the microchannels. The first case essentially concerns liquid–solid or gas–liquid–solid reactions where the solid is a catalyst. A high interfacial area is thus required, and different solutions can be proposed: If directly covering the walls by a catalyst layer is not sufficient, microposts can be erected in a staggered way by deep reactive ion etching, for example, enabling the flow to develop between the microposts with a limited pressure drop [35]. Packed beds are generally not recommended due to the technical difficulty of obtaining regular packing in microchannels, which results in high-pressure drop variations from one device to another.

The second case is of particular interest for industrial applications because the laminar regime and/or the small specific dimensions of the channels are used to create solid particles with specific properties (size, shape, surface properties, etc.). Two examples are presented in [36] and [37]. Köhler et al. [36] formed metal nanoparticles by the reduction of solutions containing metal ions. They obtained particles with diameters of less than 500 nm and pointed out the important influence of process parameters on particle properties. In particular, flow conditions are responsible for the aspect of the outer part of the particles, which determines their optical absorption capacity. Gröss et al. [37] used the microsegmented flow principle to form polymer minirods. The channel width enables the control of the diameter of the minirods (as small as 250 μm) and the segmentation in the immiscible solvent of the monomer flow controls the length of the minirods. Clogging has never occurred. Moreover, the introduction of dyed nanoparticles allows a controlled surface functionalization, which minimizes product losses.

These results illustrate how promising micromixers are for the generation of particle-controlled properties.

4.6   PROBLEMS AND SOLUTIONS

First, we must take into account that micromixers belong to a category of very recent devices. Consequently, some of the problems still being encountered when using micromixers will certainly be solved as one is able to better understand and control what happens within these devices.

Specific problems are typically linked to the size of the channels, but others result from a lack of pertinent data concerning the exact geometry of the micromixer and the physico-chemical parameters of the fluid(s). Some of the possible difficulties encountered in micromixers may be the following:

- *Problem of Connection and Feeding* In the case where several microchannels are in parallel, the well-balanced distribution of fluid must be ensured. Simplistic systems for distribution of the flow exist, but do not ensure a regular feeding of all channels. Indeed, some distribution systems have been inspired from fractal or constructal analysis enabling an equi-flow distribution [38, 39].

- *Problem of Control* In some cases, it may be necessary to determine the operating conditions—such as the temperature or flow rate—online. This may lead to two kinds of problems. The first is relative to the nature of the probe—What size thermocouple should be used? How should it be inserted into the device? What is its sensitivity and precision? etc.). The second is relative to the price of the measurement equipment because the range of the measured parameter requires very precise systems. The solutions to these problems are of different degrees. First of all, the systematic implementation of sensors is perhaps not useful. At the macroscale, the multiplication of sensors is not a problem and the common attitude is often to overestimate needs. However, special care must be taken with microdevices in order to correctly decide whether and where sensors are, in fact, useful. The second step in solving this problem will be the development of new integrated sensors. For the moment, they are primarily encountered in scientific research centers. However, one can certainly expect their commercial distribution in the future.

- *Problem of Solids Handling and Clogging* Solids handling is evidently a problem. When microchannels are not designed in order to handle solid particles, special care must be taken to prevent them from entering the device. For this, purpose filters can be placed before the feeding point. Despite the precautions taken, clogging or fouling may still occur, which leads to the problem of robustness and modularity of the system: The smaller the channels, the less robust the system. The user must also verify the consequences of complete clogging of a channel (the security aspects in particular). Furthermore, the easy replacement of a clogged channel must be considered during the design phase of the micromixer.

- *Problem of the Nature of the Surfaces* As previously mentioned, the surface-to-volume ratio in micromixers is very high and therefore the role of surfaces is very important. The problem of fouling has already been mentioned, but in addition, the degree of fouling may be more or less important depending on the construction material of the micromixer. The roughness of the surface is also an important parameter because the relative depth of this layer is much higher than in a macroscale device. Wettability is, of course, of major interest as it controls the hydrodynamic boundary conditions of zero-velocity at the walls, but also defines the nature of an emulsion (e.g., a hydrophobic surface will facilitate the creation of a water-in-oil emulsion). Consequently, special attention must be paid to the nature of the materials used, including those surrounding the active part of the microchannels (junctions, feeding parts, etc.).

BIBLIOGRAPHY

1. Hessel, V., Löwe, H., Müller, A., and Kolb, G. (eds.) (2005). *Chemical Microprocess Engineering: Processing and Plants*, Weinheim: Wiley-VCH.

2. Hessel, V., Löwe, H., and Schönfeld, F. (2005). Micromixers—a review on passive and active mixing principles. *Chem. Eng. Sci.*, 60: 2479–2501.

3. Engler, M., Kockmann, N., Kiefer, T., and Woias, P. (2004). Numerical and experimental investigations on liquid mixing in static micromixers. *Chem. Eng. J.*, 101: 315–322.

4. Wong, S. H., Ward, M. C. L., and Wharton, C. W. (2004). Micro T-mixer as a rapid mixing micromixer. *Sens. Act.* B 100: 359–379.

5. Nguyen, N.-T., and Wu, Z. (2005). Micromixers—a review. *J. Micromech. Microeng.*, 15: R1–R16.

6. Wong, S. H., Bryant, P., Ward, M., and Wharton, C. (2003). Investigation of mixing in a cross-shaped micromixer with static mixing elements for reaction kinetics studies. *Sens. Act.* B 95: 414–424.

7. Ehrfeld, W., Hessel, V., and Löwe, H. (eds.), (2000). *Microreactors: New Technology for Modern Chemistry*. Weinheim: Wiley-VCH.

8. Gad-El-Hak, M. (2002). The MEMS handbook. In F. Kreith, (ed.), *The Mechanical Engineering Handbook Series*, Boca Raton, FL: CRC Press.

9. Santiago, J. G., Wereley, S. T., Meinhart, C. D., Beebe, D. J., and Adrian, R. J. (1998). A particle image velocimetry system for microfluidics. *Exp. Fluids*, 25: 316–319.

10. Rouge, A., Spoetzl, B., Gebauer, K., Schenk, R., and Renken, A. (2001). Microchannel reactors for fast periodic operation: The catalytic dehydration of isopropanol. *Chem. Eng. Sci.*, 56: 1419–1427.

11. Günther, M., Schneider, S., Wagner, J., Gorges, R., Henkel, T., Kielpinski, M., Albert, J., Bierbaum, R., and Köhler, J. M. (2004). Characterisation of residence time and residence time distribution in chip reactors with modular arrangements by integrated optical detection. *Chem. Eng. J.*, 101: 373–378.

12. Aubin, J., Fletcher, D. F., and Xuereb, C. (2005). Design of micromixers using CFD modelling. *Chem. Eng Sci.*, 60: 2503–2516.

13. Stroock, A. D., Dertinger, S. K. W., Ajdari, A., Mezic, I., Stone, H. A., and Whitesides, G. M. (2002). Chaotic mixer for microchannels. *Science*, 295: 647–651.

14. Hoffmann, M., Schlüter, M., and Räbiger, N. (2006). Experimental investigation of liquid-liquid mixing in T-shaped micro-mixers using μ-LIF and μ-PIV. *Chem. Eng. Sci.*, 61: 2968–2976.

15. Dankwerts, P. V. (1952). The definition and measurement of some characteristics of mixtures. *Appl. Sci. Res.*, 3: 279–296.

16. Rauline, D., Le Blévec, J.-M., Bousquet, J., and Tanguy, P. A. (2000). A comparative assessment of the performance of the Kenics and SMX static mixers. *Chem. Eng. Res. Des.*, 78A: 389–396.

17. Hobbs, D. M., and Muzzio, F. J. (1997). The Kenics static mixer: A three-dimensional chaotic flow. *Chem. Eng. J.*, 67: 153–166.

18. Hobbs, D. M., and Muzzio, F. J. (1998). Reynolds number effects on laminar mixing in the Kenis static mixer. *Chem. Eng. J.*, 70: 93–104.

19. Zalc, J. M., Szalai, E. S., Muzzio, F. J., and Jaffer, S. (2002). Characterization of flow and mixing in an SMX static mixer. *AIChE J.*, 48: 427–436.

20. Aubin, J., Fletcher, D. F., Bertrand, J., and Xuereb, C. (2003). Characterization of the mixing quality in micromixers. *Chem. Eng. Technol.*, 26: 153–166.

21. Bothe, D., Stemich, C., and Warnecke, H.-J. (2006). Fluid mixing in a T-shaped micromixer. *Chem. Eng. Sci.*, 61: 2950–2958.

22. Hessel, V., Hardt, S., Löwe, H., and Schönfeld, F. (2003). Laminar mixing in different interdigital micromixers: I. Experimental characterization. *AIChE J.*, 49: 566–577.

23. Hobbs, D. M., and Muzzio, F. J. (1998). Optimization of a static mixer using dynamical systems techniques. *Chem. Eng. Sci.*, 53: 3199–3213.

24. Fournier, M. C., Falk, L., and Villermaux, J. (1996). A new parallel competing reaction system for assessing micromixing efficiency—experimental approach. *Chem. Eng. Sci.*, 51: 5053–5064.

25. Ehrfeld, W., Golbig, K., Hessel, V., Löwe, H., and Richter, T. (1999). Characterization of mixing in micromixers by a test reaction: Single mixing units and mixer arrays. *Ind. Eng. Chem. Res.*, 38: 1075–1082.

26. Panic, S., Loebbecke, S., Tuercke, T., Antes, J., and Boskovic, D. (2004). Experimental approaches to a better understanding of mixing performance of microfluidic devices. *Chem. Eng. J.*, 101: 409–419.

27. Aubin, J., and Falk, L. (2006). Micromélangeurs. In Xuereb, C., Poux, M., and Bertrand, J. (eds.), *Agitation et Mélange: Aspects fondamentaux et applications industrielles*, Paris: Dunod, pp. 306–307.

28. Haverkamp, V., Ehrfeld, W., Gebauer, K., Hessel, V., Löwe, H., Richter, T., and Wille, C. (1999). The potential of micromixers for contacting of disperse liquid phases. *Fresenius J. Anal. Chem.*, 364: 617–624.

29. Löb, P., Pennemann, H., Hessel, V., and Men, Y. (2006). Impact of fluid path geometry and operating parameters on 1/1 dispersion in interdigital micromixers. *Chem. Eng. Sci.*, 61: 2959–2967.

30. Hessel, V., Ehrfeld, W., Haverkamp, V., Löwe, H., and Schiewe, J. (1999). Generation of dispersions using multilamination of fluid layers in micromixers. In R. H. Müller (ed.), *Dispersion Techniques for Laboratory and Industrial Production*. Stuttgart: Wissenschaftliche Verlagsgesellschaft.

31. Ganan-Calvo, A. M., and Gordillo, J. M. (2001). Perfectly monodisperse microbubbling by capillary flow focusing. *Phys. Rev. Lett.*, 87: 274501/1–274501/4.

32. Hessel, V., Ehrfeld, W., Golbig, K., Haverkamp, V., Löwe, H., and Richter, T. (1998). Gas/liquid dispersion processes in micromixers: The hexagon flow. In *Proceedings of IMRET2*, New Orleans, LA, AIChE, pp. 259–266.

33. Coleman, J. W., and Garimella, S. (1999). Characterization of two-phase flow patterns in small diameter round and rectangular tubes. *Int. J. Heat Mass Trans.*, 42: 2869–2881.

34. Hessel, V., Ehrfeld, W., Herweck, V., Haverkamp, H., Löwe, H., Schiewe, J., Wille, C., Kern, T., and Lutz, N. (2000). Gas/liquid microreactors: Hydrodynamics and mass transfer. In *Proceedings of IMRET4*, Atlanta, GA, AIChE, pp. 174–181.

35. Losey, M. W., Jackman, R. J., Firebaugh, S. L., Schmidt, M. A., and Jensen, K. F. J. (2002). Design and fabrication of microfluidic devices for multiphase mixing and reaction. *J. Microelectromech. Syst.*, 11: 709–715.

36. Köhler, J. M., Held, M., Hübner, U., and Wagner, J. (2007). Formation of Au/Ag nanoparticles in a two-step micro flow-through process. *Chem. Eng. Technol.*, 30: 347–354.

37. Gröss, G. A., Hamann, C., Günther, M., and Köhler, J. M. (2007). Formation of polymer and nanoparticle doped polymer minirods by use of the microsegmented flow principe. *Chem. Eng. Technol.*, 30: 341–346.

38. Tondeur, D., and Luo, L. (2004). Design and scaling laws of ramified fluid distributors by the constructal approach. *Chem. Eng. Sci.*, 59: 1799–1813.

39. Luo, L., Tondeur, D., Le Gall, H., and Corbel, S. (2006). Constructal approach and multiscale components. *Appl. Therm. Eng.*, doi:10.1016/j.applthermaleng.2006.07.018

CHAPTER 5

MICROCHANNEL HEAT EXCHANGERS AND REACTORS

MARK GEORGE KIRBY and SVEND RUMBOLD

5.1 MICROCHANNEL HEAT EXCHANGERS

The origins of the microchannel heat exchanger can be traced to a desire on the part of the process engineer to achieve very high thermal effectiveness within a simple and compact device, with the option of including more than two streams and without the need for multiple units in series.

Compact heat exchangers have been available for many years, the principal early design being the gasketed plate exchanger, pioneered by Alfa Laval and others for use in the food and drink industry. A wide range of variants have subsequently evolved, including partially and fully welded designs, constructed from a range of high-performance alloys, to meet the more demanding temperature, pressure, and corrosion-resistance requirements of other industries. However, the gasketed plate exchanger is unsuited to high pressures and temperatures, or multistream duties.

The brazed aluminium plate-fin exchanger was initially developed for aircraft applications, where its lightweight status was the primary attraction. The concept proved ideally suited to cryogenic air separation processes, because a close approximation to countercurrent flow (and hence very close temperature approaches) can be achieved, and aluminium exhibits excellent thermal conductivity at cryogenic temperatures. The aluminium plate-fin has subsequently been widely adopted for cryogenic hydrocarbon processes, including LNG, NGL, and ethylene manufacture. Although the range of available materials of construction has been extended, material limitations still constrain the use of plate-fin designs: The use of aluminium for very-high-pressure duties is precluded by its mechanical strength, whereas stainless steel exhibits limited fin efficiency at cryogenic conditions.

Due to the weak nature of aluminum and particularly brazed joints, for a typical geometry of brazed plate-fin exchanger under steady-state conditions, the maximum

Microchemical Engineering in Practice. Edited by Thomas R. Dietrich
Copyright © 2009 John Wiley & Sons, Inc.

permissible temperature difference between streams is approximately 50°C (90°F), although some manufacturers suggest even lower figures. In more severe cases, such as two-phase flows, transient and/or cyclic conditions, this temperature difference should be lower, typically 20 to 30°C (36–54°F). It is very important to limit the cyclic or frequently repeated temperature fluctuations of any stream to $\pm 1°C$ ($\pm 1.8°F$) per minute; otherwise, leakage can occur. The printed circuit heat exchanger (PCHE) was developed by researchers at the University of Sydney to provide very high thermal effectiveness even for duties involving high temperatures and pressures, and was commercialized by Heatric in 1985. PCHEs include a matrix of flow passages on the micro/macro scale in a single-material diffusion-bonded core.

Compact lightweight equipment is of obvious benefit in many circumstances, but the benefits extend beyond size and weight savings: The ability to incorporate complex fluid flow geometries offers the opportunity for further temperature optimization, using close approaches and deep temperature crosses, while process layout can be simplified by combining complex multistream heat exchange within a single unit. Heatric has pioneered the use of the compact, diffusion-bonded PCHE, with its high-pressure and high-temperature capability, in a wide range of arduous offshore and onshore gas processing applications.

The notion of channel dimensions scale. For example, whether a heat exchanger is micro-, meso-, compact, etc., a very important consideration, is neither clear nor uniform. Here we would like to introduce the classification as proposed by Mehendale that has received wide acceptance and can be easily applied:

Microscale: $1-100\ \mu m$ (microstructured exchanger)
Meso-scale: $100\ \mu m-1\ mm$ (meso- or milli-structured)
Macroscale: $1-6\ mm$ (compact exchanger)
Macroscale: $>6\ mm$ (conventional exchanger)

Microchannel heat exchangers such as diffusion-bonded PCHEs are highly compact, typically being about one-fifth the size and weight of conventional heat exchangers for the same thermal application and pressure drops.

Passage dimensions are typically determined by the cleanliness of the fluids handled and can span the range of "microchannels" to "macroscale channels" (up to 10 mm). PCHEs can be constructed out of a range of materials, including austenitic stainless steels suitable for design temperatures up to 800°C and nickel alloys such as Incoloy 800HT suitable for design temperatures beyond 900°C.

Currently, there are thousands of tons of such microchannel matrices in hundreds of services—many of them involving arduous work on offshore oil and gas platforms where the size and weight advantages of microchannel heat exchangers are of obvious benefit. Although these matrices are predominantly involved in thermally simple two-fluid heat exchange, albeit at pressures up to 550 bar, PCHEs have also been used for many multistream counterflow heat exchangers. In addition, the field of applications is very varied, including specialized chemicals processing, with PCHEs even found orbiting the earth in the International Space Station.

Due to the inherent design flexibility of the etching process, the basic construction may be applied readily to both a wider range and the more complex integration of process unit operations than other compact exchanger surfaces.

The compact nature of microchannel exchangers precludes their use with streams that have an uncontrollable fouling tendency, whether from particulate deposition or scaling. Although there are a few straightforward issues specific to microchannel exchangers that arise in systems design, generally the application of good engineering practice will assure successful operation.

In most applications, no routine maintenance is expected to be necessary for microchannel exchangers constructed in a corrosion-resistant alloy (CRA). These exchangers are simple to operate, have low maintenance and high availability. However, various measures are available for those situations when the unexpected happens. Chemical reaction, rectification, stripping as well as boiling and condensation, can be incorporated into compact integrated process modules. Crucially, printed circuit technology has been the enabling technology in certain applications, as a result of the modules' compactness.

Techniques for chemical coating onto the surfaces of channels have been developed, with applicability to both protective coatings and catalytically active coatings. We will describe a selection of innovative printed circuit technology examples. Beyond the mainstream hydrocarbon processing industry, printed circuit technology is being developed to address the challenges of emerging energy opportunities such as nuclear power generation and fuel cell systems. These applications perhaps represent two extremes of both size and process integration, and thus aptly serve to demonstrate the range of industrial use of microchannel devices.

5.1.1 Construction

Diffusion-bonded exchangers are high-integrity compact plate-type and extended surface heat exchangers. Fluid flow channels—which can be of unlimited variety and complexity—are chemically etched (Fig. 5.1a). Flow geometries are always designed for each stream in order to optimize overall heat transfer and individual pressure drop performance. As an alternate construction method, flow channels are produced from fins that are die- or roll-formed and sandwiched between parallel plates, also called parting sheets (Fig. 5.1b). To produce the exchanger core, the layers are typically stacked using an alternating hot and cold design (Fig. 5.1c). Fluid contact can be counterflow, crossflow, coflow, or a combination of these to suit the process requirements.

Once stacked, the assembly is diffusion-bonded together in a bonding furnace to form a solid, all-metallic block (Fig. 5.1d). No melting or other gross deformation occurs, and no fluxes, fillers, or gasket-type material are added. Passage configurations are faithfully preserved in the bonded blocks, with the bonds reaching parent metal strength and ductility.

To suit the application, multiple bonded blocks are easily welded together, to form the overall exchanger core. Header shells, nozzles, and flanges are then welded to complete what is essentially a relatively simple exchanger assembly. See Figs. 5.2 and 5.3.

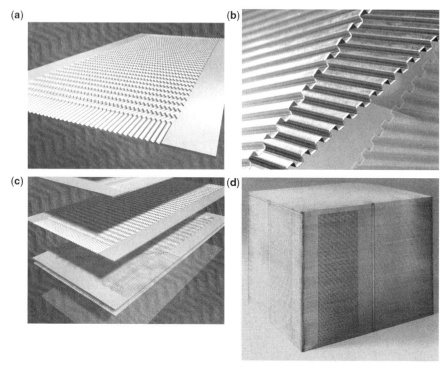

FIGURE 5.1   (a–d) Fluid flow channels.

FIGURE 5.2   Blocks welded to make core.

FIGURE 5.3 Fluid distribution headers welded to core.

Although diffusion-bonded exchangers like the PCHE may be compact, they are not necessarily small: Single units of up to 100 tons have been manufactured, providing very substantial benefits over the 500-ton shell-and-tube alternatives. See Figs. 5.4 and 5.5.

To construct the high specific surface area and high-integrity joint of the diffusion-bonded exchanger, one of the following methods is used.

FIGURE 5.4 Completed 23-ton microchanneled exchanger.

FIGURE 5.5 Microchanneled exchanger installed.

Plate Forms

Etched Plate: Printed Circuit Heat Exchanger Type Flow channels are developed along flat metal plates using a photochemical milling technique analogous to that used for the manufacture of electronic printed circuit boards, giving rise to the "printed circuit" exchanger name or PCHE.

Plate thickness used to carry passages can vary between 0.5 to 5 mm, but are commonly 1.6 or 2 mm thick. The passages are typically semicircular in crosssection after etching, having a diameter that is varied to achieve the desired performance characteristics, typically 2 mm. See Figs. 5.6 and 5.7.

Formed Plate: Plate-Fin Heat Exchanger Type This form of construction has corrugated fins (secondary surface) sandwiched between parallel plates, also called parting sheets, the primary heat transfer surface. Fins are die- or roll-formed and attached to the plates either by diffusion bonding, as practiced uniquely by Heatric, or by brazing at a multitude of companies. Other techniques used in the industry include soldering, adhesive bonding, welding, mechanical fit, or extrusion.

In the case of diffusion-bonded construction, there is no braze alloy involved. Instead, the peaks and troughs of each corrugated fin are diffusion-bonded to the adjacent separating plates. The result is a solid block of metal with engineered flow channels passing through it. Brazed assembly provides a similar result, but with joints that have somewhat lower strength compared with the plate and fin material in an industrial-size unit or similar configuration. (See Fig. 5.8.) Many

FIGURE 5.6   Etched cross-flow microchannel core.

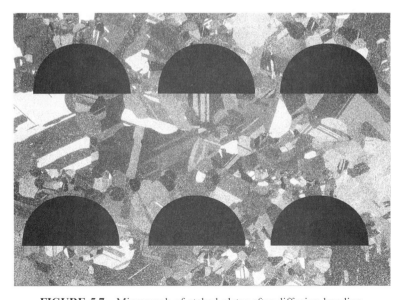

FIGURE 5.7   Micrograph of etched plates after diffusion bonding.

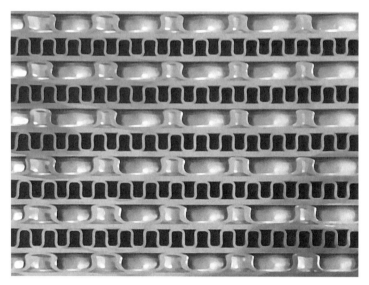

FIGURE 5.8   Fin corrugations after diffusion bonding.

different corrugations are available with various fin densities, thicknesses, and corrugation styles (plain, perforated, and herringbone).

Different flow geometries are designed for each stream in order to optimize heat transfer and pressure drop performance. The resultant layers are then stacked, ready for joining. Plate-fin style cores are generally more suited to lower-pressure services: around 110 barg in the case of brazed aluminium and up to approx. 259 barg for diffusion-bonded stainless steel. Printed circuit style cores may be used for substantially higher design pressures—sometimes in excess of 550 bar.

Diffusion Bonding

Introduction   The Welding Institute (TWI) describes diffusion bonding as a "solid-state joining process capable of joining a wide range of metal combinations to produce both small and large components." Variants of the diffusion bonding technique are available, offering the possibility of the joining of many new materials and configurations.

The technique typically involves joining together high-value materials: This, together with capital equipment costs, surface preparation requirements, and long bonding times, means that the process is traditionally associated with the manufacture of high-value, relatively low-production-volume components. However, its versatility for unusual materials and its relative ease of use for stainless steel and other corrosion-resistant alloys (CRAs) will ensure that the process continues to be developed for specific application requirements. See Fig. 5.9(a–e).

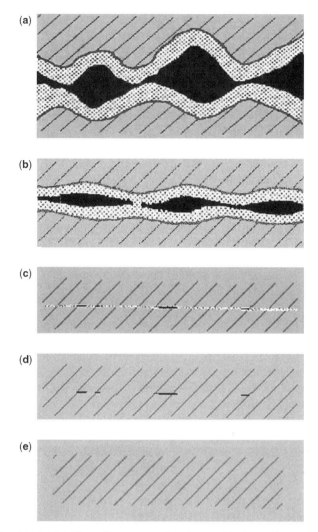

FIGURE 5.9 (a) Initial "point" contact, showing residual oxide contaminant layer. (b) Yielding and creep, leading to reduced voids and thinner contaminant layer. (c) Final yielding and creep; some voids remain with very thin contaminant layer. (d) Continued vacancy diffusion eliminates oxide layer, leaving few small voids. (e) Bonding is complete (by permission of TWI Ltd.).

The bonding process is dependent on a number of parameters—in particular, time, applied pressure, bonding temperature, and method of heat application. Other examples of solid-state joining processes include cold pressure welding, friction welding, magnetically impelled arc butt (MIAB) welding, and explosive welding.

Diffusion bonding itself can be categorized into a number of variants, depending on the form of pressurization, the use of interlayers, and the formation of a transient liquid phase. Each finds specific application for the range of materials and geometries that need to be joined.

Solid-Phase Diffusion Bonding   In its simplest form, diffusion bonding involves holding components under load at an elevated temperature usually in a protective atmosphere or vacuum. The loads used are generally below those that would cause macrodeformation of the parent material(s) and temperatures of 0.5 to 0.8 Tm (where Tm = melting point in K) are employed. Times at temperature can range from 1 to 60+ minutes, but this depends on the materials being bonded, the joint properties required, and the remaining bonding parameters.

An examination of the sequence of bonding above emphasizes the importance of the original surface finish. To form a bond, it is necessary for two, clean, and flat surfaces to come into atomic contact, with microasperities and surface layer contaminants removed from the bonding faces during bonding. Various models have been developed to provide an understanding of the mechanisms involved in forming a bond. First, they consider that the applied load causes plastic deformation of surface asperities, reducing interfacial voids. Bond development then continues by diffusion-controlled mechanisms including grain boundary diffusion and power law creep.

Bonding in the solid phase is mainly carried out in a vacuum or protective atmosphere, with heat being applied by radiance, induction, direct or indirect resistance heating. Pressure can be applied uniaxially or isostatically. In the former case, a low pressure (3–10 MPa) is used to prevent macrodeformation of the parts (i.e., no more than a few percent). This form of the process therefore requires a good surface finish on the mating surfaces as the contribution to bonding provided by plastic yielding is restricted. Typically, surface finishes of better than 0.4 μm RA are recommended; in addition, the surfaces should be as clean as practical to minimize surface contamination.

In hot isostatic pressing, much higher pressures are possible (100–200 MPa) and therefore surface finishes are not so critical, so finishes of 0.8 μm RA and greater can be used. A further advantage of this process is that the use of uniform gas pressurization allows complex geometries to be bonded, against the generally simple butt or lap joints possible with uniaxial pressurisation.

5.1.2 Design Methodology

These thermal calculations are intended to demonstrate how an approximated thermal design for a microchannel exchanger can be found. For a more detailed analysis, refer to Incropera et al. [12]

	Description	Calculation	Units
A	= heat exchange area	$= nL(\pi D/2 + D)$	m^2
A_c	= free flow area	$= n\pi D^2/8$	m^2
c_p	= heat capacity		J/kg K
D	= passage diameter		mm
D_h	= hydraulic diameter	$= 4A_c L/A$	mm

f	= friction factor	—
G	= mass flux $= W/A_c$	kg/m²s
h	= heat transfer coefficient $= jGc_p/Pr^{2/3}$	W/m²K
j	= j factor $= StPr^{2/3}$	—
L	= passage heat transfer length	mm
L_{ee}	= distributor passage lengths (entrance/exit)	mm
n	= no. of passages	passages
ΔP_f	= frictional pressure drop $= 4f \cdot VH(L + L_{ee})/D_h$	—
Pr	= Prandtl number	—
Re	= Reynolds number $= GD_h/\mu$	—
R_w	= wall resistance	—
St	= Stanton number $= h/Gc_p$	
U	= overall heat transfer coefficient $= 1/(1/h_A + R_w + 1/h_B)$	W/m²K
VH	= velocity head $= 0.5G^2/\rho$	Pa
W	= flow rate	kg/s
μ	= bulk viscosity	cP
ρ	= density	kg/m³

Hot Side Fluid

Design details

FLUID: STAGE 1 RESIDUAL

$W = 22.5$ kg/s	$D = 2.0$ mm approx.
$n = 64{,}464$	$L_{ee} = 56$ mm
$L = 1{,}284$ mm	Pr = 0.90
$c_p = 2{,}701$ J/kg K	$\rho = 68$ kg/m³
$\mu = 0.014$ cP	

By calculation:

$A = 426$ m²	$A_c = 0.101$ m²
$D_h = 1.22$ mm	$G = 223$ kg/m²s
Re = 19,430	VH = 365 Pa

THERMAL CALCULATION

Referring to fig. 10-75 in Kays and London [9] (surface: $17.8 - 3/8W$), we obtain for an order-of-magnitude indication of likely performance:

$$j = 0.0039 \quad \text{and} \quad f = 0.0190$$

Therefore,

$$h = 2{,}515 \text{ W/m}^2\text{K} \quad \text{and} \quad \Delta P_f = 30 \text{ kPa}$$

Cold Side Fluid

Design details

FLUID: FEED GAS

$$W = 28.8 \, \text{kg/s} \qquad D = 2.0 \, \text{mm approx.}$$
$$n = 67,308 \qquad L_{ee} = 50 \, \text{mm}$$
$$L = 1,230 \, \text{mm} \qquad \text{Pr} = 0.89$$
$$c_p = 2439 \, \text{J/kg K} \qquad \rho = 81 \, \text{kg/m}^3$$
$$\mu = 0.015 \, \text{cP}$$

By calculation

$$A = 426 \, \text{m}^2 \qquad A_c = 0.106 \, \text{m}^5$$
$$D_h = 1.22 \, \text{mm} \qquad G = 272 \, \text{kg/m}^2\text{s}$$
$$\text{Re} = 22,169 \qquad \text{VH} = 456 \, \text{Pa}$$

THERMAL CALCULATION

Referring to fig. 10-75 in Kays and London [9] (surface: $17.8-3/8\text{W}$), we obtain for an order-of-magnitude indication of likely performance:

$$j = 0.0036 \qquad \text{and} \qquad f = 0.0180$$

Therefore,

$$h = 2,578 \, \text{W/m}^5\text{K} \qquad \text{and} \qquad \Delta P_f = 34 \, \text{kPa}$$

Overall Performance

$$U = 1/(1/2,515 + 6.25e - 05 + 1/2,578)$$
$$= 1,179 \, \text{W/m}^5\text{K} \quad (\text{see } 944 \, \text{W/m}^5\text{K})$$

5.1.3 Characteristics of Microchannel Devices

The construction method gives rise to some obvious, and some not so obvious, characteristics of microchannel exchangers.

Compactness   The aim of the microchannel construction technique is to create a compact heat exchange core, as illustrated in Figs. 5.2 through 5.5. Although compact relative to shell-and-tubes, microchannel exchangers need not necessarily be small in absolute terms, as Fig. 5.10 illustrates.

High Efficiency   The flexible thermal design achievable with microchannel exchangers ensures high heat transfer coefficients, contributing to their compactness.

FIGURE 5.10 Relative compactness between microchanneled PCHE vs. conventional shell-and-tube.

Flexible Configuration The flexible thermal design also permits such features as pure countercurrent contact between streams, multistream contact, fluid mixing, and two-phase heat transfer (boiling, condensation, absorption, and rectification). This can give rise to considerable process enhancement. For example, the exchanger shown in Fig. 5.11 has four process streams integrated into a single unit. This multistream capacity provides space and weight advantages through reduced exchanger and piping weight, as well as simplifying process control. See Fig. 5.12.

FIGURE 5.11 Multi streamed exchanger.

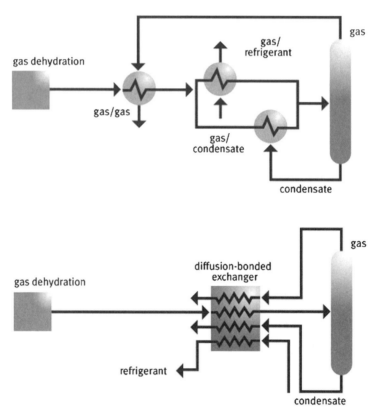

FIGURE 5.12   Original and simplified flow scheme for the multistreamed exchanger.

Mechanical Robustness   An etched diffusion-bonded microchannel exchanger such as a PCHE combines the thermal capabilities and size of compact exchangers with the temperature and pressure capabilities of tubular exchangers. Their cores are an assembly of relatively thick-walled, small-bore passages, and they therefore can withstand very high pressures. Units with design pressures of 500 bar have been manufactured.

Extreme temperatures (cryogenic to hundreds of degrees) can be accommodated using appropriate materials such as stainless steel or aluminium, as there are no gaskets or brazed metals to limit temperature. However, if start-ups and shut-downs are common and temperatures are extreme, then control strategies for start-up and shut-down require consideration.

Just as the core matrix construction protects against catastrophic failure, it also provides constraints that can induce thermal stresses and hence thermal fatigue under strongly cyclic conditions, as the mutually supporting metal sections tend to restrain the thermal expansion and contraction of each other. The importance of sound process control cannot therefore be underemphasized.

Materials The selection of materials is reasonably flexible, subject to the following principal constraints:

- Negligible general or localized corrosion in operation and shut-down
- Highly ductile, as differential thermal expansion in the heat exchange core can give rise to high local stresses that must be safely relieved by yielding
- Compatibility with manufacturing processes such as diffusion bonding, brazing, welding, etc.
- Pressure vessel codes.

Dependent on the joining technique used, microchannel matrices can be manufactured in a range of high-performance corrosion-resistant materials. See Fig. 5.13.

Diffusion-bonded exchangers have an unrivalled capability to meet mechanically, chemically, and thermodynamically demanding applications. Cores manufactured using this type of joint are available in 300 series austenitic stainless steels, 22Cr duplex steel, copper, nickel, titanium, Incoloy 800, and other metals, permitting use in a wide range of applications. They are readily designed for operating pressures of more than 550 bar and temperatures ranging from deep cryogenic to over 900°C. A brazed alloy such as aluminium is both compact and lightweight. The material is both cheaper and more thermally conductive than other materials such as 300 series stainless steels used in diffusion-bonded exchangers. Hence, if brazed aluminum is considered fit for the application, then it will usually provide a cheaper exchanger solution than stainless steel. The exception exists when moderately high operating pressure coupled with high gas flow results in the need to extensively manifold a brazed aluminium exchanger that is more onerous and expensive to fabricate than

FIGURE 5.13 Selection of diffusion-bonded alloys and passage diameters (left to right): 1 mm titanium, 1 mm copper, 600 μm SS, 100 μm SS, and 2 mm SS.

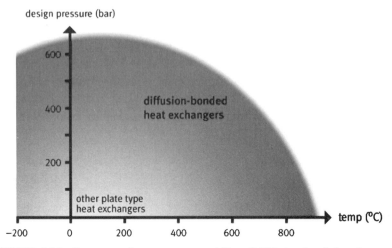

FIGURE 5.14   Pressure and temperature capability of diffusion-bonded exchangers.

a simpler single manifolded diffusion-bonded exchanger. See Fig. 5.14. Let us first consider the limitations of aluminum exchangers.

Design Temperature   In theory, heat exchangers manufactured in aluminum can be designed for temperatures as high as 204°C (400°F). In practice, however, except for very-low-pressure applications, this must be limited to 65°C (150°F), which is the maximum ASME Section VIII Division 1 Boiler and Pressure Vessel code design temperature for the preferred high-strength alloy 5083.

Design Pressure   The low mechanical strength of aluminum limits its maximum design pressure to about 100 bar according to manufacturers' current literature (although this is significantly reduced if design temperatures exceed the 65°C noted above). Fin choices are severely limited at this pressure and multiple nozzles are often required per header.

Corrosion   Mercury is a common contaminant found in hydrocarbon feed gases, and in its liquid form it attacks most aluminium alloys, particularly the high-strength Al/Mg alloys used for headers or nozzles. Manufacturers talk of mercury-tolerant construction, but in reality this only reduces the rate of destruction by limiting the potential sites likely to be attacked. Aluminium exchangers are most susceptible to mercury corrosion during plant shut-downs when mercury liquid beads collect in the nondrainable dead areas.

   Dimethylmercury (DMM) has been discovered to be one of the organomercury compounds in natural gas, besides mercury metal. The effect of DMM on the corrosion of carbon steel and aluminium metal was rather frightening, as compared to that of elemental mercury or mercuric chloride, which is well known as a severely corrosive agent. The results showed that DMM in methanol or petroleum ether as a solvent yielded corrosion patterns similar to those for the elemental Hg solution. It revealed uniform characteristics on carbon steel and a pitted appearance on aluminium. The higher the DMM concentration or higher the reaction temperature, the more severe the corrosion that occurred because of the higher elemental Hg concentration generated. A trace amount of hydrogen chloride or hydrogen sulfide, which is also present in large quantity in natural gas, remarkably increased the corrosion potential of DMM on metal. The corrosion rate was approximately 700 times faster than the rate for one containing only DMM and 40 times faster than the rate for one containing only acid.

Fatigue   Due to the weak nature of aluminium and particularly the brazed joints, it is very important to limit the cyclic or frequently repeated temperature fluctuations of any stream to $\pm 1C$ ($\pm 1.8°F$) per minute. Otherwise, leakage can occur.

Temperature Difference Between Streams   The ALPEMA guide states that for a typical geometry of brazed plate-fin exchanger under steady-state conditions, the maximum permissible temperature difference between streams is approximately 50°C (90°F), although some manufacturers suggest even lower figures. In more severe cases such as two-phase flows, transient and/or cyclic conditions, this temperature difference should be lower, typically 20 to 30°C (36–54°F).

5.1.4   Selection Criteria

Fouling Characteristics   The compact nature of microchannel exchangers precludes their use with streams that have an uncontrollable fouling tendency, whether from particulate deposition or scaling. Hydrocarbon processing—for example, the handling of gas, condensate, and cooling medium streams—yields satisfactory results, but crude oil processing does not, as reflected in the absence of reliable data supporting low fouling characteristics. Figure 5.15 shows the typical condition of passages after cooling high-pressure hydrocarbon gas for over 9 years.

It is often possible to turn "fouling" streams into adequately "clean" streams. For example:

- Strainers remove particulates that are too large to pass through exchanger passages.
- The sound design of separator internals reduces liquid and solid carry-over to acceptable levels.
- Periodic cleaning of the exchangers controls fouling deposits. Treatments can be undertaken online—for example, periodic injection of solvents to remove wax or hydrates. Offline cleaning might be as straightforward as periodic hydro-jetting or back-puffing to remove fines and dust, or chemical cleaning to remove more adhesive deposits.
- Other possibilities such as periodic online warming may be used to remove meltable solids such as hydrates, frozen glycol (from dehydration towers), and wax.

FIGURE 5.15 Clean condition of approx. 2-mm semicircle after considerable time in operation.

Operating Profile A June 1999 article [13] in *Chemical Engineering*, "Refurbishing Worn Out Heat Exchangers," made the following sound point with reference to shell-and-tube exchangers:

> Properly designed exchangers, made from corrosion-resistant materials, in steady-state operation, at design point conditions, require the least maintenance and have the longest life. Moderate deviations from design point do not materially affect service life or require extra maintenance. Large continual deviations above or below the design point can cause more-rapid-than-expected deterioration. Big excursions of short duration beyond the design point can cause unforeseen failures.

Experience shows that this also holds true for microchannel exchangers such as the PCHE. Relative to conventional exchangers:

- Microchannel exchangers are not subject to specifically "shell-and-tube" problems such as tube vibration; thermally induced tube, shell, or tube-plate overstress; tube buckling; girth flange distortion, tube impingement erosion; baffle plate erosion; tube-to-tubesheet joint leakage; expansion joint failure; etc.
- Stainless steel exchangers are not subject to low-temperature embrittlement, such as occurred in the carbon steel shell-and-tube in the Longford gas plant explosion in Victoria, Australia. Copper exchangers are not subject to ignition in oxygen service, as occurred in the aluminium plate-fin explosion at Bintulu, Sarawak, Malaysia.

- Process modifications that induce changes in the temperature profile through a compact microchannel core can give rise to thermal fatigue if sufficiently severe. Diffusion-bonded exchangers are no doubt much more resistant than aluminium plate-fins to thermal fatigue, but their high efficiency, low mass, and matrix core construction probably result in greater susceptibility than shell-and-tubes—although shell-and-tubes are by no means immune, as noted earlier.

Thermal Fatigue   The greatest threats of thermal fatigue for any exchanger type lie with malfunctioning (untuned, missized, or broken) control systems and boiling coolant. Due to the weak nature of aluminum and particularly brazed joints, it is very important to limit the cyclic or frequently repeated temperature fluctuations of any stream to ± 1C ($\pm 1.8°$F) per minute. Otherwise, leakage can occur.

In general, higher-temperature fluctuations can be accommodated by diffusion-bonded joints. Temperature fluctuations giving rise to rapid core temperature changes of less than $10°$C are not normally of concern, whereas cycling $100°$C changes certainly require immediate attention.

Here are some guidelines to help distinguish between potentially damaging and benign conditions:

- For the brazed plate-fin exchanger, the ALPEMA guide states that for a typical geometry under steady-state conditions and single phase, the maximum permissible temperature difference between streams is approximately $50°$C ($90°$F), although some manufacturers suggest even lower figures.
- In more severe cases such as two-phase flows, transient and/or cyclic conditions, this temperature difference should be lower, typically 20 to $30°$C (36–$54°$F).
- For the diffusion-bonded exchanger, under steady-state, single- or two-phase conditions, temperature differences between streams over hundreds of degrees are permissible, because the exchanger has superior resistance to thermal shock than brazed joints. There is limited opportunity for damaging thermal cycling when the difference between the stream inlet temperatures is small—say, less than $50°$C—unless conditions are remarkably unstable over prolonged periods.
- In low-efficiency applications with large differences in inlet temperatures between the streams—gas compression inter-coolers, for example—the temperature difference between the streams at any point within the exchanger can be quite wide, and hence the changes to metal thermal profiles through the exchanger are potentially large. Most commonly, it is substantial changes in fluid flow rates that can give rise to damaging cycling thermal stresses. Fluctuations in a flow of 10% or so are not normally of consequence, but rapid variations by a factor of 2 or 3 require immediate attention. The most damaging situation in gas coolers is repetitive on/off control valve action.

  Wide temperature differences also create the potential for highly unstable operation if the cooling medium is not under sufficient pressure to preclude boiling in compression inter- and after-coolers. Boiling itself is not inevitably unstable in appropriately designed systems and heat exchangers, but in the presence of temperature differences of 100 to 200°C it must be assumed to be so.

- In high-efficiency applications, such as feed/effluent heat exchange, for example, the difference in temperature between the two streams at any point within the exchanger is small, and therefore potential variations in the metal temperature profile are small. Also, it is often the case in such applications that the flows on both side of the exchanger tend to change together, again limiting the potential for significant thermal cycling. Normal control approaches are typically compatible with microchannel exchangers, provided that controllers are tuned appropriately. However, some situations may require special attention. For example:

   If the flow of a process stream is inherently unstable, feed-forward control might be required to stabilize the temperature profile in the exchanger. Alternatively, elimination of automatic control might give rise to better stability than poor, lagging control in this situation.

   Inherently on/off heat exchange duties with wide temperature differences between the stream inlet temperatures should be avoided as they effectively expose the exchanger to severe start-up on a frequent basis.

5.1.5 System Design

Although there are a few straightforward issues specific to microchannel exchangers that arise in systems design, generally the application of good engineering practices will assure successful operation.

Optimization Considerable benefits can accrue to systems designed with microchannel characteristics in mind. The implications of using a compact diffusion-bonded exchanger offshore can be profound, as there will be a direct impact on the platform sizing and structure, and reduced overpressure relief requirements. Onshore, plant modularization, miniaturization, and transportation will be simplified, and safety will be enhanced, by the low hold-up volumes. In compressor inter-cooler and after-coolers, advantage is often taken of counterflow contact in PCHEs to permit deep temperature crosses.

This allows much lower cooling medium recirculation rates, reducing piping costs and weight, and fewer pumping requirements. Tight temperature approaches are also feasible, and greater precooling of the gas enhances compressor efficiency. Careful specification of the cooling medium inlet temperature can also avoid hydrate and wax formation. The ability of certain compact exchangers to handle multistream contact can greatly simplify piping and control, and enhance process efficiency. An example is shown in Fig. 5.14.

Control Control requirements for microchannel exchangers are not arduous. In reasonably steady processes, normal PID temperature control systems are perfectly adequate so long as they are appropriately tuned and not malfunctioning. As noted above, eliminating control on some exchanger services such as hydrocarbon gas inter- and after-coolers could cut costs and improve efficiency.

Since the most damaging thermal cycles occur in heat exchangers when the flow of one stream is repetitively stopped, a good safety measure is to use a minimum stop on the control valve. Preferably, these would be physically applied stops at the valve, as they are more foolproof than software stops that can easily be overwritten. See Fig. 5.16.

Limitations on normal PID temperature control can arise in the presence of persistent process instability as the response time of compact exchangers to changing process conditions is the same order of magnitude as the dead time and response of temperature measurement instruments. The response time of the temperature measurement can be improved by measures such as the use of spring-loaded thermocouples rather than resistance temperature detectors (RTDs), and stepped vertical thermowells closely fitting the sensing elements and filled with heat transfer fluid.

If the speed of response is still not adequate in the presence of persistent and severe process instabilities, then feed-forward control may be employed to stabilize the temperature profile through the exchanger.

Instrumentation and Monitoring A complete analysis of heat exchange performance generally requires inlet and outlet temperatures, flows, compositions, pressures, and pressure drops. Pressure drops across strainers give warning of excessive particulate accumulation. More limited data may be adequate to disclose any tendency to foul, or to disclose process stability problems.

Straining It is fundamental that outsized particles be strained from streams entering PCHEs, in order to avoid plugging. Strainer apertures one-third of the minimum clearance are recommended for most circumstances. Once-off commissioning debris

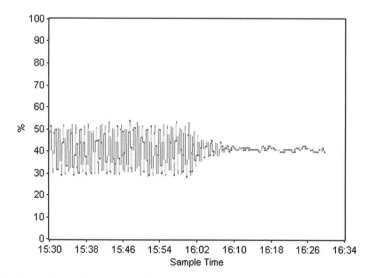

FIGURE 5.16 Damaging control valve action taken from a compressor inter-cooler showing high frequency and amplitude cycles that are readily rectified by tuning (valve position percentage vs. 24-h time).

FIGURE 5.17   Conical or top-hat strainer.

might simply require a commissioning (top hat) strainer (Fig. 5.17), whereas large and continuous particulate loads may require duplex (Fig. 5.18) or online cleanable strainers, avoiding the need to shut down for strainer cleaning.

5.1.6   Training Operators

Despite the relatively widespread use of microchannel heat exchangers, there are still many operating personnel who do not have firsthand experience with them. From

FIGURE 5.18   Duplex strainer system (by permission of Barton Firtop).

the outside they look familiar enough, with normal flange connections and heavy welding, but this apparent familiarity can induce operational errors such as removal of protective strainers or failure to tune controllers appropriately, despite clear instructions in the available operating manuals.

A good supplier provides a brief training program for operating personnel prior to commissioning, so the nature of the system's internals is fully understood. End users will find such a program greatly helpful in their avoiding basic operating errors. Basic training can reduce by more than half the number of reported incidents.

5.1.7 Commissioning and Operation

Good preparation in exchanger selection, specification, and systems design will simplify commissioning and operation. Systems do not always operate at their design conditions. To lessen the risk of fouling, it is important to consider the effect of the following circumstances:

1. Shut-down may settle particulate and increase wall temperatures.
2. Blow-down may freeze coolant not containing sufficient antifreeze, such as glycol.
3. Turn-down will lower velocities that may settle particulate and increase wall temperatures that could possibly scale.
4. Irregular cooling of medium water temperatures, such as seasonal changes in an ambient or seawater temperature, may indirectly allow gas hydrating or waxing, reducing exchanger performance.
5. Poor monitoring and maintenance of the quality of the cooling medium may allow corrosion and subsequent fouling of exchanger surfaces.
6. Variation in fluid characteristics may affect the performance of protective vessels such as knock-out pots—for example, oil wells producing higher than anticipated oil-to-gas ratios.
7. Plant modifications can unexpectedly introduce, promote, or exacerbate fouling or corrosion.

When these have not received appropriate consideration, the top three complications experienced in microchannel exchangers are:

- Blockage
- Thermal fatigue of the construction material
- Corrosion.

Therefore, critical areas for attention are:

- Strainer maintenance
- Controller action
- Coolant pressure and chemistry
- Monitor fouling trend.

Strainer Maintenance Inevitably, an abnormal quantity of particulate matter is present in pipework during commissioning, and hence strainers require close attention to ensure that they do not burst and release their contents into the exchangers. Although this is not fatal, it is certainly easier to clean a strainer than a heat exchanger.

Typically, there is a natural tendency to resist the shut-down of a processing plant such as compressors in order to service utility strainers. Therefore, thorough clean up of the cooling medium system prior to introducing process streams to coolers is recommended. This provides the opportunity for orderly servicing of coolant strainers at a period when particulate load is high. Provided that the cooling medium piping is reasonably corrosion-free, the clean-up by the strainers will then be rapid, although heavily corroded pipework will shed scale for some months. Do not allow a strainer to be removed because it persistently requires cleaning. This is a good indication that the strainer is essential or the system needs proper cleaning. Ensure that the strainer element sits snugly in the strainer body. Poorly fitted elements allow fluid and particulate to be bypassed; 100% straining of the fluid is required.

Controller Action Unstable controllers, particularly if they are running in on/off mode, can damage exchangers, and hence an early check on controller operation is important. Even if tuning is basically sound, control valves may have broken linkages or deposits may cause sticky action. Such problems need to be investigated and addressed. Unaveraged trends of flow or pressure drop normally provide the most reliable guide to successful control. Temperature measurement tends to be subject to lags that mask cycling.

Cooling Medium Pressure and Chemistry Utilities can be neglected during commissioning and operation, but they are critical to satisfactory long-term operation. Earlychecks are recommended for:

- Coolant pressure at the outlet of the PCHEs. If control valve back-pressure is relied on to suppress boiling, continuous monitoring may be safer.
- Coolant salinity, hardness, pH, oxygen, and chlorides. In carbon steel piping systems, corrosion inhibitor levels also require routine monitoring; a weekly inspection is usually sufficient.

Monitor Fouling Trend Fouling is likely to be a long-term matter and adequately monitored with infrequent data. Depending on the specific process and instrumentation, fouling might most easily be discerned from deterioration in thermal performance, disclosed by temperatures, or from pressure drops.

Control or process stability problems can only be reliably identified from trends generated with data collected at less than 1-min intervals. Readings averaged over a minute or more may mask potentially damaging instabilities, as the response time of microchannel exchangers is very rapid, and process instabilities can also be rapid. The cycling exhibited in Fig. 5.16, for example, would probably not show if subject to 1-min averaging.

Electronic communications can readily direct data streams to the supplier's office, where detailed analysis may be undertaken and informed feedback provided to operators. Monitoring of the success of the operators' installation and operating practices is a potentially powerful safeguard against future operating problems.

For plant observations to be of value, they should be properly recorded. Changes in operating conditions, no matter for how short a time, must be accurately

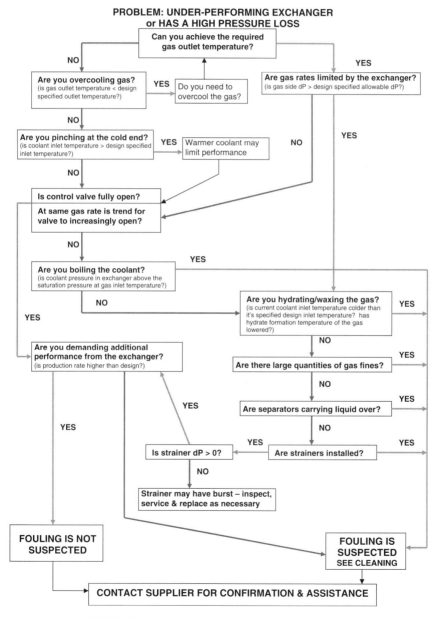

FIGURE 5.19 Problem: Cooler is underperforming.

documented because it could well be that these apparently unimportant factors are precisely those that caused an accelerated fouling problem. To determine if poor performance should be suspected, consider using something like the diagnostic guide flow scheme in Fig. 5.19, which was used to evaluate possibly underperforming compression inter- and after-coolers.

5.1.8 Maintenance

In most applications no routine maintenance is expected to be necessary for microchannel exchangers constructed in a corrosion-resistant alloy (CRA). These exchangers are simple to operate, have low maintenance and high availability. However, various measures are available for when the unexpected happens.

Strainers Strainer elements in the connecting pipework (where fitted) provide local protection and are removable for inspection and cleaning.

Gas Puffing Clearing restriction can often be quickly and successfully achieved by means of "puffing." Probably the most common maintenance procedure required is clearing from the inlet face particulates resulting from the absence of an appropriate strainer or the bursting of a strainer element due to overloading. Early cleaning of the exchanger following the mishap will help avoid particles becoming aged and adherent to exchanger surfaces. See Fig. 5.20.

Gas puffing involves installing a bursting disc at the inlet flange to the exchanger and pressurizing the exchanger and downstream piping with a gas such as nitrogen until the bursting disc bursts and releases the built-up pressure. The sudden, extensive back-flow of gas carries particulate matter out of the exchanger. Pressures up to the design pressure can be used, provided that the expelled gas and debris may be safely directed and the exchanger is suitably restrained (preferably on its supports).

FIGURE 5.20 Gas puffing through coolant inlet nozzle.

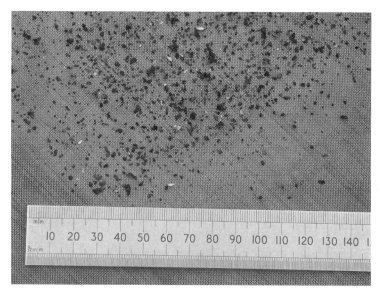

FIGURE 5.21 Coolant side fouling removed during gas puff.

See Figs. 5.21 and 5.22. To catch expelled material, we recommend placing a strainer mesh (around 200 μ) beside the PVC sheet for the first gas puff.

Hydrojetting Jetting is effective for removing adherent liquid, scale, and particulate foulant. The method is especially successful when cleaning headered chambers

FIGURE 5.22 Hydrocarbon gas side fouling removed during gas puff.

where access, usually through small service connections, allows the inlet and outlet faces of the exchanger core to receive treatment.

Chemical Cleaning This may be required when particulates have agglomerated on and/or adhered to inlet faces, or become stuck in passages, such that organic or aqueous solvents are required to release them. The task is best undertaken by chemical cleaning contractors with appropriate equipment and procedures. Once again, the earlier the procedure is undertaken, the more likely it is to be fully effective, and puffing can sometimes provide useful precleaning of bulk particulates, minimizing chemical usage. See Figs. 5.23 and 5.24.

Generally, it is helpful to degrease the core as a first step, to allow access of the acids or alkalis to any inorganic deposits. High recirculation rates through the core will assist with penetration and deposit removal, with flow reversals and pulsations sometimes found to be helpful. The inlet to the exchanger should always be protected by a strainer to avoid recontamination.

The cleaning method available is a function of the geometric complexity of the exchanger, the type of contamination present, the degree of cleanliness required, and cost. Treatments must be customized, but they should consist of a number of stages:

1. Determining level of restriction to indicate the extent of fouling. When appropriate, flow test using air or water. Testing must occur in the opposite direction to normal flow.
2. Rinsing before and after each chemical step to physically remove loose foulant that would otherwise have to be chemically dissolved, adding to downtime and the additional cost of chemicals and their disposal.

FIGURE 5.23 Coolant- side inlet passages before treatment.

FIGURE 5.24 Coolant side inlet passages after treatment.

This stage employs appropriate mechanical cleaning methods in isolation or combined together, such as gas puffing followed by water flushes. Rinse in the opposite direction to normal flow to take the foulant out of the exchanger through the shortest path. Never use seawater or poor-quality water for rinses.

3. Analyze effluent to verify that materials were removed and, if necessary, determine their composition if they appear different from any samples taken prior to treatment. Sometimes, this is the first opportunity to take such samples. If so, it is important to send them for laboratory analysis.

4. Alkali or solvent treatment is used to remove the organic portion of the deposit. This is necessary to make the subsequent acid treatment effective.

5. Acid treatment is used to soften or dissolve the deposit. The fluid can contain inhibitors to prevent attack by the acid of the metal surface.

6. Determine the level of cleanliness through further flow testing. If appropriate, repeat steps 2 through 5.

7. Final rinsing to purge residual chemicals. Confirm that all chemicals have been removed through pH or chemical testing of the effluent.

8. File a summary report of all testing, cleaning treatments, and their results.

Fouling Control and Cleaning Treatment A wide variety of methods have been successfully used for fouling abatement, treatment, and control. These are summarized in Table 5.1.

TABLE 5.1   Summary of Fouling Abatement, Treatment, and Control Methods

Treatment	Method	Addresses
System	Filtration	Particulate from commissioning
	• Side stream filtration ($<5\,\mu$)	debris or corrosion
	• Process and coolant straining ($<300\,\mu$)	
	Coolant chemistry	Stress corrosion
	• Low chlorides, typically 150 ppm	Cracking $>60°$C
		Fouling
	• Low salinity and hardness	Enhances CSCC resistance
	• Alkaline pH 9–10	Corrosion and fouling
	• Control oxygen <0.2 ppm using additives	
	Coolant circuit	Boiling: fatigue and corrosion
	• Operating pressure $>= T_{sat}$	Boiling: as above
	• Pressurized expansion tank	Corrosion and additive
	• Inert gas blanket expansion tank	degradation
Exchanger	Mechanical cleaning	Fouling
	• Gas puffing	
	• Back-flushing	
	• Hydrojetting	
	• Hydrokinetics	
	• Steaming	
	• Warm air blowing meltable solids	
	• Ultrasonic	
	• Process changes	
	Chemical cleaning	Fouling
	• Alkali	
	• Solvent	
	• Chelant	
	• Acid	

Preservation During Storage   Once fouled exchangers have been treated, they are either immediately placed back into service or stored for a period awaiting reuse. No treatment can guarantee removing every particle of contamination. Without adequate protection, risk of underdeposit corrosion during storage exists, even when corrosion-resistant alloys such as stainless steel are used.

To provide added corrosion protection:

• Before storage, ensure that cleaning fluids are adequately neutralized and purged.

• During storage, prevent oxygen, necessary for certain types of corrosion, from entering the exchanger by preserving all internal surfaces. Use either an inert gas blanket after drying, or suitable preserver liquid containing an oxygen scavenger.

5.1.9 Preventing and Treating Fouling

Fouling is generally defined as the accumulation of unwanted materials on the surface of equipment. It is unwanted in microchannel exchangers for the following reasons:

- Deposits increase surface roughness and reduce and/or block channel cross-sectional flow area, increasing pressure drop across the exchanger.
  Higher-pressure drops can restrict production throughput, requiring additional pumping, and could introduce flow maldistribution that reduces exchanger effectiveness.
- Fouling layers increase resistance to heat transfer, reducing exchanger effectiveness.
- Fouling at the exchanger wall can initiate localized corrosion.
- Deposits themselves attract, bind, or allow other particulate to impinge, further exacerbating fouling.

The following section outlines recommendations and the best practices necessary to reduce the risk of fouling and treat stainless steel microchannel exchangers, by giving consideration to the design of the heat exchanger and any attached systems, when it is desired to avoid or remove surface contaminants that:

- Impair thermal or hydraulic performance of the exchanger
- May result in later contamination of the surface
- Impair corrosion resistance of the surface.

Reduce Fouling Risk by Design   The design aim for any system should be to maximize availability and minimize life-cycle costs. More microchannel exchangers suffer unsatisfactory performance on the coolant side vs. the process side. This may result from the abundance of vessels, instrumentation, and operating procedures controlling the flow, quality, and temperature of process streams.

Since closed-loop circuits are not once-through, they will pick up, create, concentrate, and redistribute any contaminants present. There is no separator vessel or blow-down system to eliminate or lessen particulate levels. Therefore, provision should be made to filter the fluid and ensure, that it does not become corrosive or scale-forming.

The risk of fouling can be reduced by giving consideration to the design of the exchanger and corresponding systems and the means to treat fouling if it does indeed occur. To reduce the risk of fouling and suppress its effect on exchanger and system performance:

- *Consider the equipment layout.* Although sound system design will eliminate foreseeable problems, it is good practice to lay out the equipment to allow a reasonable level of access for unexpected maintenance. For example, to allow access to exchanger connections for cleaning, ensure that pipe spools can be elevated or lowered sufficiently.

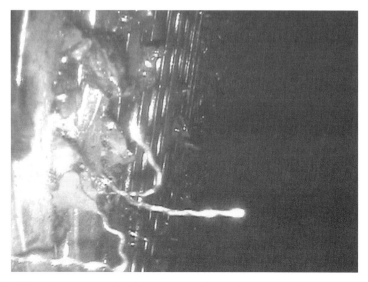

FIGURE 5.25 Passages restricted by weathered strainer and particulate as a result of bursting.

- *Provide means to determine the onset and location of fouling.* It is essential that operations staff have the ability to monitor and record data on exchanger fouling trends. See Figs. 5.25 and 5.26. When the timely onset of fouling is apparent:

  1. Quick early steps can be taken in operation to arrest or even reduce its effect on exchanger performance.

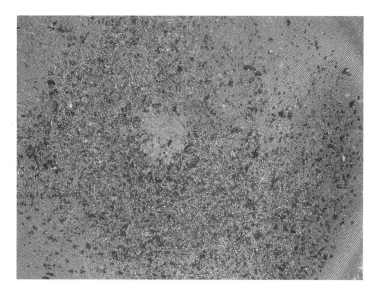

FIGURE 5.26 Strainer contents after appropriate service interval.

2. Having the information to identify whether the process side, utility side, or both involve fouling makes it possible to better plan a more effective cleaning treatment. To locate restrictions, the measured flow and pressure drop are compared against "as-built" clean calculations.

3. Early detection and intervention will improve the effectiveness of any subsequent cleaning treatment. Every deposit is subject to aging. Aging may increase the strength of the deposit through polymerization, dehydration, recrystallization, etc. Aged deposits can be considerably more difficult to treat.

• *Provide means to monitor, record, and alarm pressure drop across strainers.* Strainers may require regular cleaning depending on how clean the system is prior to use. To determine maintenance intervals, the recorded trend of the strainer pressure drop should be monitored. Action should be taken when this level reaches the low-level alarm setting. This is usually 0.5 bar. Figure 5.25 shows how poor attention to strainer pressure drop can result in its bursting and the subsequent fouling of downstream exchanger passages.

• *Use correctly sized strainers.* It is fundamental that outsize particles should be strained from streams entering passages, in order to avoid plugging. Strainer apertures on the order of one-third of the minimum clearance are recommended for most circumstances. Once-off commissioning debris might simply require a commissioning (top hat) strainer, whereas large and continuous particulate loads may require duplex or online cleanable strainers, avoiding the need to shut down for strainer cleaning.

• *Site strainers close to the exchanger they are intended to protect.* To ease maintenance, install strainers that are readily serviceable.

• *Install a polishing system.* A common coolant fluid passing through PCHEs is a closed-loop cooling medium, which in principle should be completely benign following the removal of commissioning debris.

However, problems have arisen when heavily corroded carbon steel piping sheds pipe scale into the system for months following commissioning, requiring on-going maintenance of the cooling medium strainers, something that should not be necessary. Thorough pipe cleaning and good preservation practice would eliminate this problem.

To avoid circulation with high levels of fine suspended solids, polish the coolant in carbon steel piping systems using a simple side-stream filter (Fig. 5.27). Typically, this filter employs small lines and fine filters (5 μ), as it only takes 2 to 5% of the total coolant circuit flow.

Side-stream filtration is not an alternative to 100% straining of streams entering the exchanger. It is a measure to improve the general quality of the coolant.

Alternatively, the use of light-walled stainless or GRP piping that connects to the exchanger might not only avoid carbon steel corrosion problems but also give rise to useful weight reductions.

• *Prevent fatigue, corrosion, or fouling through coolant boiling in the exchanger, during low flow or turn-down conditions.*

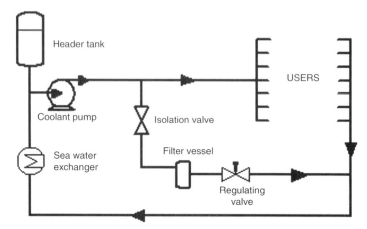

FIGURE 5.27 Closed-loop coolant side-stream filtration scheme.

Appropriate Coolant Pressure

It is essential that the cooling medium be sufficiently high in pressure within gas coolers to avoid boiling. Boiling can be highly unstable, damaging the exchanger through corrosion, thermal fatigue, or fouling through scaling.

The safest approach to having sufficient pressure is to ensure that the cooling medium cannot boil even if stagnant, as the required operating pressures are generally moderate—and well within the lowest pressure rating of most flange specifications. Thus, system costs remain largely unaffected.

Preclude this risk; deliver coolant to the exchanger at a pressure that is above the saturation pressure at the gas inlet temperature. For example, if the hottest gas temperature in the system were 180°C, ensure that coolant enters exchangers at 9 barg.

Ensure that pumps and system (pipework and other vessels using the coolant) are all rated to take and deliver coolant at the appropriate pressure to the exchanger. The opportunity to prevent boiling is lost if the exchanger is designed to take the pressure and the system is not.

Many installations successfully rely on the back-pressure generated by the control valve to pressurize the cooling medium in the exchanger, but a safer the recommended one, approach, is to ensure that pumps deliver the pressure and an inert gas is applied as trim in the cooling medium expansion tank.

Appropriate Minimum Flow

Provide a physical minimum stop on coolant control valves to prevent shutting off the coolant flow. Ensure that the minimum flow delivered is sufficient to provide adequate cooling at start-up.

• *Prevent contamination of closed-loop coolant with scale/corrosion-forming fluids.* If leakage were to occur in the sea water/closed loop coolant exchanger, corrosive and scale forming sea water is prevented from leaking into the less corrosion resistant closed-loop coolant circuit owing to its higher pressure.

Material used for the sea water side are typically titanium or glass reinforced plastics. Closed loop coolant circuits (i.e., inhibited water or glycol streams) are carbon or stainless steel.

- *Prevent ingress of oxygen into closed-loop coolant.* Oxygen can promote corrosion and degrade glycol when used in the coolant. Design the system to blanket the cooling medium expansion tank with an oxygen-free gas—for example, fuel gas or nitrogen. Nitrogen makers can allow oxygen to persist in the product stream. Therefore, check the specification of this type of equipment to see that oxygen levels are acceptably low. Avoid expansion tanks at or open to atmosphere. Design a pressurized system with the tank at a pressure sufficient to prevent leakage of air from vent or discharge lines and to provide top-up of the coolant circuit operating pressure. Problems involving poor expansion tank design vis-à-vis cooling medium circuits have occured. It is best practice to treat the expansion tank as a nonflowing header tank connected to the cooling medium circuit by a generously sized riser to allow the ascent of any gas bubbles in the circuit. A poor example is allowing the cooling medium to free-fall from the top of the expansion tank, entraining blanket gas and creating foam. When blanket gas is not oxygen-free, it can aerate the coolant, increasing its tendency toward corrosion.

- *Design to use nonscaling coolant low in chlorides.*

- *Design to provide good coolant quality and routinely monitor and record this.*

  1. Design to ensure that coolant chemistry and solids content can be measured. This will give early warning of potential problems and enable corrective action to be taken.

  2. Cooling medium is intended to cool hot walls in heat exchangers. Therefore, water salinity should be low to avoid scaling and maintained essentially free of oxygen, to avoid corrosion. Demineralized water is ideal.

  3. The prevention of chloride-induced stress corrosion cracking (CSCC) requires consideration when either stream inlet temperature exceeds 60°C. The appropriate coolant composition is a matter for specialist advice, but in general either oxygen or chloride levels must be limited. Of these alternatives, oxygen removal is likely to be the more convenient and reliable option.

     Chlorides should be maintained as low as practicable. Typically, these are below 500 ppm in closed-loop circuits. To avoid CSCC of the steel at these levels, maintain dissolved oxygen content below 0.2 ppm. Maintaining an alkaline pH between 9 to 10, up to a maximum of 12, will enhance general corrosion and CSCC resistance.

  4. Maintain oxygen scavengers and other additives within the vendor's specifications at all times.

  5. Biological growth is rare in properly maintained closed-loop circuits. If necessary, ensure that biocides are compatible with all materials in the system, including coolant chemistry.

  6. Avoid coolant additives that are prone to thermal breakdown or can form scale below the hottest heat exchanger design temperature in the system.

7. Minimize coolant losses. This will assist in the maintenance of coolant chemistry and reduce make-up costs and inconvenience.

• *Reduce particulate settling and scaling.* Compact microchannel heat exchangers foul less than shell-and-tube heat exchangers under typical operating conditions, and are much less prone to silt fouling. However, the equipment is not immune, so:

1. Design the system to provide for the maintenance of coolant circulation during a short period of plant shut-down. This should reduce the likelihood of settling and concentrating fouling deposits in the system.

2. Side-stream polishing of the cooling medium to remove fines will assist in eliminating silting in low-velocity regions of the cooling medium system.

3. Ensure the appropriate use of process and fouling design margins. Excessive margin in exchanger design leads to oversized equipment when first used. The amount of heat duty new or cleaned exchangers are capable of removing can be considerably higher compared to the design point. If the process requires control, to reduce this heat load, the water velocity must be reduced. This can result in two potentially unfavorable consequences: increases the chance of settling particulate and increases the temperature at the coolant wall and its likelihood to scale. Scales may not be removed completely even if later on the flow velocity is increased.

4. Design the exchanger to ensure a minimum flow velocity of 1 m/s. See Figs. 5.28 and 5.29.

FIGURE 5.28   Coolant scaling of outlet passages.

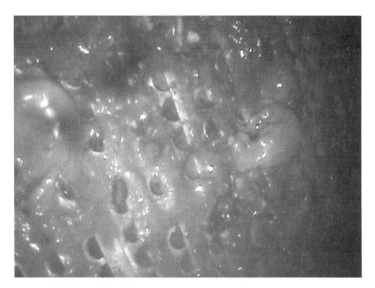

FIGURE 5.29   Biological growth on coolant inlet passages.

• *Prevent fouling by hydrates.* Hydrates were discovered in 1810 by Sir Humphrey Davy and were considered to be a laboratory curiosity. This curiosity became a major problem in the 1930s, clogging pipelines during the transportation of gas under cold conditions. Gas hydrates are crystalline solids that look like ice, occuring when water molecules form a cagelike structure around smaller "guest molecules." The most common guest molecules are methane, ethane, propane, isobutane, normal butane, nitrogen, carbon dioxide, and hydrogen sulfide; of these, methane occurs most abundantly.

Generally, hydrates should be assumed to be a risk whenever the entry temperature of the cold stream is below the hydrate temperature of the warm stream, and appropriate preventative measures should be taken. It is worth noting in this context that the hydrate resistance of shell-and-tubes is commonly subject to undue optimism, in defiance of experience.

Hydrates can be inhibited in microchannels exchangers with methanol or glycol injection, as with shell-and-tubes. Alternatively, a provision for the periodic warming of sections of the core can be included, which is not generally attractive with multiple shell-and-tubes.

Design the system to ensure that coolant inlet temperature is above the gas hydrate formation temperature at all times, including start-up, or the danger of fouling will exist.

• *Prevent fouling by waxes.* The possibility of wax solidification in the heat exchanger arises if wax constituents in hydrocarbon streams carry over from separators. The extent of carry-over should be minimized by the sound design of separator internals, which might well avoid wax problems altogether. Quantities of condensate generated during cooling in the exchanger can maintain waxy constituents in solution. If wax does form, then wax dispersants or periodic warming of the exchanger might be necessary. In one installation, resetting the cooling medium inlet temperature higher by 4°C for a few minutes was sufficient to clear wax deposits from a hydrocarbon gas inter-cooler.

Increase Cleaning Effectiveness by Design   Many factors influence fouling, including appropriate velocities, temperatures, and coolant composition. Even if all are addressed properly at the design stage, fouling may still occur. Therefore, the ability to clean and its effectiveness depend on whether adequate provisions have been made during the system's design or added as an after-thought.

- *Minimize or eliminate the presence of undrainable pockets and crevices.* These can trap particulate, solids, or cleaning solutions and reduce the effective circulation necessary for their removal. If unavoidable, add drain connections to the exchanger.
- *Minimize or eliminate unvented cavities.* These may reduce the effective circulation of cleaning fluids into such areas, making their removal during purging and equipment preservation difficult. When appropriate, separate vents should be provided.
- *Provide multipurpose connections to allow inspection and treatment of the exchanger.* Where fouling occurs, it will be heavier in regions of lowest comparable velocity. Therefore, where access permits it, we recommend positioning multipurpose connections on the exchanger adjacent to areas of lowest comparable velocity. Such connections usually range from 2 to 6 in.

Design the exchanger and/or system to provide the means to (1) inspect the condition of exchanger surfaces, (2) attach temporary cleaning fluid fill and circulation hoses, and, (3) aydrojet the core.

5.1.10 Industrial Examples

Diffusion-bonded exchangers such as the microchannelled PCHE have been supplied to hundreds of major projects around the world for a variety of applications. Applications include:

- Offshore gas compression cooling
- Gas dewpoint control
- LNG liquefaction, regassification, and superheating
- NGL extraction
- Caustic soda
- Sulphur trioxide

- Ethylene oxide
- Gas turbine fuel gas superheating
- CO_2 recuperators
- High-temperature helium feed/effluent exchangers.

Some specific examples are described below.

Gas–Utility Coolant Exchange

Gas Compression Cooling: BG Armada, Offshore North Sea

PCHEs, used as compressor coolers, made a significant contribution to improving the economic viability of this North Sea gas project, which has been on-stream since October 1997. Designed to export a maximum of 450 mmscfd from two 50% trains, the original concept, using a seawater-cooled shell-and-tube, required three separate process modules and showed marginal economic viability. Studies suggested that using PCHEs (as the HP cooler, flash gas cooler, and export coolers) could offer significant advantages.

PCHEs, in stainless steel and using a closed-loop coolant system, reduced power requirements by 15%, due to reduced seawater lift. More important, by reducing the equipment footprint and eliminating the space needed for "bundle pulling," 3 m was removed from the length of each of the three decks, resulting in considerable savings in structure.

In fact, this reduction in structure was sufficient to eliminate the need for separate modules, enabling a single lift to be used for all the integrated topsides, saving in load-out and lifting, and reducing hook-up activities.

With these results taken into account, the overall savings in project cost through the use of PCHEs was estimated at £10 million (in 1997 terms).

Gas–Gas Exchange

ARCO: Pagerungan Island, Onshore Indonesia

This onshore gas plant, with two trains delivering a total of 350 mmscfd, has been operational since 1993. Three PCHEs are used in each train: two in gas–gas duty and one smaller gas–TEG exchanger.

The plant, in a remote location, was constructed on a modular basis. Close temperature approaches were required in the gas–gas exchangers to eliminate the need for a refrigeration system in the early years of operation.

Both PCHE and shell-and-tube options were evaluated for the inlet gas–gas exchanger, upstream of the glycol contactor. For this application, the shell-and-tube design required three shells in series that would have weighed 108 tons. The single PCHE for the same application weighed only 15 tons. Savings in purchase price resulted, with the PCHE being more than 25% cheaper, but the major savings derived from installation costs, whereby the reduction in space and weight saved more than \$3 million for this unit alone. Overall capital expenditure for the project, using four large gas–gas PCHEs, was reduced by more than \$10 million (on a 1993 basis).

5.2 CHEMICAL REACTION AND MICROCHANNEL HEAT EXCHANGERS

PCHE's potential as a plug flow reactor was recognized by Heatric at an early stage [1]. Since then, the mixer/heat exchanger/reactor concept has been widely investigated [2]. By integrating heat exchange, fluid mixing, and reaction into one printed circuit block, the number of components in the system are reduced. More important, however, short mixing paths can lead to vastly accelerated reaction, reduced residence times, and thereby improved selectivity and product yields, when compared with traditional stirred tank (batch) reactors. Noncatalytic and homogeneous catalytic reactions may be carried out in devices that are, to all intents, the same as a compact heat exchanger. If desired, reactants may be introduced and mixed in a PCHE on a passage-by-passage basis, using the same fluid injection techniques that have been utilized for many years to inject glycol for hydrate suppression in offshore gas dewpoint control services.

5.2.1 Reactors for Heterogeneous Catalytic Processes

A wide variety of different reactor concepts have been used in heterogeneous catalytic processes, including fixed bed, moving bed, fluid bed, and bubble column. Within each broad category, there are further variants. For example, a fixed bed may be arranged as one or more large packed beds, or may be packed within multiple tubes (with external cooling/heating, or with feed quench), or may be packed within the annulus of a double-wall multitubular reactor. Each arrangement has been devised to address specific process requirements—especially temperature control. It is in relationship to this challenge of temperature control that microchannel heat exchange may be of particular interest.

The concept of using microchannel heat exchangers in reactors is aimed at processes that are either highly exothermic or endothermic, where their inherent compactness can both reduce the size of reactor systems and permit novel arrangements that would be infeasible using traditional heat transfer arrangements. Furthermore, PCHEs in particular are well suited to extremes of temperature and pressure.

Catalysts are commonly supplied on a porous support material such as alumina, inevitably resulting in poor thermal conductivity within the catalyst bed. Consequently, large fixed beds tend to operate adiabatically. Even in multitubular fixed bed reactors, commonly used for reactions involving large thermal loads—especially highly exothermic reactions, a very significant temperature gradient between the tube wall and center of the tube can arise. This poor in-tube heat transfer also frequently results in a highly nonisothermal axial temperature profile, with a pronounced hot spot close to the tube inlet. Loss of selectivity and product degradation are typical consequences of such uncontrollable hot spots, whereas thermal runaway and even the risk of explosion may occur in certain circumstances.

Such difficulties are sometimes addressed by the adoption of some form of fluidized bed reactor, but these can involve their own drawbacks, including high capital cost, catalyst loss (through attrition), possible flow instability or poor fluidization, and risk of line plugging. They are also substantially back-mixed, which may be detrimental to yield selectivity in certain reaction systems.

5.2.2 In-Passage Reactor

The concept of coating a heat transfer surface with a catalyst has been quite widely discussed: For example, one such study was presented at the 4th International Conference on Process Intensification for the Chemical Industry in 2002 [3]. However, one of the big challenges is how to provide sufficient catalyst surface at reasonable capital cost. The IP PCR, being structurally analogous to a PCHE, offers the desired high surface density, over the full range of operating conditions, at demonstrably competitive cost.

Nevertheless, perhaps the biggest challenge is the availability of suitable catalysts. Substantially improved catalyst activity is needed if we are to take full advantage of the opportunity offered by a coated passage reactor. Catalyst life and resistance to poisoning or deterioration are also of paramount importance when considering layers only a few microns thick. Techniques for applying catalyst coatings to the passages within a microchannel heat exchanger are being developed, and it appears that robust and renewable coatings can be applied in a cost-effective manner. The key benefits of a coated microchannel heat exchanger reactor include the following:

- The reactant flow contacting the catalyst surface has no dead-spots or recirculating flows, which commonly compromise selectivity.
- The level of turbulence in the reactant flow can be readily adjusted to minimize gas-phase mass transfer resistances, which would become increasingly important at higher reaction rates.
- Process pressure drop can be readily optimized.
- By skillful application of counter-, cross-, and coflow, the cooling medium flow pattern can be configured to ensure tight control over catalyst temperature profiles.
- Microchannel heat exchangers are cost-effectively manufactured in high-performance material.
- The microchannel structure is inherently small-scale and, in the case of potentially explosive reactants, can be designed to quench any tendency toward explosion.
- Preheaters and postcoolers may be conveniently close-coupled to the reactor within the same microstructured core, thereby minimizing piping and structure costs.

Inevitably, it will not always be possible to match (or even approximate) the required catalyst surface with the required heat transfer surface. In this case, the coated microchannel heat exchanger reactor risks simply becoming a rather expensive catalyst support, and an alternative approach is needed. A logical alternative is then to decouple the catalyst and heat transfer surfaces in some way.

Although such an approach may be expedient in certain situations, it clearly imposes limitations on passage geometry and reactor size. However, we should remember that there is no reason to confine ourselves to small semicircular passages: Ribbon-shaped passages are also possible, and the etched plates may be oriented face to face, resulting in a symmetrical cross section—one perhaps more suited to certain reactor applications.

5.2.3 Multiple Adiabatic Bed (MAB) Reactor

When the required catalyst surface area is very large and substantially exceeds the required heat transfer area, a better balance between capital cost and performance may be achieved by approximating the in-passage reactor with a large number of shallow adiabatic beds, with heat exchange between each bed: the multiple adiabatic bed or MAB PCR.

Thus far, the feasibility of such an approach has been constrained by the cost of successive reactor vessels, heat exchangers, and interconnecting piping. Even in the most sophisticated integrated reactor designs, the number of reaction and heat exchange steps rarely exceeds three or four, since the volume required for conventional heat exchange within a reactor vessel becomes increasingly costly with increasing overall reactor vessel size. However, by making use of the very compact nature of PCHEs, it is possible to devise cost-effective reactor layouts with many tens of adiabatic beds, with intermediate heating or cooling between each.

One way of constructing a MAB PCR is to have an array of shallow catalyst beds interposed between thin PCHE panels. Alternatively, a very large number of small adiabatic beds can be incorporated within a single PCR block, together with heat exchange zones to adjust the initial feed temperature, to add or remove the heat of reaction between each catalyst bed, and to adjust the product temperature. The heating or cooling medium may simply be a utility stream, or it too can undergo a separate reaction within a separate sequence of catalytic beds. One such unit has been manufactured to investigate this concept for a petrochemical process. It has four catalyst beds, with staged reactant addition and interstage cooling in integral microchannel exchangers.

BIBLIOGRAPHY

1. Johnston, T. "Miniaturized Heat Exchangers for Chemical Processing." *The Chemical Engineer*, December 1986, pp. 36–38.
2. Wille, C. H., Johnston, N. M., Pua, L., Rumbold, S. O., Unverdorben, L., and Wehle, D. "Modern Technologies for Optimised Speciality Chemicals Production Processes."

Presented at Switching from Batch to Continuous Processing Conference, November 22–23, 2004, London.

3. Babovic, M., Gough, A., Leveson, P., and Ramshaw, C. "Catalytic Plate Reactors for Endo- and Exothermic Reactions." Presented at 4th International Conference on Process Intensification for the Process Industries, September 2001, Bruges, Belgium.

4. Burn, J., Johnston, A. M., and Johnston, N. M. "Experience with Printed Circuit Heat Exchangers." GPA General Meeting, Budapest, Hungary, 27th August 1999.

5. BS EN ISO 15547–2:2005, *Petroleum, Petrochemical and Natural Gas Industries – Plate-type Heat Exchangers – Part 2: Brazed Aluminium Plate-fin Heat Exchangers (ISO 15547-2:2005)*, European Committee for Standardization, 28 October 2005.

6. Pua, L. M., and Rumbold, S. O. "Industrial Microchannel Devices—Where Are We Today?", Presented at 1st International Conference on Microchannels and Minichannels, April 2003, Rochester, New York.

7. Heatric Recommended Practice, RP 305, *Monitoring of Heatric Compact Diffusion Bonded Heat Exchangers*, Rev 1, February 2007.

8. Dunkerton, S. "TWI Knowledge Summary—Diffusion Bonding." The Welding Institute, Cambridge, UK, 2001. http://www.twi@twi.co.uk.

9. Kays, W. M., and London, A. L. *Compact Heat Exchangers*, 3rd ed. New York: McGraw-Hill, 1984.

10. Mehendale, S. S., Jacobi, A. M., and Shah, R. K. "Heat Exchangers at Micro and Meso-scales." Proceedings of the International Conference on Compact Heat Exchangers and Enhancement Technology for the Process Industries, Banff, Canada, 1999, pp. 55–74.

11. Wongkasemjit, S., and Wasantakorn, A. "Laboratory Study of Corrosion Effect of Dimethyl-Mercury on Natural Gas Processing Equipment." *J. Corr. Sci. Eng.*, 1(12) (1999).

12. Incopera, F. P., DeWitt, D. P., Bergman, T. L., Lavine, A. S., *Introduction to Heat Transfer*, 5th Edition, John Wiley & Sons Inc., 1 Sep 2006.

13. Silverberg, P., "Refurbishing Worn-out Heat Exchangers." *Chemical Engineering*, June 1st 1999, http://www.che.com.

CHAPTER 6

SEPARATION UNITS

ASTERIOS GAVRIILIDIS and JOHN EDWARD ANDREW SHAW

6.1 INTRODUCTION

Separations play a significant role in many chemical and biochemical processes at all scales. In industrial chemical processes, they account for a significant fraction of plant costs. They are utilized to purify reactor feeds and products to meet required specifications, and to recover reactants, solvents for recycling, and waste by-products for disposal. During the 20th century the petrochemical industry drove the technology forward. Future separation needs are primarily related to the pharmaceutical, microelectronics, aerospace, water, energy (e.g., hydrogen) and life sciences industries [1]. Microprocess engineering is well placed to serve many of these needs, due to its small production scale and unique advantages offered by point of use and intensified operation.

Preparative separations can be applied to obtain small quantities of product for direct application or as raw materials for reactions to produce useful quantities of derivatives. However, much work on microfluidic separators has thus far been directed at analysis, where separations produce material sufficient only for identification and quantification and products may be discarded or destroyed. Analytical microfluidic operations and assays, especially very-small-scale electrophoretic and chromatographic processes, are well covered in a number of papers and reviews, especially those by Manz [2, 3].

This chapter focuses on processing (i.e., nonanalytical) microfluidic separations, where a product or derivative is at least potentially collectable. However, it must be recognized that there is considerable overlap between processing and analytical operations. In general, quantities for preparative separations may be somewhat larger with the use of parallel structures. In analytical procedures such as HPLC and capillary electrophoresis, action is applied to a small discrete sample, the

Microchemical Engineering in Practice. Edited by Thomas R. Dietrich
Copyright © 2009 John Wiley & Sons, Inc.

components of which become separated along the length of a channel, or equivalently eluting at different rates. Microfluidic separation processes designed to generate product streams from a substantial or continuous input may involve cross-flow transfer processes in which the transport of material across a main flow allows products to be directed to selected effluent streams.

Separation can be achieved by phase creation or addition, and by introducing a barrier, solid agent, force field, or gradient, through the use of energy and/or mass separating agents [4]. During separation, the rate of mass transfer of the species to be separated is enhanced relative to the mass transfer of all species. The driving force and direction of mass transfer are governed by thermodynamics. Thus, both transport and thermodynamics considerations are crucial in separation operations. The rate of separation is generally governed by mass transfer, although kinetics, for example, a slow chelation reaction related to partition, may mean that mass transfer rate limits are not attained. The extent of separation is limited by thermodynamic equilibrium. Microseparation units can impact mass transfer limited processes. Fast transfer processes can be engineered due to short transport distances, large gradients that can be applied, and high surface-to-volume ratios. In this way, thermodynamic equilibrium may be approached rapidly.

There is a plethora of macroscale separation operations available at various levels of maturity. So far, a rather limited number have been investigated employing microstructured devices, especially if one excludes separations limited to analytical applications. This chapter presents information on microseparation units that have been realized to date. They are described along with their modes of operation and, when possible, comparisons with conventional counterparts are offered.

Microfluidic separation processes may be classified on the basis of the type and number of immiscible fluid phases present, for example, processes involving two immiscible fluid phases with material transfer between phases, single fluid phase processes where the material dissolved is to be separated or concentrated, etc. One may also subdivide microfluidic separation units on the basis of structural elements such as membranes or filters, and applications of fields.

Multiple phase separations involving gas–liquid and liquid–liquid phases have much in common, but the lower density, electrically insulating nature, and higher diffusive transport processes in a gas phase give rise to some differences. In general, for gas–liquid processes, the material transport limitations tend to be confined to the liquid phase [5] or to any membrane structures involved.

Single-phase separations include the case of selective transfer through a membrane, where the membrane transport properties constitute the enabling feature, and separations employing a field acting across at least part of the flow. The latter may be regarded as versions of field flow fractionation (FFF). FFF processes, including split flow FFF, are essentially microfluidic separation processes established before that term was widely applied. These methods, originally described and much developed by Giddings [6–8], have formed the basis of much later work in which microengineering techniques were applied to device construction and a range of fields applied across the main flow direction.

6.1.1 Transport in Microfluidic Separation Devices

The performance of microfluidic separations is commonly dependent on material transport within fluid flows in microchannels. In almost all cases, at least for liquids, the flows are laminar with viscous forces dominating inertial forces, as exemplified by Reynolds number (Re $= lv\rho/\mu$) values well below those for transition to turbulent flow. For laminar flows, the transfer of dissolved or suspended material across stream-lines occurs by migration and not convective processes. Various fields may drive migration, but where these are not applied or effective, cross-stream transfer proceeds by diffusion.

The interaction of laminar convective flow and diffusion can, in principle, be evaluated using CFD techniques. However, in many cases, reasonable estimates may be provided by analytical expressions for simpler systems; for example, in a limited plane sheet, the value of the Fourier number (Fo $= Dt/l^2$) characterizes a system [9]. Generally, Fo $\rightarrow 1$ corresponds to diffusive transport evolving to a steady-state condition over length l. This allows the estimation of suitable device dimensions when designing fluidic systems where diffusion is a controlling process. The liquid diffusion coefficients D for species whose separation is desired depends on their size, approximately according to the Stokes–Einstein equation ($D = kT/6\pi r\mu$) [10]. Knowledge of solute or particle diameter may be lacking, but diffusivity for many cases relates to molecular weight. Examination of diffusion data indicates an approximate dependence on MW$^{-0.5}$, as shown in a plot derived from a compilation of data from a variety of sources (Fig. 6.1). Whereas the trend with molecular weight is clear, the divergence between different macromolecule types probably reflects differences in shape. A consideration of these values, and the relationship between

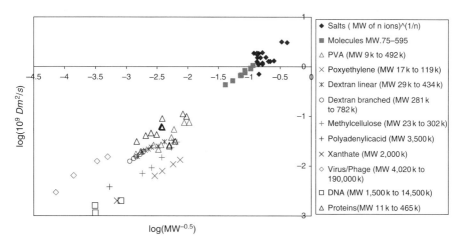

Figure 6.1 Variation of diffusion coefficients (in water) with particle mass (data from a range of sources, including *CRC Handbook of Chemistry and Physics* [11] and *Polymer Handbook* [12]).

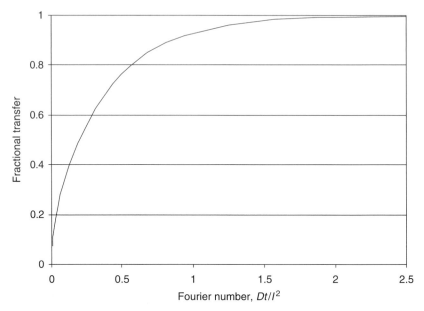

Figure 6.2   Fractional diffusive transfer of a solute with diffusivity D into a plane sheet of thickness l with fixed concentration at one surface and no flux conditions at the other.

the Fourier number and progress in the diffusive process indicated in Fig. 6.2, suggests that effective separation, based on diffusion alone, requires that the species to be separated are substantially different in size (MW). For example, the equilibration time t indicated by $Dt/l^2 = 1$ for diffusion into a 50-μm-thick layer is ~2 s for glycine (MW 75), ~5 s for sucrose (MW 434), and ~40 s for serum albumin (MW 67000).

6.2   MEMBRANE SEPARATION OF GASES

Membrane separation utilizes a semipermeable membrane through which one or more species move faster than another or other species to separate a feed consisting of two or more components. The feed mixture is separated into a retentate (the part that does not pass through the membrane) and a permeate (the part that does pass through the membrane). The membrane is porous or solid (or even a liquid), and made out of polymer, ceramic, or metal materials. Even though the technology is not new, it is still evolving and being progressively used in new applications. Depending on the components to be separated and the membrane utilized, different physicochemical principles control the separation. Microtechnology offers the opportunity to prepare very thin, strong, and uniform membrane layers, which is a key issue for separation efficiency and productivity. Microengineered membranes have been developed for gas and particle separation. In this section, we concentrate on gas separation, in which membranes have been developed from palladium and zeolites.

6.2.1 Palladium and Palladium–Silver Micromembranes

Hydrogen is extensively used in petroleum refining, fertilizers, metals, and electronics processing, although one of its emerging uses is as a clean energy carrier, particularly in relation to fuel cells. Palladium is a unique material in its ability to absorb and allow the fast diffusion of hydrogen through its bulk. Other molecules, even as small as helium, cannot permeate it. This property has made possible the use of palladium membranes for the separation of hydrogen from gas mixtures. Palladium is often alloyed with silver to avoid the problem of hydrogen embrittlement when the latter is exposed to thermal cycling and to improve its permeability. Conventional technology is limited by the need to have robust membranes, which necessitates large (self-supported) membrane thickness on the order of 50 μm [13]. This results in not only low hydrogen fluxes but also expensive membranes due to the high cost of palladium. To address this problem, various researchers have fabricated thin Pd or Pd alloy membranes supported onto ceramic and other inorganic porous supports utilizing slip casting, electroless plating, sputtering, and chemical and electrochemical vapor deposition techniques [14]. These methods have their limitations, because if very thin palladium layers are deposited, the pores of the support are not well covered, leading to pinholes in the membrane and subsequent gas leakage. To prevent pinhole formation, the membrane thickness should be larger than the diameter of the pores of the immediate support layer. Furthermore, mass transfer resistances in the porous support, which has to be relatively thick for sufficient mechanical strength, decrease overall hydrogen permeability.

These problems can be circumvented through microfabrication technology by depositing the membrane in the environment of a dust-free, clean room and on wafers with smooth, nonporous surfaces. Apertures are etched on the backside of the wafer *after* membrane deposition to provide access to the thin film membrane. In this way, the membrane film covers the support completely, leading to defect-free membranes. Two slightly different approaches have been utilized in the microfabrication of Pd and Pd–Ag membranes. In the simplest one, free-standing membranes of Pd or Pd–Ag alloy are deposited on a silicon wafer coated with a Ti adhesion layer [15–17]. Long narrow slits are etched on the opposite side of the wafer until the backside of the Pd–Ag film is revealed. The thickness of these membranes is 0.5 to 1.2 μm.

An approach that can lead to even thinner membranes is to support the Pd or Pd–Ag film on microsieves that provide additional mechanical stability. Tong and coworkers [16–19] fabricated SiO_2/SiN microsieves ~ 1 μm thick with 5-μm pore size on a silicon wafer containing large apertures to provide immediate Pd–Ag membrane support (Fig. 6.3). Strong membranes with 500-nm thickness were thus fabricated. Wilhite et al. [20] fabricated SiO_2/SiN microsieves ~ 500 nm thick with 4-μm pore size to support Pd and Pd–Ag membranes 200 nm thick. Karnik et al. [21] fabricated $Cu/Al/SOG$ (spin-on-glass) microsieves with varying thickness to support 50- or 200-nm Pd membranes. Copper was employed because it could act as a catalyst in the water gas shift reaction, while aluminium was used for mechanical strength and to aid adhesion of the Pd layer.

Deposition of the Pd or Pd–Ag film can be carried out by sputtering or cosputtering using two pure metal targets [15, 17]. This allows sputtering of Pd and Ag

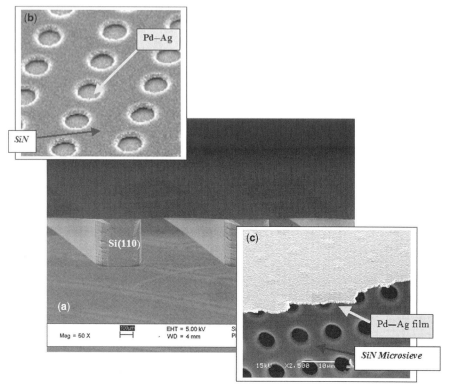

Figure 6.3 SEM image of microfabricated Pd–Ag membrane. (a) An overview of the membrane on Si<110>. Pd–Ag film on the SiN microsieve: (b) as seen from the top of (a); (c) as seen from the bottom of (a) (reprinted with permission from [18]; copyright 2004 American Chemical Society).

simultaneously with a homogeneous composition. Alternatively, Wilwhite et al. [20] deposited by electron beam deposition alternate layers of Pd and Ag, which upon annealing provided an almost uniform composition of the two metals. Assembly of the membrane can be performed by bonding the silicon-supported membrane by four-electrode anodic bonding to two glass wafers containing manifold and fluidic channels [15]. This bonding method can, in principle, be used to construct a stack of silicon wafers separated by glass plates. Other researchers formed the second channel by bonding a structured silicon wafer to the wafer containing the membrane [20, 21]. Heating of the membranes can be performed by either placing the membrane in an oven or by incorporating microfabricated heaters on the membrane chip.

The mechanical strength of microfabricated membranes is excellent. The rupture strength at room temperature of a 1-μm-thick, free-standing Pd–Ag alloy film supported on silicon with narrow slits of 23-μm width was found to be in excess of 4 to 5 bar [15], whereas that of 0.2- to 0.5-μm-thick Pd and Pd–Ag films supported on SiN microsieves was 4 to 5 bar [18, 22]. Typical porosity of the supporting microsieves, and therefore the accessible metal film, is about 20% but in principle it can be increased to 50% [18, 19].

In most of the microfabricated membranes presented above, the hydrogen flux dependence on hydrogen partial pressure was close to first order, deviating from Sievert's law and indicating that surface adsorption and desorption of hydrogen are the transport limiting steps [20]. However, Karnik et al. [21] found a 0.5 order, while Gielens and coworkers [17] determined 1.2 order dependence. The latter also observed that hydrogen flux increased faster with increasing temperature as compared to conventional membranes. This was attributed to the different transport limiting step (hydrogen surface reactions), which results in higher activation energy than if the transport is limited by diffusion through the membrane. The membranes in [21], for which the transport limiting step seemed to be hydrogen diffusion, showed activation energy significantly lower than that for conventional membranes, possibly due to the different microstructure of the sputtered Pd films. These membranes showed exceptionally high flux (up to 4.5 mol/m^2 s at a hydrogen partial pressure of 0.5 bar) even at room temperature. Overall, microfabricated Pd and Pd–Ag membranes exhibit hydrogen fluxes one order of magnitude higher than those of conventional membranes, while maintaining high hydrogen selectivity [19].

Membrane stability in hydrogen mixtures was very good. In long-duration permeation studies, hydrogen flux and selectivity did not show any significant reduction for $\sim$1,000 h, during which period hydrogen concentration and temperature cycling were enforced [18]. The good stability of the membrane was attributed to the prevention of Pd and Ag diffusion into the support due to intermediate layers of SiN and SiO$_2$, materials that are excellent diffusion barriers. The presence of ammonia and carbon dioxide in the feed mixture has only a minimal effect on hydrogen permeability, while partially recoverable loss of H2 permeation was observed for exposure to carbon monoxide, for both Pd and Pd–Ag membranes [20].

Although the thickness of Pd–Ag membranes can be substantially reduced by microfabrication technology, the membrane cost is still high compared to that of conventional membranes due to the use of clean rooms, specialized equipment, and silicon and glass substrates. Tong et al. [19] estimated that to manufacture a 500-nm-thick Pd–Ag membrane on a 6-in. {110} Si wafer costs $340, with less than 3% of this cost attributed to the Pd and Ag materials. Such a membrane can provide 600 L/h of high-purity hydrogen, enough to produce power of 1 kW from a fuel cell.

6.2.2 Zeolite Micromembranes

Zeolites with their molecular sieving properties and tunable pore structure can separate molecules based on their size, shape, and polarity [23]. In addition, they are thermally stable and resistant to most acids, bases, and organic solvents. Due to their high cost, the best opportunities for the industrial implementation of zeolite membranes are expected in the fields of small and microscale applications [24].

Made of inorganic crystals, zeolite membranes accumulate stress during thermal treatment and high-temperature operation that can lead to mechanical failure. The mechanical stress increases in proportion to membrane area, making large membranes more susceptible to cracks and defects. Hence, an important advantage of micro-scale zeolite membranes is preventing stress-related failure, by numbering up small individual membrane areas [25].

Free-standing zeolite microstructured membranes fabricated on a silicon wafer were first reported by den Exter et al. [26]. Wan et al. [25] investigated various fabrication strategies for zeolite films and prepared microstructured silicalite membranes that separated a chamber and channel etched in opposite sides of a silicon chip. Yeung and coworkers [27, 28] fabricated free-standing MFI zeolite membranes by arranging them in a 7×7 grid pattern, as shown in Fig. 6.4. The excellent intergrowth of zeolite crystals ensured that interstitial transport pathways were kept to a minimum. Localized growth within the grid pattern was achieved through selective seeding. Membranes were grown onto the seeded recesses by hydrothermal synthesis of appropriate solutions. Seeding induced secondary growth of the zeolite membrane and could directly affect crystal size and orientation. The silicalite and ZSM-5 membranes had 0.6 to 4.4 μm crystal size and were 5 μm thick. In this way, 49 unconnected membrane areas were created. The mechanical strength of the membrane was improved not only by keeping individual, unconnected membrane areas small but also by using an oxygen plasma method to remove the template at low

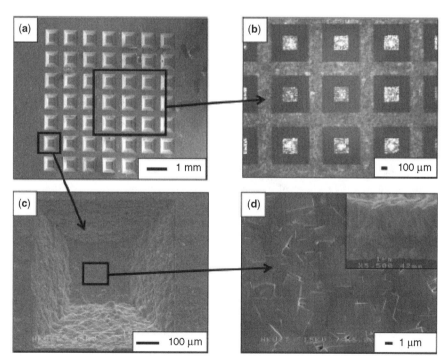

Figure 6.4 (a) Optical micrograph of silicalite micromembrane. (b) A higher-magnification image showing the size of the individual micromembranes. SEM images of (c) a single micromembrane and (d) its surface microstructure and cross section (inset) (from [27]; reproduced by permission of the Royal Society of Chemistry).

temperature. Membranes before template removal could withstand a 5-bar gas pressure difference.

The diffusion of small gas molecules through zeolite pores and the interstitial transport pathways such as grain boundaries and imperfections is primarily Knudsen-type. However, the diffusion of larger molecules is influenced by interaction with the pore wall. The microfabricated membranes displayed an order of magnitude higher permeability compared to traditional supported membranes, while at the same time they possessed excellent permselectivity. In gas mixtures, heavier hydrocarbons and carbon dioxide selectively permeated through the silicalite membrane, resulting in a hydrogen-enriched retentate stream. This is advantageous when a concentrated hydrogen stream is required at high pressure. The addition of Al in the ZSM-5 membranes weakened the adsorption of methane and carbon dioxide; as a result, hydrogen was concentrated at the permeate side.

6.3 ABSORPTION OF GASES

Absorption is utilized to selectively remove a certain component (or components) from a gas mixture by mass transfer from a gas to a liquid phase that has high affinity for it. Absorption is employed in various cases, such as purifying gas streams (that need to be used as raw materials for chemicals/energy production or to be released to the atmosphere) from undesired components, such as CO, CO_2, H_2S, SO_2, etc. Some of the absorption processes are reactive, that is, the absorbing component reacts with another component in the solution. Absorption is similar to extraction, the main difference being that an immiscible liquid phase replaces the gas phase in the latter. Passing of the component molecules through the interface is normally assumed to be so fast that at the interface equilibrium concentrations are considered. The step limiting the separation rate is usually diffusion in the bulk phase of the liquid (or liquids in the case of extraction). For this reason, separation rates can be increased by improving mass transfer, which can conveniently be achieved by microchannel contactors. There are principally two approaches to bringing into contact two phases: The first is to keep both phases continuous and use the contactor to create an interface between them (*continuous-phase microcontactors*), whereas the second is to disperse one phase into the other by using an appropriate inlet or a micromixer upstream of the contactor section (*dispersed-phase microcontactors*) [29].

6.3.1 Continuous-Phase Microcontactors

In continuous-phase microcontactors, the gas and liquid phases form two streams that are fed separately in the liquid and gas region of the contactor and are also withdrawn separately at the contactor outlet. The advantage of such contactors is that the phases are not intermixed, and the gas–liquid interfaces are well defined. When interdispersion processes are applied, this can result in foams or mists that may be difficult to break in order to separate the phases. The crucial design issue of continuous-phase contactors is the way the interface and/or liquid film is stabilized.

In the falling film microreactor [30], which is basically a *falling film micro-contactor*, gravity in conjunction with a plate that contains microchannels (typically 300 μm wide, 100 μm deep, and 78 mm long separated by 100-μm-wide walls) creates and stabilizes the liquid film, which can be less than 100 μm thick. The microchannels prevent break-up of the liquid film at low flow rates. Due to the combination of capillary forces and small channel widths, liquid being pulled up along the sides of the channels takes up a significant portion of the channel width and the surface of the liquid film assumes the form of a flowing meniscus [31].

Carbon dioxide absorption in sodium hydroxide solution has been studied both experimentally and theoretically [32]. The concentration of OH^- did not decrease to zero near the interface for any of the cases investigated, while the CO_2 dissolved in the liquid phase was consumed, at most, within 25% of the liquid thickness. As a result, the liquid film thickness did not affect significantly CO_2 absorption. However, thinner liquid films led to better utilization of NaOH. The transverse concentration profiles of CO_2 in the gas phase were not as steep as those in the liquid phase. The amount of CO_2 absorbed improved by increasing gas residence time and NaOH concentrations. This was primarily due to the direct effect of concentration on reaction rate. Complete CO_2 absorption could be obtained for gas residence times between 10 to 60 s, depending on the CO_2 gas feed concentration, NaOH liquid concentration, and microchannel and gas chamber dimensions. For example, all CO_2 was removed from a stream containing 26% CO_2 in 9.5 s using a 1-M NaOH solution and a 2.5-mm gas chamber depth. This gives for CO_2 $Dt/l^2 = (0.139^*9.5/0.25^2) = 21$, whereas $Dt/l^2 = 1$ yields $t = 0.45$ s, confirming that absorption is not controlled by gas-phase diffusion.

In *overlapping microchannel* and *mesh microcontactors*, the two immiscible fluids flow through separate channels. To provide stable operation, the fluid interface is immobilized by well-defined openings obtained by partial overlapping of the (open on one side) channels where the two fluids flow [33] (see Fig. 6.5), or by a thin mesh [34] (see Fig. 6.6). Interfacial forces help to stabilize the fluid interface within the openings. The meniscus shape defines the available area for mass transfer and is a function of the contact angle, opening geometry, and pressure difference between phases. Operation of such a channel or mesh devices requires that the pressures used to drive flow do not produce excessive differential pressures across the opening. The breakthrough of one phase into the other is broadly predictable by the Young–Laplace expression ($\Delta P = \gamma \cdot \cos \theta / R$ for a slot and $\Delta P = 2\gamma \cdot \cos \theta / R$ for a circular opening). For more precise evaluation of breakthrough limits, advancing and receding contact angles, and meniscus contortion at the opening mouth(s) must also be considered [35]. In practice, while narrower openings certainly provide greater stabilization, observed pressure differentials for breakthrough tend to be somewhat lower than predicted.

An overlapping microchannel contactor manufactured by anodically bonding etched silicon and glass substrates was developed by Shaw et al. [36, 37]. The contactor configuration containing 120 sets of overlapping channels 14 mm long could handle liquid flow rates 1 to 10 mL/h, corresponding to liquid residence times of 2 to 20 s. Experiments with NH_3 absorption into dilute acid solutions containing cresol blue indicator confirmed rapid absorption, consistent with diffusion control

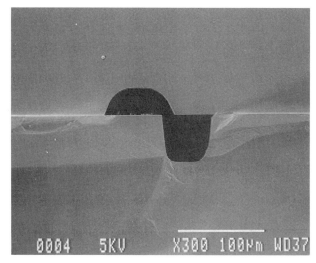

Figure 6.5 SEM image of cross-section of Si/glass-bonded structure with overlapping channels to form microcontactor (from [33]; reproduced with permission of Springer Science and Business Media).

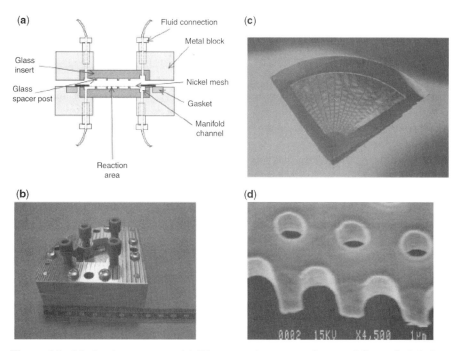

Figure 6.6 Mesh microcontactor. (a) Diagrammatic cross-section, partially exploded view. (b) Assembled device. (c) Photo of complete mesh showing frame and struts. (d) SEM image of mesh pores (from [34]; reproduced by permission of the Royal Society of Chemistry).

in the liquid phase. The generation of constricted openings, set to values between 2 and 30 μm, by overlapping channels required high alignment accuracy. This was achieved by the use of a system based on an optical mask aligner [33].

Due to geometric and fabrication constraints, the above contactors have limited stability and small working volume and interfacial area (2–15% of the total). The stability is particularly an issue when uneven pressure gradients are obtained, such as during countercurrent operation. These problems can be alleviated by using mesh or porous plate microcontactors [34, 38, 39]. Wenn et al. [34] implemented a mesh structure to separate planar chambers containing the two fluids (see Fig. 6.6). The mesh-to-wall distances could be set generally to $\sim$100 μm, whereas the pore widths were typically $\sim$5 μm and provided adequate stability. The ratio of pore length (mesh thickness) to width was $\sim$1 : 1 and ensured low diffusive transport resistance through the mesh. The open area of the mesh was about 20 to 25%, which led to a gas–liquid interfacial area of 2,000 m^2/m^3. It was placed between two glass layers that formed the chambers for the two fluids. Struts fabricated on the mesh were aligned with pillars on the glass to provide the necessary channel depth and additional support to the mesh. The mesh was fabricated in nickel using photolithography and a two-stage electroplating method. However, nickel meshes are limited to noncorrosive media. This contactor was used for the absorption of oxygen into an aqueous alkaline pyrogallol solution under stop-flow conditions for the liquid and 1 mL/h flow of air or air diluted with nitrogen. Over a 60-s contact time, depletion of oxygen from the gas stream, expressed as a fraction of that supplied, was 31% when the feed was air.

Calculations using diffusivity values in the literature indicated that significant mass transfer resistances still existed [40]. Other methods for manufacturing meshes include photochemical machining and laser micromachining, direct or through a mask. However, the thickness of these meshes is usually larger, which can introduce additional mass transfer resistance. Martin et al. [41] fabricated meshes with 10- to 15-μm diameter holes, laser-micromachined in polyimide or 130-μm diameter holes, photochemically machined through stainless steel shims.

Using porous plate microcontactors, TeGrotenhuis et al. [38] investigated the removal of carbon dioxide from a mixture of CO_2 and N_2 using diethanolamine solutions as the absorbent, and for scrubbing CO from a hydrogen-rich stream using copper ammonia formate solutions. Over 90% of the carbon dioxide was removed in less than 10 s from a stream containing 25% CO_2 with a 40% diethanolamine solution. For the copper ammonia formate system, carbon monoxide was reduced from 1% to less than 10 ppm in 1.3 s gas residence time.

6.3.2 Dispersed-Phase Microcontactors

In dispersed-phase contactors, a gas–liquid dispersion is created by an inlet that induces merging of the gas and liquid streams. The inlet contains a dual- or multiple-feed structure that splits the phases in thin lamellae, which eventually form a dispersion. After the inlet section, both phases are delivered to one or more pipes or microchannels. The hydrodynamics obtained depend on various parameters, such as inlet design, channel dimensions, viscosity, density, and surface tension, but one of the most critical is the ratio of gas-to-liquid flow rates. As the ratio increases, the flow obtained ranges from bubble flow, where bubbles have smaller diameter than the contactor channel, to slug flow (or segmented or Taylor flow), where bubbles are large enough to fill the channel width. Unlike conventional bubble columns, a relatively uniform bubble size distribution can be obtained in these contactors. At very high values of gas-to-liquid flow rate ratio, annular flow can be obtained: a thin liquid annular film surrounds a gas core and hence the contactor becomes a continuous-phase one. Varying the wetting conditions of the channel wall can affect the flow patterns formed or even result in new ones, such as the pattern seen in 100-μm quartz capillaries where the gas bubbles in slug flow are connected via gas stems (so-called Yakitori flow) [42].

A key dimensionless number for dispersed flow is the Capillary number (Ca = $\mu v / \gamma$, where μ is the liquid viscosity, v the bubble velocity, and γ the interfacial tension). During Taylor flow at high Ca, a complete bypass flow pattern with no flow reversal is observed, whereas at lower Ca toroidal vortices form inside the slug [43], as shown in Fig. 6.7. For relatively low flow rates, mass transfer by diffusion superimposed on the vortices results in well-mixed conditions inside the slugs [44]. This allows rapid transfer of solute in a moving slug that can far exceed the rates expected for diffusion from slug ends. When a micromixer is used upstream of the contactor channel and when the latter has a much larger diameter than the fluid lamella created in the micromixer, a foam is created [45]. These contactors

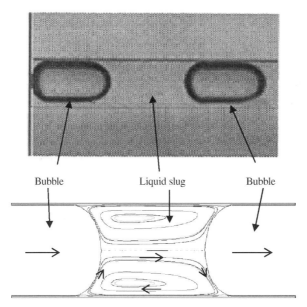

Figure 6.7 Image of Taylor (slug) flow in a glass capillary and streamlines (obtained by CFD) within the liquid slug (reprinted with permission from [29]; copyright 2005 American Chemical Society).

can provide a large gas-phase void fraction. However, coalescence may occur especially at large residence times.

A scaled-out version of a single-channel dispersed flow contactor was presented by Hessel et al. [30]. In this device, termed a *microbubble column*, the gas and liquid feeds were split into a number of substreams and subsequently brought into contact in the main channels, so that one gas and one liquid substream were introduced into one main channel. In order to achieve flow equipartition, the gas and liquid inlet channels were designed with very different hydraulic diameters (7 μm and 20 μm, respectively). The channels were of 50 μm $\times$ 50 μm or 300 μm $\times$ 100 μm cross section. Interfacial areas up to 19,000 m^2/m^3 for isopropanol–nitrogen and 18,000 m^2/m^3 for water–nitrogen were observed for sufficiently high gas flow rates for the 300 μm $\times$ 100 μm channels [46]. This is much larger than those observed for conventional bubble columns and agitated tanks (up to 300 m^2/m^3) [47].

The microbubble column was used for CO_2 absorption from nitrogen mixtures in a NaOH solution. Nearly 100% absorption was observed for mean residence times of the fluid mixture below 0.1 s for the 50 μm $\times$ 50 μm channels, and 1 s for the 300 μm $\times$ 100 μm channels. A maximum in absorption with residence time was observed. This was attributed to the fact that decreasing residence time increases the specific interfacial area, due to flow regime change from bubbly to slug to annular, while it decreases the amount of contact time between the two fluids. These short residence times resulted in high space time yields—higher than that for the falling film microcontactor at similar inlet conditions (227–816 mol/m$^3 \cdot$ s vs. 56–84 mol/m$^3 \cdot$ s). Similar improved performance for CO_2 absorption by decreasing channel size was observed by Sato and Goto [48].

Separation of phases at the end of a dispersed flow contactor can be problematic as gravity, generally effective for large-scale operations, may become ineffective as dimensions are reduced. Body forces driving separation, arising from gravity or acceleration and fluid density difference, become less significant than surface tension effects at small dimensions. Capillary forces tend to dominate as the Bond number becomes sufficiently small ($Bo = \Delta \rho g R^2 / \gamma < 1$), which happens at $\sim$2 mm for immiscible water–organic liquid combinations at terrestrial gravity [35, 49, 50]. Similar pore sizes may be encountered for gas–liquid systems where higher values of density difference may be partially countered by greater interfacial tension.

Phase separation structures or devices may therefore need to be incorporated downstream of the contactor region, if the phases need to be collected separately. The approaches developed so far for gas–liquid mixtures make use of capillary and surface forces to induce such a separation. TeGrotenhuis and Stenkamp [51] designed a device that contained a wicking structure for the separation of water from water–air mixtures. The gas–liquid mixture entered the device and flowed through a slit microchannel, on the bottom of which wicking material was located. The liquid was sorbed on the wicking structure that provided a path for the liquid to flow to a liquid outlet while preventing gas intrusion. Suction was applied to take the liquid out of the wicking structure. The gas phase exited from a separate outlet. The microchannel could be open when the phases were relatively continuous, such as during annular and slug flow. Two mechanisms were required for effective operation: a wicking mechanism and a mechanism for excluding gas. The first one was accomplished by a porous structure that was wetted by the liquid in order to cause preferential sorption. The second was accomplished by using a pore throat layer (adjacent to the wicking layer), whose pore size determined the maximum allowable pressure difference to avoid gas breakthrough (as discussed earlier for the mesh contactor).

In a similar fashion, Günther et al. [52] separated liquid during slug flow by a separator, placed at the end of the microchannel contactor, cotaining 16 capillaries, each one ∼20 μm wide. Hsieh and Yao [53] used a separator consisting of two parallel silicon wafers, both containing an array of etched holes. One wafer was coated with a hydrophilic film (liquid separator that facilitates the removal of the liquid) and the other with a hydrophobic film (gas separator that prevents the liquid from leaking through it).

6.4 STRIPPING OF VOLATILE COMPONENTS

Stripping is essentially the opposite of absorption and involves the removal of volatile components from a liquid stream into a gas stream. As such, microcontactors used for absorption can also be used for stripping. However, the optimal design characteristics and operating conditions will, in general, be different than when they are used for absorption.

Turner et al. [39] used both an overlapping channel and a mesh microcontactor for stripping acetone from an organic stream (1 : 2 : 2 by weight phenol, acetone, xylene) using air. The overlapping channel contactor consisted of silicon with 95 μm × 50 μm (width × depth) channels bonded to glass with 97 μm × 46 μm channels. Each wafer had a set of 120 overlapping channels and for the whole device four wafers were manifolded. Channels were aligned with an overlap of 4 μm. The mesh microcontactor had a mesh area of 16 mm × 4 mm with 10-μm pores at 15-μm pitch. On either side of the mesh were glass channel structures 500 μm × 100 μm (width × depth) at 200-μm pitch. The geometry of the mesh microcontactor allowed operation at a gas–liquid flow rate ratio much larger than the overlapping channel one, while maintaining stability. This resulted in ∼50% acetone removal in the mesh device; almost none was removed in the channel device.

Cypes and Engstrom [54] used a microfabricated stripping column (MFSC) made from two silicon and one glass wafer, similar in design to the mesh microcontactor. The gas–liquid interface was pinned at holes 50 μm × 50 μm, separating the gas and liquid streams that were flowing in 330-μm-deep channels. The stripping of toluene from toluene-saturated water was investigated at room temperature in the MFSC and a conventional packed tower. The number of theoretical transfer units was, in general, greater for the packed tower, but the overall capacity coefficient was nearly an order of magnitude larger for the MFSC. The MFSC was also ∼10 times more volumetrically efficient than the packed tower. This was attributed to lower mass transfer resistance in the liquid film due to its smaller thickness, which is not feasible with traditional packed towers even near flooding. In both cases, mass transfer resistance in the gas phase was determined to be negligible, which is expected for volatile components with a large Henry constant. The MFSC has other advantages as well, such as lower pressure drop and no danger of flooding. However, care is required to control the pressure difference between gas and liquid streams to ensure stable operation and to control the quality of the fluid streams to avoid fouling the openings.

6.5   DISTILLATION OF BINARY MIXTURES

Distillation is one of the most widely used separation techniques. It relies on differences of vapor pressures of components to separate high from low boiling ones. A number of consecutive stages are required, each operating at a progressively higher temperature, if a high degree of separation is necessary. Microchannel distillation technology has the potential to decrease the height equivalent of a theoretical stage by an order of magnitude [55].

Sotowa and Kusakabe [56] investigated the separation of a water–methanol mixture on a chip, which consisted of a single stage operating at 345 K. The chip had three sections: the mixture inlet, where the liquid mixture entered the chip; a meandering channel equilibration section, where the vapor and liquid were generated; and a phase separation chamber for separation of the two phases. A 70% methanol mixture yielded 81.5% methanol-containing vapor and 65% methanol-containing liquid.

Wootton and de Mello [57] used a carrier gas to effect evaporative transport in a glass device consisting of three sections. In the first one, gas and liquid streams were brought together and rapidly heated. The streams were then split, so that liquid and some carrier gas were diverted back to the feedstock. Vapor-saturated carrier gas passed through a long condensation channel, which allowed heat transfer to encourage condensation. The final section consisted of a separation module allowing depleted carrier gas and condensate to be separated, and the condensate to be collected. By heating the first section to 60°C, a 9-fold enrichment of acetonitrile concentration was achieved for a 50% mixture in DMF, which was equivalent to 0.72 theoretical plates.

Kockmann and Woias [58] fabricated a meandering channel distillation chip by KOH etching of silicon; it contained a heating and a cooling channel at its ends for temperature control. Preliminary testing demonstrated the need for better thermal insulation and better control of flow and phase contacting within the device.

6.6   IMMISCIBLE PHASE LIQUID–LIQUID EXTRACTION

In liquid–liquid extraction, a liquid feed consisting of the *carrier* and *solute* (or solutes) is contacted with the *solvent* that is immiscible with the carrier. The solute is miscible in the solvent and, as a result, it transfers from the carrier to the solvent, effecting at least partial separation of the feed. Promotion of the transfer of solute between contacting immiscible liquids is enhanced by increasing the contact area and decreasing mass transfer distances within the liquids. In large-scale systems, such as mixer/settlers and packed columns, this is achieved by turbulent mechanical mixing. Microsystems emulate this but with constrained dimensions largely eliminating turbulence. The means for achieving liquid–liquid contact within microfluidic formats include phase mixing to create droplet dispersion, segmented flow with the formation of slugs flowing sequentially along a channel, and contactors where the flow of each fluid forms a continuous layer laterally contacting the other (as for gas/liquid systems in Section 6.3.1). Falling-film-type devices described for gas–liquid contacting are generally not suited to liquid–liquid operation at microfluidic scales due to unfavorable density and capillarity effects.

6.6.1 Droplet Dispersion Microcontactors

The principles and much of the microfluidic equipment used for dispersed-phase liquid–liquid contacting are similar or identical to those described for gas–liquid systems in Section 6.3.2. Although many microfluidic mixers have been described, the literature on their application to immiscible phases to generate dispersions is dominated by the output from IMM [59, 60]. Examination of the dispersions generated in compact interdigitated channel mixers indicates that decreasing channel width decreased droplet size, but that the effects of increased absolute flow rates and flow ratios were also effective in generating droplets much smaller than the channel dimensions [59]. The high interfacial surface areas generated favor fast extraction, and this appears to be confirmed for several extraction chemistries by a closer approach to equilibrium with increasing flow rate, as described in [60] directed at evaluating a static micromixer as an alternative to stirred apparatus in a mixer settler set-up. Some uncertainty exists about true extraction times as breaking of the emulsion in a settler was not immediate and for some systems at least transfer during settling may be significant.

The phase dispersion method appears to yield relatively high throughputs compared with other microfluidic liquid–liquid extraction formats, but it requires structures and time for subsequent phase separation. The phase-separating structures of TeGrotenhuis and Stenkamp [51], and Hsieh and [53] referred to in the earlier section on gas–liquid systems, should be applicable, in principle, to liquid–liquid systems. Separating structures applied to liquid–liquid systems have been described and evaluated. Okubo et al. [61] passed oil–water droplet dispersions through thin cells formed between glass and PTFE plates and showed that droplets with a size approaching the cell height agglomerated, resulting in sequential slug flow. However, droplets significantly below cell height passed through intact. Kralj et al. [62] described an electrocoalescence cell. Quite moderate AC fields (~ 10 V over $547\ \mu$m) appeared to effectively cause agglomeration, and operation of the system for extraction was demonstrated for a dispersed-phase, surfactant-aided extraction of phenols from aqueous solution into hexane. Although both Okubo et al. [61] and Kralj and coworkers [62] only converted droplet dispersions to slug flows, the relatively large slugs may be further separated into simpler settler, hydrophylic/hydrophobic channel, or filter arrangements.

6.6.2 Segmented Flow Microcontactors

Contacting immiscible fluids as sequential slugs passing along a capillary, with each filling the width, has been applied relatively infrequently in microfluidic operations despite the relative simplicity of the structure. The same hydrodynamic effects of recirculating flow within slugs present in gas–liquid Taylor flow (Fig. 6.7) applies to liquid–liquid systems and may be observed using simple dye transfer. Wetting of the wall material by the combinations of liquids involved affects the contact area between phases, which may range from little more than the capillary cross section to much larger values if one phase wets the wall sufficiently to form a continuous wall coating. Burns and Ramshaw [63] demonstrated the extraction of acetic acid from kerosene with an aqueous alkaline solution containing pH dye indicator by slug flow using 380-μm channels. Both flow speed and slug length were varied but not independently. Thus, an empirical relationship for transfer time was derived, but it may not be general. Although theoretical considerations of slug flow have mainly concentrated on the capillaries of a circular cross section, the use of channels with corners is likely to affect transfer as the wall wetting phase will tend to be retained there during the passage of slugs of the other phase.

Reproducible generation of slug lengths is not facile and subject to variations in the junction geometry characteristics and flow delivery systems. Interfacial tension and viscosity are key parameters affecting slug formation and transition to parallel flows [64]. Control of phase volume ratios for slug contactors by control of slug lengths with the aspirated formation of slugs has been employed in an application directed at determining organic/aqueous partition coefficients [65]. Microextractions at a volume ratio of 100 : 1 (aqueous/organic) were performed in 320-μm quartz capillary by repeatedly shuttling a slug chain over a length ~ 1.5 times that of the longer (aqueous) slug as this produced effective axial mixing.

6.6.3 Continuous-Phase Microcontactors

The advantage of extraction operated with immiscible continuous-phase streams, having lateral contact without mixing, is that phase separation involves only connection to separate outlets. A variety of strategies have been employed in microfluidic devices to generate and maintain such lateral streams. These include the use of restricted openings through meshes [38, 66, 67] overlapping channels [33, 36, 37, 68, 69], channels containing continuous or discontinuous ridges [70, 71], selective positioning of hydrophobic coatings [72, 73], and interfacial tension and flow rate control [64].

Without stabilizing structures in the lateral flow of immiscible fluids, the tendency exists for interfacial energy minimization to cause the streams to break up, in some cases converting to slug flow or more irregular forms. This has been addressed by Kuban and coworkers [64] who demonstrated that some control could be exercised by the use of relatively low interfacial tension fluids and flow speeds sufficiently high to provide body forces counteracting the interfacial tension effects. Such an approach can result in parallel flows without stabilizing structures, but is severely limiting on available extraction chemistries and contact times. Similar results have been reported by Reddy and Zahn [74], as applied to microfluidic DNA purification systems, who used SDS as a surfactant to modify aqueous to phenol/chloroform interfacial tension. For suitable chemistries, this method has the attraction of compatibility with simple channels. However, in general, methods employing stabilizing structures have wider application.

TeGrotenhuis et al. [38] described expressions and protocol to guide the design of microfluidic contactors and predicted that for the same output, the volume of such microfluidic hardware would be an order of magnitude lower than that of conventional extraction equipment employing high-efficiency structured packing materials. During extraction of acetone from water using 1,1,2-TCA, equilibrium effluent concentrations were approached in cocurrent operation, while countercurrent flow was able to achieve greater than one theoretical stage. TeGrotenhuis et al. [66] performed the extraction of cyclohexanol from a water to cyclohexane stream using micromachined contactor plates fabricated by laser drilling a matrix of holes through Kapton films separating 300-μm-deep channels. These films, which were 25 and 50 μm thick, provided $\sim$35% fractional transfer at 2.4 min, while a Teflon composite microporous membrane with a total thickness of 180 μm resulted in $\sim$25% fractional transfer in the same timeframe. This indicates that thinner membranes can improve performance by decreasing mass transfer resistance.

As discussed in Section 6.3, thinner mesh layers ($\sim$5 μm thick) may be fabricated (see Fig. 6.6). These have been used in a contactor with 100-μm-deep channels for the extraction of 3-hydroxy-nitrobenzoic acid from an ethyl acetate solution to a an aqueous sodium carbonate solution, achieving 30% fractional transfer in 1.5 s [67]. Overlapping channel contactors have been shown to achieve 20% fractional transfer in $\sim$5 s in the extraction of Fe^{3+} from an aqueous acid to a TPB/xylene solution [37, 69], but apart from the chemical robustness of silicon and glass, these contactors suffer from fabrication and low throughput issues, as discussed in Section 6.3.

A series of mainly coflow microchannel extractions have been described by Kitamori and coworkers, mostly aimed at providing platforms for colorimetric analysis. These do not require the collection of separate effluent streams, although provision for this is made in some structures. In some cases, the units described have ridges along the channels. Tokeshi et al. [70] indicated that without the ridges (see Fig. 6.8, where the ridge height is 5 μm in a 20-μm-deep channel) stabilization was not possible. Such ridge-type devices produced for an earlier study [68] yielded problematic results because of the difficulties in controlling ridge height due to fast etching at the ridge tip during fabrication. Tokeshi et al. [75] reported that extraction of a complex ion into a chloroform stream was not observed during flow, but occurred within the expected diffusion time under stopped flow conditions. The relatively fast flows used, giving short contact times ($\sim$0.5 s), may have been required to achieve stability.

Maruyama et al. [71] reported on a contactor in which a discontinuous ridge was formed between contacting channels. Images show that curved liquid–liquid interfaces form at the openings, and rates for a metal ion extraction were observed to be elevated above those for an open interface device of similar geometry.

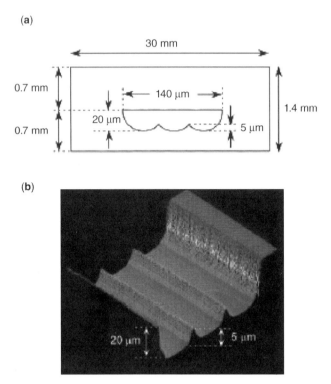

Figure 6.8 (a) Schematic cross-section of microchannel showing ridges and (b) depth profile image of structure (reprinted with permission from [70]; copyright 2002 American Chemical Society).

Microscopic observation of beads and CFD modeling indicated a flow component crossing the main flow trajectory. The elevation of transfer rates reported by Ueno et al. [76] to be associated with nonsymmetric, zigzag, side-walled channels showing a sinusoidal liquid–liquid interface may result from such a cross-flow component as well as the cited increase in interfacial area.

Hibara et al. [72] described a contactor device in which channels in a pair of substrates cross to form liquid–liquid interfaces and stabilization of the phase separation is aided by pretreating one substrate with octadecylsilane. Nitrobenzene and water flowed in 250-μm-wide channels and interacted only in the small region where the two channels passed across each other. Aota et al. [77] reported that a similar octadecylsilane-treated system could be used for countercurrent extraction. A novel method for generating such localized hydrophobic coating in an assembled microfluidic device has been described by Maruyama et al. [73]: by passing laminar streams of octadecylsilane in toluene and pure toluene through the structure. A device with three contacting channels, in which the central one had been rendered hydrophobic in this way, was used for metal ion selective transfer between two aqueous streams through an organic extractant stream acting as a liquid membrane. Fifty percent of Y^{3+} present in the feed phase (which also contained Zn^{2+}) transferred selectively to the receiving phase within 2.4 s. It should be noted that the operation of all devices discussed in this section is governed by phase breakthrough issues, which have been discussed in detail in Section 6.3.

6.7 PARTICLE SEPARATION FROM LIQUIDS

The surface tension effects that apply for immiscible liquid–liquid and gas–liquid systems are absent for systems involving a single phase or miscible phases. Although there is no fully defined boundary surface where miscible fluids meet, under laminar flow conditions streamlines are maintained and material transfer between streams occurs by migration processes only. Devices where transfer occurs by a diffusive process only have been described by Yager [78], whereas other migration drivers (inertial, electrophoretic, magnetic, etc.) are covered in a range of studies and structures often related to FFF and split flow FFF.

6.7.1 Separation by Diffusive Transfer Across Laminar Flow

The theory and practice of diffusion-based separations have been presented in a series of papers by Yager. The operation of a T-sensor cell with contacting streams with indicator dye and different pH buffers was described by Weigl et al. [79], followed by variants with fluorescent dyes and beads [78, 80, 81].

"H-filter" devices, in which product streams are separated, have been developed as a platform for the diffusive separation of molecular species from cells, allowing serum components to be separated from whole blood [82]. Within each of these devices, simple diffusive separation is limited to species with substantially different diffusion coefficients ($>100:1$) corresponding to high molecular weight ratios ($>10,000:1$). However, conditions causing the generation of slowly diffusing macromolecules or particles, or the capture of smaller molecules on particles, can raise efficiency. This was demonstrated by Bowden et al. [83] in a hexane–oil extraction where the formation of colloids by polyaromatic species of limited hexane solubility effectively prevented their diffusion into a hexane stream. H-cell-type separations have for the most part been applied to aqueous solutions for which polymer laminate constructions are suited, but Bowden and colleagues [83] demonstrated that a glass device can perform well for organic solvents.

6.7.2 Field-Assisted Separations in Laminar Flows

Separations based on the application of cross-flow fields in microfluidic devices represent developments of field flow fractionation (FFF) methods introduced by Giddings [6]. Much of the early FFF work effectively involved sedimentation of particles to channel walls and subsequent elution. Of greater interest are split flow versions with multiple outlets where materials transfer across flow streams remaining in suspension and exit via selective outlets. Particle transit is controlled by fluid flow speeds and applied fields. A component of flow across the main flow stream can be provided by flows through porous membranes or frits that form the channel walls [6, 8] or by bringing together flows as parallel streams [84] in structures very similar to the Yager H-cell. In such a device, the use of unequal inlet flows allows hydrodynamic pinching, confining the incoming particles to a narrow laminar region before the field separation takes effect. Although a component of cross-channel flow is introduced in each of these systems, the field providing separation is inertial (gravitational).

Sharma et al. [85] described a FFF system incorporated in a centrifuge to improve inertial-force-driven separation, but the system was configured to separate discrete samples on the basis of elutriation times, not for continuous operation with split flows. The use of microfluidic devices for split flow FFF (referred to as SPLITT) was covered by Fuh [86] and more recently reviewed by Zhang et al. [87]. Separations at split flow outlets are described by Fuh for centrifugal, magnetic, and electrical fields. Zhang et al. provide a section on the forces governing particle and fluid flow in a channel, followed by a presentation of CFD modeling that includes the application of acoustic and magnetic fields within the channel. The modeling presented agreed well with experimental work on acoustic cell washing in a SPLITT system, as described by Hawkes et al. [88, 89].

The acoustic drive of particles across a flow stream can provide a concentration of particles at nodes [90], allowing separation of lipid particles from cells [91]. Pamme [92] discusses the use of magnetic fields to drive separations in H-cell-type separators. For biological separations, this generally requires binding selected species to paramagnetic particles, as covered in a paper by Mikkelsen and Bruus [93]. The use of inherent cell properties without need for such a binding step is described by Han and Frazier [94] for a high magnetic field gradient separating paramagnetic red blood cells from diamagnetic white cells in a microfluidic structure.

The application of electric fields acting across flows has been addressed in microfluidic separation technology. The flow channels used are generally of rectangular cross section and relatively wide ($\sim 5-100$ mm), but with low heights ($10-1,000$ μm). In electrical FFF, including SPLITT systems, the field is applied across the short channel height with potential differences in aqueous solutions restricted to less than 1.7 V to avoid unwanted electrolytic generation of gas. For free flow electrophoresis as described by Raymond et al. [95] and Zhang and Manz [96], the narrow stream bearing material to be separated enters, flanked by broader flows of clear solution, and the field is applied across the channel width. Species transport is a combination of the lateral electrophoretic movement and laminar flow along the channel, but since that flow has a parabolic profile, there is substantial broadening of the trajectories, which limits separation efficiency.

Gale et al. [97] developed the theory of electrical FFF as it applies to microfluidic devices, indicating how separation in at least the elutriation mode could be enhanced and describing the performance of a device for the separation of polystyrene particles in the 44- to 261-μm range. SPLITT operation using electrical fields was described by Fuh [86], and Merugu et al. [98] illustrate the operation of a pair of devices in series to improve separation. Electrochemical action at the electrodes can generate acid and base, and this has been used to generate pH gradients across the intervening fluid. The use of this approach to modify the electrophoretic mobility of species such as proteins and carry out isoelectric focusing in the H-cell or SPLITT system has been described by Cabrera and coworkers [99]. The issues with electrochemical action at the contacting electrodes can be dealt with by provision of intervening ion permeable membranes. Although this is the established technology for salt removal by electrodialysis [100], such arrangements have not been much used in microfluidic separation devices.

Dielectrophoresis is a process in which electric signals, normally AC, act on particles to direct them toward or away from regions of increasing electric field, depending on their polarizability and that of the surrounding medium. Applications in microfluidics mainly focus on the separation of cells and macromolecules using systems which in some cases approximate embodiments of FFF or SPLITT. The breadth and rapid growth of the literature in this area can only be covered by referring to specialist reviews and articles [101–103]. The need for field concentraton in dielectrophoresis can require that particles be brought very close to electrode structures, limiting channel height and throughput. More recent papers appear to concentrate on using asymmetric electrodes to provide lateral as well as vertical deflections [104, 105] and the use of insulating pillars to shape fields and obviate the need for complex electrode structures [106].

6.7.3 Filtering and Hydrodynamic Separation

Advances in the application of microengineering have enabled the formation of regular but sometimes complex structures that may be used for size segregation of particles in flows without the application of fields. This area has been covered in a recent review by Eijkel and van den Berg [107] that included the theoretical basis and highlighted devices with anisotropic sieving structures and the use of hydrodynamic effects on particles in near wall flows. The desire for continuous flow operation for the processing and classification of particulate suspensions is constrained by issues related to the blocking of channels and structures, especially where particle agglomeration can occur. The need therefore exists to consider the conditions compatible with cross-flow filtration and identification of structural features reducing undesired flow blocking effects.

A book by van Rijn [108] makes a substantial contribution in this area, as well as describing the practical fabrication techniques for the production of membranes, especially of silicon nitride micromeshes, and discussing mesh robustness to pressure loading and related design rules. A chapter on microfiltration discusses the forces affecting particles near pores in a cross-flow arrangement. Furthermore, expressions are derived for the required conditions for cross-flow filtration allowing fluid removal without formation of pore-blocking cake layers. Micromeshes made and operated so as to conform to these conditions have been successfully used for beer cross-flow filtration, demonstrating flux and fouling improvements over those obtained with conventional equipment [109].

A structure has been proposed in which filter pores are located on a nonplanar surface with a view to reduce blocking, possibly by enhancing lateral flow near the openings [110]. In addition to the issue of filter structure blockage and release of agglomerating particles, the problem of particle damage through wall contact, particularly relevant to plasma separation from blood, has been a driver for the development of hydrodynamic separations in microfluidic structures. Some systems have been described using body forces imposed in spinning systems or resulting from flows along curved trajectories to achieve separation [111, 112]. Ookawara et al. [113] describe a microseparator/classifier consisting of a semicircular microchannel whose downstream end bifurcates to separate particles from a slurry. Experimental work with an aqueous slurry of 1.8 to 30-μm acrylic particles using a semicircular microchannel with 200 μm $\times$ 150 μm cross section and 20 mm radius of curvature demonstrated better classification compared to a representative hydrocyclone [114]. The centrifugal forces result also in Dean vortices that tend to entrain the dispersed particles. A model formulated to analyze this system [115] indicated that such entrainment may be prevented for large particles (20 μm), but not for small ones (1.8 μm), by lift forces developing near the upper and lower microchannel walls as a result of high shear rates. The separation efficiency remains high, provided that the slurry is dilute.

A number of recent papers describe the use of hydrodynamic effects near the wall, and at channel pinches and side channels. They provide separation of fluid from particles and the size classification of particles without directly using the size exclusion effects of pores and at least potentially reducing the resultant blockage issues.

When a particle is transported in a channel under laminar flow, its center of mass is excluded from the slow-moving wall region, resulting in increased transport rate for the larger particles. This is the basic concept of hydrodynamic separation proposed by DiMarzio and Guttman [116]. A detailed model of the separation mechanism, accounting for attractive (Van der Walls) forces, repulsive (double-layer charge) forces, and gravitational forces, has been presented by Lin and Jen [117]. Yamada et al. [118] described structures and experiments with particle streams "pinched" by a second fluid stream and then entering a sharply widening channel. Good separation was demonstrated for 15 and 30-μm particles as long as the entering particle flow was fluidically pinched to a thickness less than the lowest particle diameter. The process did not degrade or change significantly with increasing total flow rates. Further work by the same group [119] described an improved pinch flow fractionation structure with multiple outlets, as shown in Fig. 6.9, which was able to separate a mixture of 1–5-μm particles, as well as red blood cells from blood.

The microfluidic plasma separation device presented by Yang et al. [120] relied on "hydrodynamic filtration" to retain blood cells in a straight main stream, while plasma was withdrawn along narrow side channels. Hydrodynamic filtration relies on similar principles as hydrodynamic separation presented above. Consider a laminar flow containing particles being introduced in a microchannel with a number of side channels, through which flow rates are significantly smaller. Particles whose diameter is larger than a specific value would never go through the side channels, even if they are flowing near the sidewalls, and even if the particle size is smaller than the cross-sectional size of the side channel [121]. This is due to the fact that the center of the particle cannot be present within a certain distance from the sidewall, which is equal to the particle radius. This principle was employed by Yamada and Seki [121] for achieving not only separation but also fractionation of particles by introducing larger side channels after a first group of channels designed for particle concentration.

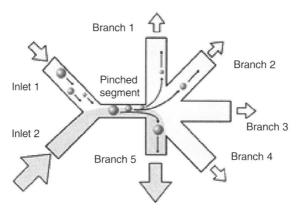

Figure 6.9 Schematic of pinched flow particle classifier (from [119]; reproduced by permission of the Royal Society of Chemistry).

6.8 CONCLUDING REMARKS

The amount of research dedicated to microchannel separation systems is significantly smaller as compared to microchannel reaction systems (if one excludes processes that have been developed primarily for analytical applications), even though these are necessary for microstructured chemical plants. A number of the separation methods applied in macroscopic systems have been converted to microfluidic formats and new variants of some of these methods have been enabled. Separation techniques relying on phenomena present to a significant extent only in microstructures have also been developed. In many cases, significant advantages can be derived from the dimensional restrictions, although when microfluidic structural dimensions remain large compared with molecular dimensions, the properties dependent on atomic and molecular interactions, for example, thermodynamic equilibria, partition coefficients, and molecular mobilities, are rarely affected. For the most part, it is behavior related to mass and heat transport, and processes dependent on the application of fields and field gradients, that can most radically be affected by suitably reducing contactor dimensions. The level of maturity that microfluidic separations have achieved is shown in Table 6.1. Significant improvements in performance have been

TABLE 6.1   Maturity Level of Various Microfluidic Separation Processes

Membrane separation of gases	√√
Absorption of gases	√√
Stripping of volatiles	√√
Distillation	√
Extraction	√√√
Particle separation from liquids	√√√

demonstrated for gas separation by membranes, but the cost of such devices is still high, mainly due to fabrication processes. Various contactors have been developed for multiphase separations (absorption, stripping, and extraction), and the area has benefited further from devices designed for multiphase mixing and reaction processes. A great deal of research on particle separation has occurred for analytical applications, and this can potentially evolve to higher-volume applications.

BIBLIOGRAPHY

1. Noble, R. D., and Agrawal, R. (2005). Separation research needs for the 21st century. *Ind. Eng. Chem. Res.*, 44: 2887–2892.

2. Manz, A. (1997). Ultimate speed and sample volumes in electrophoresis. *Biochem. Soc. Trans.*, 1: 278–281.

3. Reyes, D. R., Iossifidis, D., Auroux, P. A., and Manz, A. (2002). Micro total analysis systems. 1. Introduction, theory, and technology. *Anal. Chem.*, 74: 2623–2636.

4. Seader, J. D., and Henley, E. J. (2006). *Separation Process Pronciples*, 2nd ed. Hoboken, NJ: Wiley.

5. Danckwerts, P. V. (1970). *Gas–Liquid Reactions*. New York: McGraw-Hill.

6. Giddings, J. C., Yang, F. J., and Myers, M. N. (1977). Flow field-flow fractionation: New method for separating, purifying, and characterizing the diffusivity of viruses. *J. Virology*, 21: 131–138.

7. Giddings, J. C. (1985). A system based on split-flow lateral transport thin (SPLITT) separation cells for rapid and continuous particle fractionation. *Sep. Sci. Technol.*, 20: 749–768.

8. Liu, M., Li, P., and Giddings, J. C. (1993). Rapid protein separation and diffusion coefficient measurement by frit inlet flow field-flow fractionation. *Prot. Sci.*, 2: 1520–1531.

9. Crank, J. (1980). *The Mathematics of Diffusion*, 2nd rev. ed. Oxford, UK: Clarendon Press.

10. Cussler, E. L. (1997). *Diffusion Mass Transfer in Fluid Systems*, 2nd ed. Cambridge, UK: Cambridge University Press.

11. Lide, D. R. (ed.) (2005). *CRC Handbook of Chemistry and Physics*, 86th ed. Boca Raton, FL: Taylor & Francis.

12. Brandrup, J. Immergut, E. H. and Grulke E. A. (eds.) (1999). *Polymer Handbook*, 4th ed. New York: Wiley.

13. Gryaznov, V. M. (1986). Hydrogen permeable palladium membrane catalysts. An aid to the efficient production of ultra pure chemical and pharmaceuticals. *Plat. Met. Rev.*, 30: 68–72.

14. Paglieri, S. N., and Way, J. D. (2002). Innovations in palladium membrane research. *Sep. Purif. Methods*, 31: 1–169.

15. Tong, H. D., Berenschot, J. W. E., De Boer, M. J., et al. (2003). Microfabrication of palladium–silver alloy membranes for hydrogen separation. *J. Microelectromech. Systs.*, 12: 622–629.

16. Keurentjes, J. T. F., Gielens, F. C., Tong, H. D., et al. (2004). High-flux palladium membranes based on microsystem technology. *Ind. Eng. Chem. Res.*, 43: 4768–4772.

17. Gielens, F. C., Tong, H. D., van Rijn, C. J. M., et al. (2004). Microsystem technology for high-flux hydrogen separation membranes. *J. Membr. Sci.*, 243: 203–213.

18. Tong, H. D., Gielens, F. C., Gardeniers, J. G. E., et al. (2004). Microfabricated palladium–silver alloy membranes and their application in hydrogen separation. *Ind. Eng. Chem. Res.*, 43: 4182–4187.

19. Tong, H. D., Gielens, F. C., Gardeniers, J. G. E., et al. (2005). Microsieve supporting palladium–silver alloy membrane and application to hydrogen separation. *J. Microelectromech. Systs.*, 14: 113–124.

20. Wilhite, B. A., Schmidt, M. A., and Jensen, K. F. (2004). Palladium-based micro-membranes for hydrogen separation: Device performance and chemical stability. *Ind. Eng. Chem. Res.*, 43: 7083–7091.

21. Karnik, S. V., Hatalis, M., and Kothare, M. V. (2003). Towards a palladium micro-membrane for the water gas shift reaction: Microfabrication approach and hydrogen purification results. *J. Microelectromech. Systs.*, 12: 93–100.

22. Franz, A. J., Jensen, K. V., and Schmidt, M. A. (1999). Palladium membrane reactors. In *Proceedings of 3rd International Conference on Microreaction Technology*, Frankfurt, Germany, April 18–21, pp. 267–276.

23. Coronas, J., and Santamaria, J. (1999). Separations using zeolite membranes. *Sep. Purif. Methods*, 28: 127–177.

24. Coronas, J., and Santamaria, J. (2004). The use of zeolite films in small-scale and micro-scale applications. *Chem. Eng. Sci.*, 59: 4879–4885.

25. Wan, Y. S. S., Chau, J. L. H., Gavriilidis, A., et al. (2001). Design and fabrication of zeolite-based microreactors and membrane microseparators. *Microp. Mesop. Materials*, 42: 157–175.

26. den Exter, M. J., van Bekkum, H., Rijn, C. J. M., et al. (1997). Stability of oriented silicalite-1 films in view of zeolite membrane preparation. *Zeolites*, 19: 13–20.

27. Chau, J. L. H., Leung, A. Y. L., and Yeung, K. L. (2003). Zeolite micromembranes. *Lab Chip*, 3: 53–55.

28. Leung, Y. L. A., and Yeung, K. L. (2004). Microfabricated ZSM-5 zeolite micro-membranes. *Chem. Eng. Sci.*, 59: 4809–4817.

29. Hessel, V., Angeli, P., Gavriilidis, A., et al. (2005). Gas-liquid and gas-liquid-solid microstructured reactors: Contacting principles and applications. *Ind. Eng. Chem. Res.*, 44: 9750–9769.

30. Hessel, V., Ehrfeld, W., Herweck, T., et al. (2000). Gas/liquid microreactors: Hydrodynamics and mass transfer. In *Proceedings of 4th International Conference on Microreaction Technology*, Atlanta, Georgia, March 5–9, pp. 174–186.

31. Yeong, K. K., Gavriilidis, A., Zapf, R., et al. (2006). Characterisation of liquid film in a microstructured falling film reactor using laser scanning confocal microscopy. *Exper. Therm. Fluid Sci.*, 30: 463–472.

32. Zanfir, M., Gavriilidis, A., Wille, C., et al. (2005). Carbon dioxide absorption in a falling film microstructured reactor: Experiments and modeling. *Ind. Eng. Chem. Res.*, 44: 1742–1751.

33. Shaw, J., Nudd, R., Naik, B., et al. (2000). Liquid/liquid extraction systems using micro-contactor arrays. In *Proceedings of 4th International Conference on Miniaturized Chemical and Biochemical Analysis Systems*, MicroTAS2000, Enschede, Netherlands, May 14–18, pp. 371–374.

34. Wenn, D. A., Shaw, J. E. A., and Mackenzie, B. (2003). A mesh microcontactor for 2-phase reactions. *Lab Chip*, 3: 180–186.

35. Amador, C., Angeli, P., Gavriilidis, A., et al. (2003). Meniscus shape, position and stability in straight pores. In *Proceedings of Annual AIChE Meeting*, San Francisco, November 16–21, Paper No. 277a.

36. Shaw, J., Turner, C., Miller, B., et al. (1998). Reaction and transport coupling for liquid and liquid/gas microreactor systems. In *Proceedings of 2nd International Conference on Microreaction Technology*, New Orleans, Louisiana, March 9–12, pp. 176–180.

37. Shaw, J., Turner, C., Miller, B., et al. (1998). Characterisation of microcontactors for solute transfer between immiscible liquids and development of arrays for high throughput. In *Proceedings of 2nd International Conference on Microreaction Technology*, New Orleans, Louisiana, March 9–12, pp. 267–271.

38. TeGrotenhuis, W. E., Cameron, R. J., Viswanathan, V. V., et al. (1999). Solvent extraction and gas absorption using microchannel contactors. In *Proceedings of 3rd International Conference on Microreaction Technology*, Frankfurt, Germany, April 18–21, pp. 541–549.

39. Turner, C., Shaw, J., Miller, B., et al. (2000). Vapour stripping using a microcontactor. In *Proceedings of 4th International Conference on Microreaction Technology*, pp. 106–113.

40. Amador, C. (2006). Principles of Two-Phase Flow Microreactors and Their Scale-Out. Ph.D. thesis, University College, London.

41. Martin, P. M., Matson, D. W., and Bennett, W. D. (1999). Microfabrication methods for microchannel reactors and separations systems. *Chem. Eng. Comm.*, 173: 245–254.

42. Serizawa, A., Feng, Z., and Kawara, Z. (2002). Two-phase flow in microchannels. *Exper. Therm. Fluid Sci.*, 26: 703–714.

43. Thulasidas, T. C., Abraham, M. A., and Cerro, R. L. (1997). Flow patterns in liquid slugs during bubble-train flow inside capillaries. *Chem. Eng. Sci.*, 52: 2947–2962.

44. Salman, W., Gavriilidis, A., and Angeli, A. (2004). A model for predicting axial mixing during gas-liquid Taylor flow in microchannels at low Bodenstein numbers. *Chem. Eng. J.*, 101: 391–396.

45. Löb, P., Pennemann, H., and Hessel, V. (2004). g/l-dispersion in interdigital micromixers with different mixing chamber geometries. *Chem. Eng. J.*, 101: 75–85.

46. Haverkamp, V. (2002). Charakterisierung einer Mikroblasensäule zur Durchführung stofftransportlimitierter und/oder hochexothermer Gas/Flüssig-Reaktionen (in Fortschritt-Bericht VDI, Reihe 3, Nr. 771). Ph.D. thesis, Universität Erlangen-Nürnberg.

47. Lee, S. Y., and Tsui, Y. P. (July/1999). Succeed at gas/liquid contacting. *Chem. Eng. Prog.*, 23–49.

48. Sato, M., and Goto, M. (2004). Gas absorption in water with microchannel devices. *Sep. Sci. Tech.*, 39: 3163–3167.

49. Adam, N. K. (1941). *The Physics and Chemistry of Surfaces*. 3rd ed. Oxford, UK: Oxford University Press.

50. Adamson, A. W., and Gast, A. P. (1997). *Physical Chemistry of Surfaces*, 6th ed. New York: Wiley.

51. TeGrotenhuis, W. E., and Stenkamp, V. S. (2001). Normal gravity testing of a microchannel phase separator for in situ resource utilization. NASA Contract Rept. CR-2001-210955.

52. Günther, A., Jhunjhunwala, M., Thalmann, M., et al. (2005). Micromixing of miscible liquids in segmented gas-liquid flow. *Langmuir*, 21: 1547–1555.

53. Hsieh, C. C., and Yao, S. C. (2004). Development of a microscale passive gas/liquid separation system. In *Proceedings of 5th International Conference on Multiphase Flow*, Yokohama, Japan, May 31–June 3, Paper No. 566.

54. Cypes, S. H., and Engstrom, J. R. (2004). Analysis of a toluene stripping process: A comparison between a microfabricated stripping column and a conventional packed tower. *Chem. Eng. J.*, 101: 49–56.

55. Palo, S. R., Stenkamp, V. S., Dagle, R. A., et al. (2006). Industrial applications of microchannel process technology in the United States. In N. Kockmann (ed.), *Microprocess Engineering*. Weinheim: Wiley-VCH.

56. Sotowa, K. I., and Kusakabe, K. (2003). Design of microchannels for use in distillation devices. In *Abstracts from 7th International Conference on Microreaction Technology*, Lausanne, Switzerland, September 7–10, pp. 156–157.

57. Wootton, R. C. R., and deMello, A. J. (2004). Continuous laminar evaporation: Micronscale distillation. *Chem. Commun.*, 3: 266–267.

58. Kockmann, N., and Woias, P. (2005). Separation principles in micro process engineering. In *Proceedings of 8th International Conference on Microreaction Technology*, Atlanta, Georgia, April 10–14, Paper No. TK129a.

59. Haverkamp, V., Ehrfeld, W., Gebauer, K., et al. (1999). The potential of micromixers for contacting of disperse liquid phases. *Fresenius J. Anal. Chem.*, 364: 617–624.

60. Benz, K., Jäckel, K. P., Regenauer, K. J., et al. (2001). Utilization of micromixers for extraction processes. *Chem. Eng. Technol.*, 24: 11–17.

61. Okubo, Y., Toma, M., and Ueda, H., et al. (2004). Microchannel devices for the coalescence of dispersed droplets produced for use in rapid extraction processes. *Chem. Eng. J.*, 101: 39–48.

62. Kralj, J. G., Schmidt, M. A., and Jensen, K. F. (2005). Surfactant-enhanced liquid–liquid extraction in microfluidic channels with inline electric-field enhanced coalescence. *Lab Chip*, 5: 531–535.

63. Burns, J. R., and Ramshaw, C. (2001). The intensification of rapid reactions in multiphase systems using slug flow in capillaries. *Lab Chip*, 1: 10–15.

64. Kuban, P., Berg, J., and Dasgupta, P. K. (2003). Vertically stratified flows in microchannels. Computational simulations and applications to solvent extraction and ion exchange. *Anal. Chem.*, 75: 3549–3556.

65. Law, B., Colclough, N., Temesi, D., et al. (2001). The development and evaluation of a microfluidic device for the determination of organic-aqueous partition coefficients to support agrochemical and drug discovery. *Presented at 5th International Conference on Miniaturized Chemical and Biochemical Analysis Systems, MicroTAS2001.* Monterrey, California, October 21–25.

66. TeGrotenhuis, W. E., Cameron, R. J., Butcher, M. G., et al. (1998). Microchannel devices for efficient contacting of liquids in solvent extraction. In *Proceedings of 2nd International Conference on Microreaction Technology*, New Orleans, Louisiana, March 9–12, pp. 329–334.

67. Turner, C., Shaw, J., Miller, B., et al. (2000). Solvent extraction using micro-mesh reactors. In *Proceedings of 4th International Conference on Microreaction Technology*, Atlanta, Georgia, March 5–9, pp. 334–340.

68. Shaw, J., Miller, B., Turner, C., et al. (1996). Mass transfer of species in micro-contactors: CFD modelling and experimental validation. *Analyt. Methods Instrum.*, Basel, Switzerland, November 20–22, Special Issue, MicroTAS96: 185–188.

69. Bibby, I. P., Harper, M. J., and Shaw, J. (1998). Design and optimisation of micro-fluidic reactors through CFD and analytical modelling. In *Proceedings of 2nd International Conference on Microreaction Technology*, New Orleans, Louisiana, March 9–12, pp. 335–339.

70. Tokeshi, M., Minagawa, T., Uchiyama, K., et al. (2002). Continuous-flow chemical processing on a microchip by combining microunit operations and a multiphase flow network. *Anal. Chem.*, 74: 1565–1571.

71. Maruyama, T., Kaji, T., Ohkawa, T., et al. (2004). Intermittent partition walls promote solvent extraction of metal ions in a microfluidic device. *The Analyst*, 129: 1008–1013.

72. Hibara, A., Nonaka, M., Hisamoto, H., et al. (2002). Stabilization of liquid interface and control of two-phase confluence and separation in glass microchips by utilizing octade-cylsilane modification of microchannels. *Anal. Chem.*, 74: 1724–1728.

73. Maruyama, T., Matsushita, H., Uchida, J., et al. (2004). Liquid membrane operations in a microfluidic device for selective separation of metal ions. *Anal. Chem.*, 76: 4495–4500.

74. Reddy, V., and Zahn, J. D. (2005). Interfacial stabilization of organic–aqueous two-phase microflows for a miniaturized DNA extraction module. *J. Coll. Interf. Sci.*, 286: 158–165.

75. Tokeshi, M., Minagawa, T., and Kitamori, T. (2000). Integration of a microextraction system on a glass chip: ion-pair solvent extraction of Fe(II) with 4,7-diphenyl-1,10-phenanthrolinedisulfonic acid and tri-*n*-octylmethylammonium chloride. *Anal. Chem.*, 72: 1711–1714.

76. Ueno, K., Kim, H. B., and Kitamura, N. (2003). Channel shape effects on the solution-flow characteristics and the liquid/liquid extraction efficiency in polymer microchannel chips. *Anal. Sci.*, 19: 391–394.

77. Aota, A., Nonaka, M., Hibara, A., et al. (2003). Micro counter-current flow system for highly efficient extraction. In *Proceedings of 7th International Conference on Miniaturized Chemical and Biochemical Analysis Systems*, MicroTAS2003, Squaw Valley, California, October 5–9, pp. 441–444.

78. Weigl, B. H., and Yager, P. (1999). Microfluidic diffusion-based separation and detection. *Science*, 283: 346–347.

79. Weigl, B. H., Holl, M. R., Schutte, D., et al. (1996). Diffusion-based optical chemical detection in silicon flow structures. *Analyt. Methods Instrum.*, Basel, Switzerland, November 20–22, Special Issue, MicroTAS96: 174–184.

80. Brody, J. P., and Yager, P. (1997). Diffusion based extraction in a microfabricated device. *Sens. & Act. A: Physical*, 58: 13–18.

81. Kamholz, A. E., Weigl, B. H., Finlayson, B. A., et al. (1999). Quantitative analysis of molecular interaction in a microfluidic channel: The T-sensor. *Anal. Chem.*, 71: 5340–5347.

82. Weigl, B. H., Bardell, R. L., and Cabrera, C. R. (2003). Lab-on-a-chip for drug development. *Adv. Drug Deliv. Rev.*, 55: 349–377.

83. Bowden, S. A., Monaghan, P. B., and Wilson, R., et al. (2006). The liquid–liquid diffusive extraction of hydrocarbons from a North Sea oil using a microfluidic format. *Lab Chip*, 6: 740–743.

84. Rings, A., Lücke, A., and Schleser, G. H. (2004). A new method for the quantitative separation of diatom frustules from lake sediments. *Limnol. Oceanogr. Methods*, 2: 25–34.

85. Sharma, R. V., Edwards, R. T., and Beckett, R. (1993). Physical characterization and quantification of bacteria by sedimentation field-flow fractionation. *Appl. Environ. Microbiol.*, 59: 1864–1875.

86. Fuh, C. B. (2000). Split-flow thin fractionation. *Anal. Chem.*, 72(7): p266A–271A.

87. Zhang, Y., Barber, R. W., and Emerson, D. R. (2005). Particle separation in microfluidic devices—SPLITT fractionation and microfluidics. *Cur. Analyt. Chem.*, 1: 345–354.

88. Hawkes, J. J., Barrow, D., and Coakley, W. T. (1998). Microparticle manipulation in millimetre scale ultrasonic standing wave chambers *Ultrasonics*, 36: 925–931.

89. Hawkes, J. J., Barber, R. W., and Emerson, D. R., et al. (2004). Continuous cell washing and mixing driven by an ultrasound standing wave within a microfluidic channel. *Lab Chip*, 4: 446–452.

90. Nilsson, A., Petersson, F., and Jönsson, H., et al. (2004). Acoustic control of suspended particles in micro fluidic chips. *Lab Chip*, 4: 131–135.

91. Petersson, F., Nilsson, A., and Holm, C., et al. (2005). Continuous separation of lipid particles from erythrocytes by means of laminar flow and acoustic standing wave forces. *Lab Chip*, 5: 20–22.

92. Pamme, N. (2006). Magnetism and microfluidics. *Lab Chip*, 6: 24–38.

93. Mikkelsen, C., and Bruus, H. (2005). Microfluidic capturing—dynamics of paramagnetic bead suspensions. *Lab Chip*, 5: 1293–1297.

94. Han, K. H., and Frazier, A. B. (2006). Paramagnetic capture mode magnetophoretic microseparator for high efficiency blood cell separations. *Lab Chip*, 6: 265–273.

95. Raymond, D. E., Manz, A., and Widmer, H. M. (1994). Continuous sample pretreatment using a free-flow electrophoresis device integrated onto a silicon chip. *Anal. Chem.*, 66: 2858–2865.

96. Zhang, C. X., and Manz, A. (2003). High speed free-flow electrophoresis on chip. *Anal. Chem.*, 75: 5759–5766.

97. Gale, B. K., Caldwell, K. D., and Frazier, B. (1998). A micromachined electrical field-flow fractionation (μ-EFFF) system. *IEEE Trans. Biomed. Eng.*, 45: 1459–1469.

98. Merugu, S., Narayanan, N., and Gale, B. K. (2003). High throughput separations using a microfabricated serial electric SPLITT system. In *Proceedings of 7th International Conference on Miniaturized Chemical and Biochemical Analysis Systems*, MicroTAS2003, Squaw Valley, California, October 5–9, pp. 1191–1194.

99. Cabrera, C. R., Finlayson, B., and Yager, P. (2001). Formation of natural pH gradients in a microfluidic device under flow conditions: Model and experimental validation. *Anal. Chem.*, 73: 658–666.

100. Brauns, E., van Hoof, V., Dotremont, C., et al. (2004). The desalination of an Arthrospira platensis feed solution by electrodialysis and reverse osmosis. *Desalination*, 170: 123–136.

101. Hughes, M. P. (2002). Strategies for dielectrophoretic separation in laboratory-on-a-chip systems. *Electrophoresis*, 23: 2569–2582.

102. Gascoyne, P. R. C., and Vykoukal, J. (2002). Particle separation by dielectrophoresis, *Electrophoresis*, 23: 1973–1983.

103. Kua, C. H., Lam, Y. C., and Yang, C., et al. (2005). Review of bio-particle manipulation using dielectrophoresis. Innovation in Manufacturing Systems and Technology (IMST), Web site http://www.hdl.handle.net/1721.1/7464.

104. Park, J., Kim, B., and Choi, S. K., et al. (2005). An efficient cell separation system using 3D-asymmetric microelectrodes. *Lab Chip*, 5: 1264–1270.

105. Choi, S. and Park, J. K. (2005). Microfluidic system for dielectrophoretic separation based on a trapezoidal electrode array. *Lab Chip*, 5: 1161–1167.

106. Lapizco-Encinas, B. H., Simmons, B. A., and Cummings, E. B., et al. (2004). Dielectrophoretic concentration and separation of live and dead bacteria in an array of insulators. *Anal. Chem.*, 76: 1571–1579.

107. Eijkel, J. C. T. and van den Berg, A. (2006). Nanotechnology for membranes, filters and sieves. *Lab Chip*, 6: 19–23.

108. van Rijn, C. J. M. (2004). *Nano and Micro-engineered Membrane Technology*. Membrane Science and Technology Series, No. 10, Amsterdam: Elsevier.

109. Kuiper, S., van Rijn, C., and Nijdam, W., et al. (2002). Filtration of lager beer with microsieves: Flux, permeate haze and in-line microscope observations. *J. Membr. Sci.*, 196: 159–170.

110. Mielnik, M. M., Ekatpure, R. P., and Sætran, L. R., et al. (2005). Sinusoidal crossflow microfiltration device—experimental and computational flowfield analysis. *Lab Chip*, 5: 897–903.

111. Brenner, T., Haeberle, S., and Zengerle, R., et al. (2004). Continuous centrifugal separation of whole blood on a disk. In *Proceedings of 8th International Conference on Miniaturized Systems for Chemistry and Life Sciences*, MicroTAS2004, Malmo, Sweden, September 26–30, pp. 566–568.

112. Kim, K. C. K. (2004). Novel particle separation using spiral channel and centrifugal force for plasma separation from whole blood. In *Proceedings of 8th International Conference on Miniaturized Systems for Chemistry and Life Sciences*, MicroTAS2004, Malmo, Sweden, September 26–30, pp. 614–616.

113. Ookawara, S., Higashi, R., and Street, D., et al. (2004). Feasibility study on concentration of slurry and classification of contained particles by microchannel. *Chem. Eng. J.*, 101: 171–178.

114. Ookawara, S., Oozeki, N., and Ogawa, K. (2005). Experimental benchmark of a metallic micro-separator/classifier compared with representative hydrocyclone. In *Proceedings of 8th International Conference on Microreaction Technology*, Atlanta, Georgia, April 10–14, Paper No. TK129g.

115. Ookawara, S., Street, D., and Ogawa, K. (2006). Numerical study on development of particle concentration profiles in a curved microchannel. *Chem. Eng. Sci.*, 61: 3714–3724.

116. DiMarzio, E. A., and Guttman, C. M. (1970). Separation by flow. *Macromolecules*, 3: 131–146.

117. Lin, Y. C., and Jen, C. P. (2002). Mechanism of hydrodynamic separation of biological objects in microchannel devices. *Lab Chip*, 2: 164–169.

118. Yamada, M., Nakashima, M., and Seki, M. (2004). Pinched flow fractionation: Continuous size separation of particles utilizing a laminar flow profile in a pinched microchannel. *Anal. Chem.*, 76: 5465–5471.

119. Takagi, J., Yamada, M., and Yasuda, M., et al. (2005). Continuous particle separation in a microchannel having asymmetrically arranged multiple branches. *Lab Chip*, 5: 778–784.

120. Yang, S., Ündar, A., and Zahn, J. (2005). Biological fluid separation in microfluidic channels using flow rate control. *Presented at IMECE05 ASME International Mechanical Engineering Congress*, Orlando, Florida, November 15–11, Paper No. IMEC2005-80501.

121. Yamada, M., and Seki, M. (2005). Hydrodynamic filtration for on-chip particle concentration and classification utilizing microfluidics. *Lab Chip*, 5: 1233–1239.

CHAPTER 7

CALCULATIONS AND SIMULATIONS

DIETER BOTHE

7.1 INTRODUCTION

The large area-to-volume ratio of microreactors gives the prospect of, for example, better yield and selectivity than for conventional designs, since diffusive fluxes of mass and heat in microdevices scale with the area, while the rate of changes corresponding to sources and sinks is proportional to the volume. Most applications of microreactors aim at an intensification of chemical transformations or the design of new classes of chemical processes, especially fast and/or highly exothermic chemistry. From the very definition of intensification of a chemical process as an increase of the space/time-yield at massively reduced costs, it is obvious that residence times must be minimized and, hence, should hardly exceed the time required for the chemical reaction due to its pure kinetics. Consequently, transport processes can still limit the process performance even if the dimensions of the system are small. To tab the full potential of ever-expanding microreaction engineering technology, a fundamental understanding of the transport processes on all relevant time and length scales is therefore necessary.

When decreasing the length scale of a fluidic system, new phenomena may appear. These fall into two categories:

1. True microeffects due to physico-chemical phenomena that are negligible in macrosystems, such as electrokinetic effects that may significantly change the boundary conditions, nonlinear material properties that may become important due to possibly much higher shear rates, or influences of the solid walls because the ratio of wall area per fluid volume, of course, also increases. Furthermore, especially in gas flows within small channels, deviations from the continuum mechanical description can occur.

Microchemical Engineering in Practice. Edited by Thomas R. Dietrich
Copyright © 2009 John Wiley & Sons, Inc.

2. Scale effects that can be described with the common continuum mechanical balances for mass, momentum, species mass, and energy, but simply reflect changes in the individual time scales of the relevant mechanisms. Besides the example of diffusive fluxes mentioned above, a typical scale effect is the emergence of viscous heating even in simple channel flow.

In this chapter, the main emphasis lies on scale effects, since these are present in any microsystem, in the calculation of such scale effects and how they should be taken into account in the design of microreactors as well as in the modeling and CFD simulation of the transport and transformation processes that take place across a hierarchy of time and length scales. Concerning microeffects, an intense discussion persists in the literature about to what extent and on which scales such effects occur in microchannels [1–8]. In the case of water or aqueous solutions flowing through microchannels with wall distances above some 10 μm, there seem to be no deviations from macroscopic channels concerning the flow behavior and its description. In fact (see, e.g., [3, 7]), the transport of momentum is then adequately described by the Navier–Stokes equations with appropriate boundary conditions.

7.2 MECHANISMS AND SCALES

We concentrate on the case of a chemical process that takes place inside a fluid phase within a microreactor. Such reactors are usually operated continuously, that is, the educts enter the reactor through two or more inlets and the fluid is driven through the microsystem by means of a pressure difference between inlets and outlet where the reacting mixture leaves the system.

A prototype device that has the functionality to unify two incoming fluid streams and to host them in a mixing/reacting zone is a T-shaped microchannel, as illustrated in Fig. 7.1. Typical length scales in the cross-directions of microreactors for productional purposes fall in a (sub-)millimeter rather than true micrometer range, with hydraulic diameters of a few 100 μm. Since the mechanisms of chemical reactions act on an Ångström length scale, even in such small devices about seven orders of magnitude have to be bridged by transport processes in order to mix the initially segregated inlet streams on the molecular level. This mixing has to happen fast enough such that the chemical transformations are not masked by transport processes; see Fig. 7.2. The fact that the local state of mixing can massively affect, for instance, the selectivities of chemical reactions is well known for macrosystems [9, 10], but applies equally to microchemical processes.

The basic timescale for the design of a continuously driven chemical reactor is the characteristic time of the chemical reaction, meaning again the time for conversion due to the intrinsic kinetic. In the case of a single chemical reaction, assumed to be of first order for the sake of simplicity, this timescale is

$$\tau_R = \frac{1}{k} \tag{7.1}$$

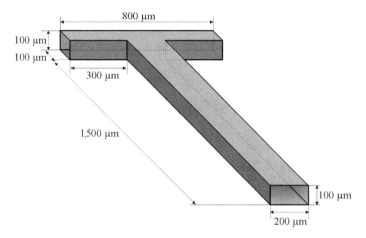

FIGURE 7.1 T-shaped microreactor.

where k denotes the rate constant. Given the value of τ_R, the hydrodynamic residence time

$$\tau_H = \frac{L}{U} \tag{7.2}$$

with L being the reactor length and U the mean axial velocity, should be a few times higher in order to allow complete conversion but not more, which gives as a first simple design rule the equation

$$\frac{L}{U} = a\tau_R \tag{7.3}$$

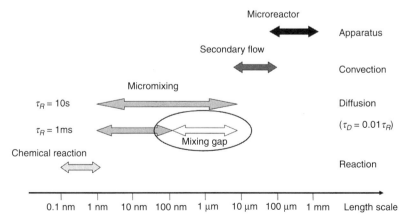

FIGURE 7.2 Hierarchy of length scales.

with $a = 3 \ldots 5$, say. Everything else depends on the geometry, flow conditions, and employed mixing mechanisms. To start with, consider a reacting flow in a microchannel of rectangular cross section at strictly laminar flow conditions. The latter implies certain restrictions on the Reynolds number

$$\mathrm{Re} = \frac{dU}{\nu} \tag{7.4}$$

of the flow, say, $\mathrm{Re} \leq 1$, where d is the hydraulic channel diameter and ν denotes the kinematic viscosity of the flow. In this case, there is no secondary flow that would promote mixing. Hence, the required mixing in a cross-direction can only be achieved by diffusion—this subprocess is usually called micromixing. The time required to homogenize the concentration profiles across the diameter d of the channel is given approximately by

$$\tau_D = \frac{d^2}{D} \tag{7.5}$$

where D is the diffusivity of the respective chemical species. Here, two comments are in order: First, it is assumed that the educts are segregated on a scale comparable to d, which is the case at least at the entrance; second, even if the educts enter the reaction channel completely mixed, the Poiseuille-like velocity profile results in significantly varying residence times along different streamlines, resulting in different degrees of conversion along a cross section, that is, it results in an apparent demixing. Therefore, unless further mixing mechanisms are active, the diameter of the channel has to be used to compute the timescale of diffusive mixing. To avoid a mixing dependence of the chemical process, τ_D should be well below τ_R, that is,

$$\frac{d^2}{D} = \varepsilon \tau_R \tag{7.6}$$

with $\varepsilon = 0.01$, say, that allows for computing the channel diameter.

The main constraint on the velocity stems from the corresponding pressure drop. In the case of a fully developed flow (Fig. 7.3), the flow field is given as the stationary solution of the Navier–Stokes equations (see Section 7.3 below) for a channel of rectangular cross section. It reads as

$$w(x,y) = \frac{16\Delta p}{\eta L \pi^4} \sum_{i,j \in N} \frac{a^2 b^2}{ij(i^2 b^2 + j^2 a^2)} \sin\left(i\pi \frac{x}{a}\right) \sin\left(j\pi \frac{y}{b}\right) \tag{7.7}$$

where w denotes the axial velocity component, a and b are the width and height of the channel cross section (with $a \leq b$, say), and N denotes the set of all odd natural numbers; see [11] or [12] for another common representation of this solution.

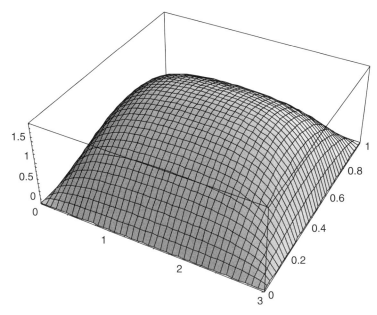

FIGURE 7.3 Fully developed channel flow (aspect ration 1 : 3).

From Eq. (7.7), the pressure drop Δp can be calculated as

$$\Delta p = \alpha \frac{\eta L U}{a^2} \quad \text{with} \quad \frac{1}{\alpha} = \frac{64}{\pi^6} \sum_{i,j \in N} \frac{1}{i^2 j^2 (i^2 + j^2 (a/b)^2)} \tag{7.8}$$

This resembles the well-known formulas for the pressure drop in a pipe or between two parallel plates, the only difference being the coefficient α that depends on the aspect ratio $a:b$; see Table 7.1.

TABLE 7.1 Pressure Drop Coefficient in a Rectangular Channel vs. Aspect Ratio

$a:b$	α
1 : 1	28.45
1 : 2	17.49
1 : 3	15.19
1 : 5	13.73
1 : 10	12.81
1 : ∞	12.00
Pipe	32.00

To fix the case, assume that the channel has a quadratic cross section of width d. Exploiting (7.2), (7.5), (7.3), and (7.6) then yields

$$\Delta p \approx 30 \, \mathrm{Sc} \frac{\tau_H}{\tau_D} \rho U^2 \approx 10^4 \, \mathrm{Sc} \, \rho U^2 \tag{7.9}$$

where

$$\mathrm{Sc} = \frac{\nu}{D} \tag{7.10}$$

is the Schmidt number. If Δp_{max} is the maximal feasible pressure drop, this leads to the velocity constraint

$$U \leq \frac{1}{100} \sqrt{\frac{\Delta p_{max}}{\rho \, \mathrm{Sc}}} \tag{7.11}$$

Specialized to aqueous systems with typical values of $\mathrm{Sc} \approx 1{,}000$, inequality (7.11) means

$$U \leq 10^{-5} \sqrt{\Delta p_{max} \times \frac{\mathrm{m}^3}{\mathrm{kg}}} \tag{7.12}$$

Hence, due to the small hydraulic diameter, a pressure difference of 10 bar generates a flow of only 0.01 m/s and an even smaller volumetric flow rate $\dot{V}$; note that $\dot{V}$ scales with d^4. To maximize the flow rate, let

$$U_{max} = \frac{1}{100} \sqrt{\frac{\Delta p_{max}}{\rho \, \mathrm{Sc}}} \tag{7.13}$$

Then $\mathrm{Re} \leq 1$ implies

$$d \leq d_{max} = 100 \sqrt{\frac{\rho \nu^3}{D \Delta p_{max}}} \tag{7.14}$$

which gives $d \leq 100 \, \mu\mathrm{m}$ for aqueous systems. With the maximum diameter, the associated time scale for diffusive mixing is

$$\tau_D = 10^4 \, \mathrm{Sc}^2 \frac{\eta}{\Delta p_{max}} \tag{7.15}$$

That is, 10 s in the specific case above. Here, some polishing is possible: $d/2$ instead of d can be used in (7.5), and in Einstein's formula for the diffusion time an additional factor of 2 appears in the nominator, which adds up to about one order of magnitude. This value then determines the admissible chemistry such that masking by mixing is avoided. Knowing the reaction time, Eq. (7.3) yields the reactor length. In the example above with a reduced $\tau_D = 1$ s, the result is $\tau_R = 100$ s, leading to a reactor length of 1 m.

The mechanisms, dimensionless numbers, and scales used above can be found, for example, in books on chemical engineering such as [13, 14]. Although recalling these basic scale considerations yields some initial design rules, it also clearly shows that purely diffusive mixing is not enough, except for slow chemistry and applications like chemical analysis where extremely small flow rates are sufficient. Therefore, efficient mixing is a fundamental demand also for microreactors. Since microchemical processes are mainly run continuously, mixing evolves along an axial direction and mixing in axial direction is often unwanted, because it reduces conversion and strongly influences selectivities.

Consequently, in contrast to mixing in batch reactors, mixing should be anisotropic, ideally combining complete and instantaneous homogenization in cross-directions with no relative axial movements. The latter would require plug flow that cannot be achieved due to no-slip at the walls because of the liquid's viscosity; recall also (7.7) and Fig. 7.3. Fortunately, various types of microreactors have been developed that promote mixing in one or the other way. Therefore, the "mixing gap" illustrated by Fig. 7.3 can hopefully be bridged. Some of the mechanisms underlying these micromixers are (see Fig. 7.4(a–c) and especially Chapter 4):

- Geometric splitting or focusing of feed streams
- Stretching and folding (Baker's map)
- Secondary flow mixing (e.g., chaotic advection)

In addition, there are active micromixers employing effects such as flow modulation (e.g., by piezo-electrical devices), crossing flow, electrowetting, ultrasound, etc. More information about passive and active mixing mechanisms and micromixers may be found in reviews [15, 16] available in the literature.

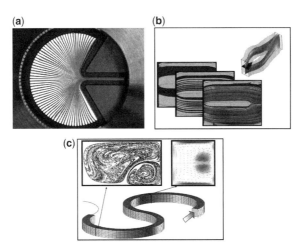

FIGURE 7.4 (a) Multilamination, (b) split-and-recombine, and (c) chaotic. Courtesy of IMM GmbH.

From the discussion thus far, it should be clear that transport processes are of great significance and that sufficient mixing cannot simply be assumed but has to be proven—if possible before designing a certain microchemical process. Here, simple scale considerations are not enough but have to be complemented by numerical simulations.

7.3 MODELING AND COMPUTATION OF REACTING FLOWS

We focus on situations where the reacting flow is governed by the standard continuum mechanical balances. If incompressibility is assumed, the balances of mass and momentum lead to the Navier–Stokes equations that in nondimensional form read as

$$\nabla \cdot \mathbf{u} = 0 \qquad \text{and} \qquad \partial_t \mathbf{u} + \mathbf{u} \cdot \nabla \mathbf{u} = -\nabla p' + \frac{1}{\text{Re}} \Delta \mathbf{u} \qquad (7.16)$$

with velocity field $\mathbf{u}$ and reduced pressure p', complemented by appropriate boundary conditions. In such a fluid flow, the transport of an ideally diluted chemical species that undergoes a chemical reaction, say, again a reaction of first order for simplicity, is governed by the species equation

$$\partial_t c + \mathbf{u} \cdot \nabla c = \frac{1}{\text{Re Sc}} \Delta c - \frac{d}{L} \text{Da}_\text{I} c \quad \text{with } \text{Da}_\text{I} = \frac{kL}{U} \qquad (7.17)$$

where c denotes the molar concentration made dimensionless using a reference value c_0.

Detailed information about the balance equations (7.16) and (7.17) as well as those to come can be found in [13, 14, 18, 20, 21, 24]. Here, the main purpose in recalling these fundamental differential equations is that they determine which dimensionless quantities are involved and the latter includes the behavior under changes of scales. Indeed, given geometric similarity, the flow behavior is solely determined by the Reynolds number, regardless of the dimensions of the device, if the in- and outlet conditions are scaled correspondingly. For the complete system (7.17) and (7.18), physico-chemical and hydrodynamical similarity cannot be achieved simultaneously for reactors of different size. This is due to the fact that the Reynolds number depends on the product Ud and the Damköhler(I) number on the quotient L/U. Hence, for the fixed ratio d/L, both numbers cannot be kept fixed while changing the size. To quantify the effect of scaling, we introduce a geometrical scaling factor λ, with $\lambda \ll 1$. Let $L' = \lambda L$, $d' = \lambda d$, and $U' = U/\lambda$. The resulting dimensionless numbers for the micro- and macroscale system satisfy the relations

$$\text{Re}' = \text{Re}, \quad \text{Sc}' = \text{Sc}, \quad \text{but} \quad \text{Da}'_\text{I} = \lambda^2 \text{Da}_\text{I} \qquad (7.18)$$

This phenomenon results from changes in the involved time scales of reaction, convection, and diffusion that are given in (7.1), (7.2), and (7.5). Under the same scaling as above, they behave as

$$\tau'_R = \tau_R, \qquad \tau'_H = \lambda^2 \tau_H, \qquad \tau'_D = \lambda^2 \tau_D \qquad (7.19)$$

Hence, compared to the time scale of the chemical reaction, the characteristic times of the transport processes are massively reduced. This leads to a shift in the range of accessible reaction times, that is, certain reactions that are too fast to be performed in a macroscale reactor, can be accomplished in microscale systems.

In the case of exothermic reactions, for example, thermal effects have to be taken into account. For a single reaction of first order, the balance of heat given for the nondimensional temperature ϑ reads as

$$\partial_t \vartheta + \mathbf{u} \cdot \nabla \vartheta = \frac{1}{\text{Re Pr}} \Delta \vartheta + \frac{\text{Ec}}{\text{Re}} \mathbf{S} : \nabla \mathbf{u} + \frac{d}{L} \text{Da}_{\text{III}} c \qquad (7.20)$$

with

$$\vartheta = \frac{T - T_0}{T_0}, \quad \text{Pr} = \frac{\nu}{\alpha}, \quad \text{Ec} = \frac{U^2}{\hat{c}_p T_0}, \quad \text{Da}_{\text{III}} = \frac{(-\Delta_R H) k c_0}{\rho \hat{c}_p T_0 U / L} \qquad (7.21)$$

Here, Pr denotes the Prandtl number, Ec is the Eckert number relating kinetic to thermal energy, and the Damköhler(III) number Da_{III} relates the maximum rate of heat of reaction to the convective energy flux into the reactor. The second term on the right-hand side of (7.20) corresponds to viscous heating with the viscous stress tensor in dimensionless form, that is,

$$\mathbf{S} = \eta(\nabla \mathbf{u} + \nabla \mathbf{u}^{\mathsf{T}}) \qquad (7.22)$$

and the double dot product $\mathbf{S} : \nabla \mathbf{u}$ is the trace of the matrix product $\mathbf{S} \cdot \nabla \mathbf{u}$.

The heat balance (7.21) has to be complemented by the boundary condition

$$\alpha \frac{\partial \vartheta}{\partial \mathbf{n}} = -\alpha_W \text{Nu} \frac{d}{d_W} (\vartheta - \vartheta_W) \qquad (7.23)$$

on the lateral wall with

$$\vartheta_W = \frac{T_W - T_0}{T_0}, \quad \text{Nu} = \frac{\partial T}{\partial \mathbf{n}} \bigg/ \frac{T_W - T}{d_W} \qquad (7.24)$$

In (7.23) and (7.24), Nu are the local Nusselt numbers, T_W is the outside wall temperature assumed to be constant, and d_W is the wall thickness. Under the scaling $L' = \lambda L$, $d' = \lambda d$, $U' = U/\lambda$ as before, the dimensionless numbers behave as

$$\text{Re}' = \text{Re}, \quad \text{Pr}' = \text{Pr}, \quad but \quad \text{Ec}' = \lambda^{-2} \text{Ec}, \quad \text{Da}'_{\text{III}} = \lambda^2 \text{Da}_{\text{III}} \qquad (7.25)$$

Equation (7.25) in particular, shows that viscous heating becomes important in microscale systems, which has also been shown experimentally and by means of numerical simulations [6, 8]. It is also evident that viscous and reactive heating scale in the opposite manner: Reducing the system's dimension at fixed Re diminishes the overall rate of heating until the viscous contribution finally leads to an increase.

In order to control the reactor thermally, both heat sources have to be balanced by heat flow through the lateral boundary. To see whether this can be achieved, the effect of the boundary condition, which acts as a sink in case of $\vartheta_W < \vartheta$, has to be quantified as an equivalent volumetric source term. For a duct with constant cross section A, it can be shown [17] that the effect of the boundary condition (7.23) may indeed be expressed by a volumetric source term, leading to

$$\partial_t \vartheta + \mathbf{u} \cdot \nabla \vartheta = \frac{1}{\text{Re Pr}} \Delta \vartheta + \frac{\text{Ec}}{\text{Re}} \mathbf{S} : \nabla \mathbf{u} + \frac{d}{L} \text{Da}_{III} c$$
$$- \frac{4}{\pi} \frac{\text{Nu}}{\text{Re Pr}_W} \frac{|\partial A|}{d_W} \langle \vartheta - \vartheta_W \rangle_{\partial A}$$

where $|\partial A|$ denotes the circumferential length, while $\langle \vartheta - \vartheta_W \rangle_{\partial A}$ is the average of the (nondimensional) temperature difference on the circumference of a cross section. If Nu only depends on Re and Pr_W, the source term scales as d/d_W. Although due to technical constraints, d_W might possibly not decrease by the same factor as d does, this still leads to a scaling of type

$$\frac{d'}{d'_W} = \lambda^{0\ldots1} \frac{d}{d_W} \tag{7.27}$$

for the heat flux through the boundary. Consequently, the overall volumetric rate of heating as a function of the scaling parameter λ is of the type

$$b_V \lambda^{-2} + b_R \lambda^2 - b_W \lambda^{0\ldots1} \langle \vartheta - \vartheta_W \rangle_{\partial A} \tag{7.28}$$

with certain coefficients b_V, b_R, and b_W, corresponding to viscous heating, reactive heating, and wall cooling, respectively. Hence, it is clear that heat control performs better for smaller system dimensions as long as viscous heating is negligible. For typical values of the viscous contribution to heating, see [8].

The models given above rely on certain simplifications. First, Sorét and Dufour effects modeling thermal or pressure diffusion have been neglected; see [18] and especially [17]. Second, and this might be too crude an approximation, the electrical field is not taken into account. The latter is, of course, important for the transport of ionic species, in which case the ions generate an intrinsic electrical field that causes additional convection-like fluxes, so-called electromigration; see [19]. Then, although the system will behave electroneutral macroscopically, all diffusive fluxes are altered by a strong coupling and the model as well as its numerical solution becomes much more complicated. Third, as it stands, the model will only apply to dilute systems. In the case of higher concentrations, as will occur especially in intensified processes, the Stefan–Maxwell equations have to be used instead of species equations of type (7.17); for more details on this topic, see [18]. Other general references concerning modeling of (reacting) flows are, for example, [20, 21]. For more information on mixing, see [22].

There are several commercial computational fluid dynamics tools (CFD tools) available that allow for the numerical solution of partial differential equations (7.16),

(7.17), and (7.20). A review of the underlying numerical schemes, their pros and cons, is outside the scope of this chapter, just as an evaluation of the commercial codes would be. Concerning the former, there are excellent books on the subject, such as [23–25], whereas the latter would only be a biased snapshot with a very limited life span. Instead, let us start with some general comments and hints. The standard CFD codes are rather mature concerning the computation of flow fields in stationary flow regimes. The latter are of interest for most microfluidic applications due to the typically moderate Reynolds numbers. Still, due to secondary flows, the resulting flow fields can show complicated small-scale structures. These can promote mixing and, hence, need to be resolved in a CFD calculation.

To estimate the smallest length scale present, we start with the balance for mechanical energy, which reads as

$$\eta \int_V |\nabla \mathbf{u}|^2 dV = \dot{V}\Delta p + \int_{\Gamma_{in}} \frac{\rho}{2}|\mathbf{u}|^2 U_{ax} dA - \int_{\Gamma_{out}} \frac{\rho}{2}|\mathbf{u}|^2 U_{ax} dA \qquad (7.29)$$

for a control volume V in the stationary case, neglecting stress contributions at the in- and outflow boundaries Γ_{in} and Γ_{out}, respectively. Here, U_{ax} denotes the component of $\mathbf{u}$ normal to the boundary in the main flow direction. A simple scaling argument as above shows that for small system dimensions the boundary terms become negligible compared to the pressure drop contribution. The pressure drop splits into a part Δp_{wall} acting against the wall friction and a possible excess part Δp_{mix} that drives secondary flow and generates small-scale velocity structures. To estimate their effects in the mixing zone V, decompose $\mathbf{u}$ according to $\mathbf{u} = \mathbf{u}_0 + \mathbf{u}'$, where $\mathbf{u}_0$ is the main flow field, say, a Poiseuille flow in simplest cases, and $\mathbf{u}'$ is a superimposed secondary flow field. Insertion into (17.29) yields

$$\eta \int_V |\nabla \mathbf{u}'|^2 dV \leq \dot{V}\Delta p_{mix} + 2\eta \int_V |\nabla \mathbf{u}_0||\nabla \mathbf{u}'| dV \qquad (7.30)$$

The length scales of the smallest structures are such that

$$Re_{local} = \frac{U' \lambda_{vel}}{\nu} \geq 1 \qquad (7.31)$$

since viscous dissipation smears out smaller scales faster than they are generated by convection.

Using this estimate on the left-hand side leads to the inequality

$$\frac{1}{\lambda_{vel}^4} \leq 4\frac{Re^2}{d^3}\frac{1}{\lambda_{vel}} + \frac{\dot{V}\Delta p_{mix}}{\nu^3 \rho V} \qquad (7.32)$$

Then, by an elementary calculation, it follows that

$$\lambda_{\text{vel}} \geq \frac{d}{\text{Re}^{2/3}} \left(1 + \sqrt{3 + \frac{\dot{V}\Delta p_{\text{mix}} d^4}{\nu^3 \rho V \text{Re}^{8/3}}} \right)^{-1/2} \tag{7.33}$$

Note that all quantities on the right-hand side are known except Δp_{mix} and the latter can be obtained from a measurement at the actual Reynolds number and a second measurement at a low reference Reynolds number where the flow is strictly laminar. In the case of the T-shaped micromixer from Fig. 7.1 with a hydraulic diameter of 133 μm and Re = 186, this estimate predicts a lower bound for the smallest length scale of about 3 μm, in good agreement with numerical simulations from [17] showing velocity fluctuations down to length scales of 5 to 10 μm. Such scales can be resolved with moderate grid size; for example, for the T-mixer from [17], a Cartesian grid of 300,000 grid cells is sufficient to resolve the velocity fields, as illustrated in Fig. 7.5.

Concentration fields can show considerably finer structures, with typical dimensions given by the Batchelor length scale [27]

$$\lambda_{\text{conc}} = \frac{\lambda_{\text{vel}}}{\sqrt{\text{Sc}}} \tag{7.34}$$

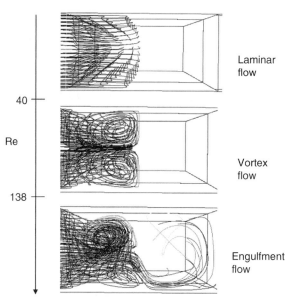

FIGURE 7.5   Streamlines at the entrance of the mixing channel inside a T-micromixer at different Re.

Since the finest structures are dissipated by diffusion, they can only persist if both dissipation and their convective generation act on the same timescale, that is, if

$$\frac{\lambda_{vel}}{U} = \frac{\lambda_{conc}^2}{D} \tag{7.35}$$

With $U \approx \nu/\lambda_{vel}$, this yields

$$\frac{\lambda_{vel}^2}{\nu} = \frac{\lambda_{conc}^2}{D} \tag{7.36}$$

which provides a simple explanation for (7.34).

While Schmidt numbers are moderate for gas flows, the above-mentioned values of about 1,000 for aqueous systems—or even much higher values for diffusion in liquids of higher viscosity—can lead to 30-fold finer structures and, hence, 30,000 times more degrees of freedom in three spatial dimensions. Due to these facts, a so-called direct numerical simulation (DNS for short) of reacting liquid flows in three dimensions, that is, a numerical solution in which all relevant length scales are fully resolved, is not feasible today. This refers to not only turbulent flows but also complex laminar flows with secondary flow fields. The main problem here is the occurrence of artificial smearing of the steep gradients: so-called numerical diffusion that is mainly due to discretization errors. Note that in the finite volume method, which is the standard method in commercial CFD codes, the solution is represented by a single constant value per grid cell. Hence, each grid cell is "ideally mixed" such that no subgrid structure of the solution can be represented. When a flux from one cell to a neighboring one has been computed, the transported quantity immediately fills the entire grid cell. If numerical tracer experiments are performed to asses the mixing efficiency of a flow field, the observed mixing might be completely due to numerical diffusion.

At least for nonreactive mixing, a way to circumvent this problem is to replace the continuous tracer concentration by a number concentration obtained from particles that are tracked during the simulation. This approach hardly suffers from artificial diffusion, because the position of the tracer particles can be resolved with a much higher accuracy than the cell size, and the velocity field at these particle positions can be obtained by interpolation from the grid values. In particular, this method is often applied to compute mixing in liquids of higher viscosity; see [28] for a recent review of this approach as well as Chapter 4. However, let us mention that the particle tracking method is not well suited for convective-diffusive transport (although there are current attempts to extend the particle tracking by solving the stochastic Langevin equation that accounts for Brownian motion, but extremely large numbers of Brownian particles need to be computed to allow for reliable statistical quantities) and even less suited to incorporate chemical reaction.

Another more computationally driven approach is to use adaptive grids with several levels of refinement, exploiting the fact that only the fine scale structures require the finest resolution. Combining this with parallel computation techniques, direct numerical simulations of reacting flows within small subregions of the reactor are possible today under certain simplification. Such simplifications are needed in the case of fast reactions because these lead to extremely stiff differential equations that then require special solution methods or very fine time steps. In the limiting case of instantaneous reactions such as neutralizations or (almost all) radical reactions, the stiffness can sometimes be removed by an old trick already used by Toor in [29]. Indeed, if A and B react to a product P and c_A, c_B denote the molar concentrations of A and B, respectively, then

$$\partial_t c_A + \mathbf{u} \cdot \nabla c_A = \frac{1}{\mathrm{Re}\,\mathrm{Sc}_A}\Delta c_A - \frac{d}{l}\mathrm{Da_I}c_A c_B \tag{7.37}$$

$$\partial_t c_B + \mathbf{u} \cdot \nabla c_B = \frac{1}{\mathrm{Re}\,\mathrm{Sc}_B}\Delta c_B - \frac{d}{l}\mathrm{Da_I}c_A c_B \tag{7.38}$$

where c_A and c_B are normalized to a common reference concentration c_0, and the Damköhler(I) number is given by

$$\mathrm{Da_I} = \frac{kl}{U}c_0 \tag{7.39}$$

here, because the reaction is now of second order. Subtracting (7.36) from (7.35) yields the convection-diffusion equation

$$\partial_t \phi + \mathbf{u} \cdot \nabla \phi = \nabla \cdot (a(\phi)\nabla \phi) \tag{7.40}$$

where $\phi := c_A - c_B$. In the special case of equal diffusivities, which is a realistic assumption for instance in the case of a binary ionic system, this further simplifies to

$$\partial_t \phi + \mathbf{u} \cdot \nabla \phi = \frac{1}{\mathrm{Re}\,\mathrm{Sc}}\Delta \phi \tag{7.41}$$

Now, if the reaction is much faster than the transport processes, the species A and B cannot coexist. Consequently, solving (7.40) for ϕ eventually yields c_A and c_B because of the relations

$$c_A = \max\{\phi, 0\} \quad \text{and} \quad c_B = \max\{-\phi, 0\} \tag{7.42}$$

This approach has been exploited to compute a neutralization reaction inside the T-micromixer shown in Fig. 7.1. Elimination of the reaction term as explained above yields a pure transport problem that can be tackled by adapted grids with parallel computation. In the stationary flow case under consideration, the grid adaptation can be done manually in a few successive steps. Figure 7.6 shows the outcome of corresponding CFD simulations done with Fluent 6.2. Underlying this computation is the neutralization reaction between hydrochloric acid and sodium hydroxide. In addition, a transport equation for the pH-sensitive fluorescence tracer fluoresceine has been solved that allows for the calculation of the signal intensity. This intensity is depicted in the top part of Fig. 7.6, and it shows good agreement with experimental measurements. The bottom part of Fig. 7.6 shows a small part of the adaptive grid having four different refinement levels with a finest spatial resolution of 0.3 μm.

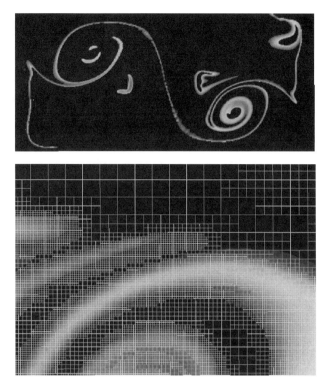

FIGURE 7.6   Simulated fluorescence intensity (top) and adaptive grid (bottom).

A general problem, especially in flow simulations, is the fact that in- and outflow boundaries are nonphysical boundaries that are introduced artificially in order to define a certain computational domain. Therefore, it is unclear what boundary conditions are appropriate at such surfaces. Although this problem can be solved rather easily for inflow boundaries, where usually the velocity profile can be prescribed—for instance, as a fully developed flow as in (7.7)—the problem at outflow boundaries is nontrivial: There, if the boundary condition is "too rigid", it can alter the flow field even far upstream. Note, for example, that the usual homogeneous Neumann boundary condition of vanishing directional derivative in the normal direction of the velocity (together with a prescribed pressure at the outflow boundary) will severely distort exiting vortices. Actually, the investigation of so-called artificial boundary conditions is an active research area; see [30] for a starting point.

7.4   EVALUATION AND VALIDATION OF CFD SIMULATIONS

Computational fluid dynamics has become a very powerful technique for obtaining deep insights into complex flow-based transport processes if the available tools are used with sufficient care and comprehension. A basic point is that commercial

CFD codes are typically very robust in the sense that the solver is likely to converge to a certain outcome which might then be viewed as a solution, although it should actually be viewed as just a possible candidate for the solution. Three types of errors will always occur: modeling errors like the ones mentioned above, errors due to discretization that replaces the original system by a discrete—actually a finite—system, and errors in the computation, for example, deviations because iterative schemes need to be stopped at some point. So, while some result is usually easy to obtain, it is unfortunately left as the user's duty—a rather difficult one—to check whether this result gives an appropriate approximation of the physical solution.

This situation results in a strong need to develop efficient means to evaluate the simulation results for increasingly complex models. In the case of reacting flows with small-scale structures that need to be resolved, the fundamental question is whether all structures inherent in the true solution are present in the computed approximation thereof. Although this query cannot be rigorously answered since the true solution is unknown, one can check if a higher resolution leads to further small-scale structures. A direct check by comparison of two numerical solutions given on millions of grid cells, say, is of course cumbersome. Hence, to check this in an efficient way, an integral measure for the total amount of resolved structures is needed. In [26] we proposed the quantity

$$\Phi(V) = \frac{1}{|V|} \int_V \|\nabla f\| dV \quad \text{with } f = \frac{c}{c_{\max}} \tag{7.43}$$

for this purpose; here, $|V|$ denotes the volume content of V and $\|\nabla f\|$ is the (Euclidian) length of the gradient of the normalized concentration. Mathematically speaking, the quantity Φ is nothing but the *total variation* of the normalized concentration profile, that is, the sum of all variations of the normalized values of c.

Indeed, the quantity Φ can be used to verify whether all relevant length scales are resolved in a numerical simulation. For this purpose, the change of Φ under a grid refinement is observed. With a finer grid, smaller length scales are resolved, and hence, if smaller scales are present in the true solution (and if the numerical scheme converges against the true solution for grid size tending to zero), the value of Φ increases until an asymptotic value is reached for a fine enough spatial discretization. This is illustrated in Fig. 7.7, which shows the total variation obtained from the concentration profile of a diffusive tracer vs. the Schmidt number of the tracer for different grid sizes and different numerical schemes. The concrete simulations again refer to the convective-diffusive mixing in the T-shaped micromixer described above.

In Fig. 7.7, the upper line corresponds to a locally adapted grid with two levels of refinement in regions of steep gradients. The figure clearly shows the increase of resolved scales until eventual convergence: If Sc is kept constant, the values for Φ increase until they end up on a straight line. If the resolution is insufficient, the resulting value falls below that line. Hence in this particular case, a Schmidt number of 500 can be handled with an adaptive grid of 17 million grid cells (see the single point in

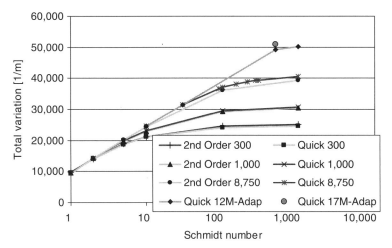

FIGURE 7.7 Total variation vs. log (Sc).

the figure); this computation has been done in parallel using 36 processors and four levels of grid refinement. The point that a linear correlation between the total variation and logarithm of the Schmidt number appears is interesting in its own right; whether this is a special feature of the particular flow or has a more universal meaning must be further investigated. The quantity Φ is also useful to define an integral length scale: Recall that the striation thickness s is a well-known quantity to characterize the length scale in the case of lamellar structures and s^{-1} then corresponds to the specific contact area. Given any segregated concentration distribution $c(\mathbf{x})$ inside a volume V, the quantity Φ coincides with the specific contact area inside V; see [31]. The reciprocal of Φ has the dimension of length and may be interpreted as the average distance between regions of high and low species concentration. This interpretation is exact in the case of lamellar structures. Hence, Φ^{-1} is then a mean striation thickness. The quantity Φ maintains these meanings for almost segregated concentration fields, whereas these interpretations lose their significance for fields close to homogeneous. In any case, physically speaking, Φ represents the *potential for diffusive mixing*, because it measures the total driving force for diffusive fluxes within the concentration field.

While grid independence is a necessary property of a numerical solution in order to be valid, this need not be sufficient. Since additional model errors are present as well, further validations against experimental data—at least under certain key conditions—are indispensable. Furthermore, local data are needed to asses the integrity of complex models and the numerical methods involved.

Fortunately, the well-defined flow conditions that can be achieved in microfluidic systems allow for accurate measurements at improved spatial resolutions. In particular, μ-PIV (micro particle image velocimetry) measurements and μ-LIF (laser induced fluorescence in combination with a confocal laser scanning microscope technique) have been developed further and are now available; see [31]. Figure 7.8

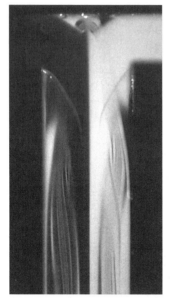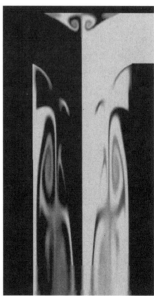

FIGURE 7.8 Experimental (left) and simulated (right) tracer distribution on a horizontal section of the T-micromixer for Re $= 186$.

compares the results of μ-LIF measurements and a CFD simulation with Fluent 6.2, showing very good qualitative agreement between the experiment and simulation. For further details concerning the experimental and numerical analysis of the T-micromixer from Fig. 7.1, see [17, 31] and the references given there.

BIBLIOGRAPHY

1. Gad-el-Hak, M. (1999). The fluid mechanics of microdevices. *J. Fluids Eng.*, 121: 5–33.
2. Gad-el-Hak, M. (2004). Transport phenomena in microdevices. *Z. Angew. Math. Mech.*, 84(7): 494–498.
3. Guo, Z.-Y. and Li, Z.-X. (2003). Size effect on microscale single-phase flow and heat transfer. *Int. J. Heat Mass Trans.*, 46: 149–159.
4. Herwig, H. (2002). Flow and heat transfer in micro systems: Is everything different or just smaller? *J. Appl. Math. & Mech.*, 82(9): 579–586.
5. Koo, J. and Kleinstreuer, C. (2003). Liquid flow in microchannels: Experimental observations and computational analyses of microfluidic effects. *J. Micromech. Microeng.*, 13: 568–579.
6. Koo, J. and Kleinstreuer, C. (2004). Viscous dissipation effects in microtubes and microchannels. *Int. J. Heat Mass Trans.*, 47: 3159–3169.
7. Lee, P.-S., Garimella, S. V., and D. Liu, (2004). Investigation of heat transfer in rectangular microchannels. *Int. J. Heat Mass Trans.*, 48: 1688–1704.

8. Morini, G. L. (2005). Viscous heating in liquid flows in microchannels. *Int. J. Heat Mass Trans.*, 48: 3637–3647.

9. Rys, P. (1992). The mixing-sensitive product distribution of chemical reactions. *Chimia*, 46: 469–476.

10. Bourne, J. R. (2003). Mixing and the selectivity of chemical reactions. *Organ. Proc. Res. & Develop.*, 7: 471–508.

11. Muralidhar, K. and Biswas, G. (1996). *Advanced Engineering Fluid Mechanics.* London: Narosa Publ. House.

12. Shah, R. K. and London, A. L. (1978). *Laminar Flow Forced Convection in Ducts.* New York: Academic Press.

13. Belfiore, L. A. (2003). *Transport Phenomena for Chemical Reactor Design.* Hoboken, NJ: Wiley.

14. Kee, R. J. Coltrin, M. E., and Glarborg, P. (2003). *Chemically Reacting Flow.* Hoboken, NJ: Wiley.

15. Hessel, V. Löwe, H., and Schönfeld, F. (2005). Micromixers—a review on passive and active mixing principles. *Chem. Eng. Sci.*, 60: 2479–2501.

16. Nguyen, N.-T. and Wu, Z. (2005). Micromixers—a review. *J. Micromech. Microeng.*, 15: R1–R16.

17. Bothe, D. Stemich, C. and Warnecke, H.-J. (2006). Fluid mixing in a T-shaped micromixer. *Chem. Eng. Sci.*, 61: 2950–2958.

18. Bird, R. B. Stewart, W. E. and Lightfoot, E. N. (2002). *Transport Phenomena*, 2nd ed. New York: Wiley-Interscience.

19. Newman, J. S. (1991). *Electrochemical Systems*, 2nd ed. Englewood Cliffs, NJ: Prentice Hall.

20. Probstein, R. F. (2003). *Physicochemical Hydrodynamics*, 2nd ed. New York: Wiley-Interscience.

21. Slattery, J. C. (1999). *Advanced Transport Phenomena.* Cambridge, UK: Cambridge University Press.

22. Ottino, J. M. (1989). *The Kinematics of Mixing, Stretching, Chaos, and Transport.* Cambridge, UK: Cambridge University Press.

23. Ferziger, J. H. and Peric, M. (2002). *Computational Methods for Fluid Dynamics*, 3rd ed. Berlin: Springer-Verlag.

24. Warsi, Z. U. A. (1999). *Fluid Dynamics*, 2nd ed. Boca Raton, FL: CRC Press.

25. Wesseling, P. (2001). *Principles of Computational Fluid Dynamics.* Berlin: Springer-Verlag.

26. Bothe, D. Stemich, C. and Warnecke, H.-J. (2006). Mixing in a T-shaped microreactor: Scales and quality of mixing. In W. Marquardt, and C. Pantelidis (eds.), *16th European Symposium on Computer Aided Process Engineering and 9th International Symposium on Process Systems Engineering*, Garmisch-Partenkirchen, Germany: Elsevier, pp. 351–357.

27. Ottino, J. M. (1994). Mixing and chemical reactions: A tutorial. *Chem. Eng. Sci.*, 49: 4005–4027.

28. Phelps, J. H. and Tucker, C. L. (2006). Lagrangian particle calculations of distributive mixing: Limitations and applications. *Chem. Eng. Sci.*, 61: 6826–6836.

29. Toor, H. and Chiang, S. (1959). Diffusion-controlled chemical reactions. *AIChE J.*, 5: 339–344.

30. Sani, R. L. and Gresho, P. M. (1994). Resume and remarks on the open boundary condition minisymposium. *Int. J. Num. Meth. Fluids*, 18: 983–1008.

31. Hoffmann, M. Schlüter, M., and Räbiger, N. (2006). Experimental investigation of liquid–liquid mixing in T-shaped micromixers using μ-LIF and μ-PIV. *Chem. Eng. Sci.*, 61: 2968–2976.

PART III

PERIPHERIC EQUIPMENT

CHAPTER 8

DOSAGE EQUIPMENT

ASIF KARIM and WOLFGANG LOTH

8.1 CONCEPT AND REQUIREMENTS

8.1.1 Basics of Metering for Micromixing (Volumetric Conveyance)

The praxis of microreaction engineering (μRE) is a focus on flowing systems and streams, that is, continuous processes. The density of fluids divides the field into two areas: dense fluids and gaseous streams. Because of gases' compressibility, their metering with pumps and syringes is problematic. Thus, gaseous streams are mostly delivered from pressurized reservoirs and pressurized gas cylinders, and metering valves and flow controllers control the flow.

The motion of dense fluids is driven by pressure difference. It can be realized by placing the reservoir on a higher level, pressurizing the reservoir, or by pumps. What are the necessary basic conditions for producing exact feeds in μRE systems?

Nonpulsating, exact feed is necessary for keeping the flow and pressure drop of microstructured devices at a constant level. This is also the basic condition for mixing fluid streams in micromixing devices, because the ratio between the feeds must be exactly defined and precisely adjusted for the mixture and reaction conditions. Micromixing, pulsating feeds can cause an alternating chain of fluid segments instead of proper mixing because of the small interfacial area that is available. In microreactor flow, pulsation or drifting flow may result in composition and reaction conditions that are not constant. Even if the feed is varied around an average, the mixing ratio is inconsistent, so the correlation between the time and concentration of the feed components is not a constant. Process, reaction conditions and product properties cannot be reproduced or defined this way.

Metering with constant pressure is important when direct control of the feed is absent. Otherwise, a strong increase in pressure drop cannot be compensated for in microscopic channels, where viscosity is increasing.

Microchemical Engineering in Practice. Edited by Thomas R. Dietrich
Copyright © 2009 John Wiley & Sons, Inc.

TABLE 8.1   **Typical Dimensions for Scale-up from Laboratory to Production**

Scale	Form	Feed (mL/h)	Reactor Volume (L)	Pressure (bar)	Form of Dosage
Laboratory scale	Discontinuously	1–5,000	0.05–1	0–6	1 shoot or stepwise feeding into reactor, i.e., batch/ semibatch
Miniplant	Continuously	10–2,000	0.5–5	0–50	Mixing of multiple feeds with additional dosing
Microprocess engineering	Continuously	0.1–2,000	0.01–0.1	0–100	Mixing of multiple feeds with additional dosing

The dosing of fluids from pressurized reservoirs via check valves and flow controllers might be a simple and desirable solution for delivery to μRE systems when the dilution of gas is low and creates no problems in the μRE set-up by out-gassing and forming a second unwanted phase, such as bubbles. When a gas phase does occur, gas volume and interfacial tension effects lead to maldistribution in the microstructures and undefined residence time behavior. Thus, exact pulse-free, metering, with constant pressure, and feed control via a mass flow controller are a must for realizing a reliable microprocess engineering experimental set-up.

8.1.2   Comparison of Miniplant Laboratory Scale and Microprocess Engineering

While developing processes for chemical reactions, different scales are used. For first-time experiments with a low cost for ingredients and for the experimental set-up, laboratory scale usually is the right choice. This scale normally works under discontinuous conditions, that is, in a batch vessel. After initial success in the laboratory scale, the scale-up experiments may be run optionally in a miniplant under continuous conditions and finally scaled up to production scale. Typical dimensions for that equipment are listed in Table 8.1.

8.2   TYPE OF PUMPS

8.2.1   Design

For realizing μRE-compatible feeds, a variety of different pump types are available. Most of them follow the principle of displacement. These are syringe, plunger, and

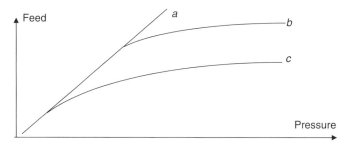

FIGURE 8.1 Pressure delivery characteristics.

gear pumps. Some utilize an impulse-driven principle like centrifugal pumps. They deliver a smooth, pulse-free stream, but are available only for large-volume streams. To obtain small streams with the desired pressure, they have to be staggered and throttled, which generates a lot of energy dissipation that has to be removed as heat. That is why this type of pump is inappropriate for a μRE set-up in the lab and at the miniplant scale.

8.2.2 Metering Characteristics

Pressure Stability Ideally, pumps follow a linear correlation between pressure and feed (see Fig. 8.1 line a). For syringe, piston, and plunger pumps such as high pressure liquid chromatography (HPLC) pumps, a slight deviation occurs with increasing pressure and feed. Because of the compressibility of solvents under real conditions, this correlation runs into a limit value (line b). Even less stiff performance may be found with gear pumps and comparable types of pumps due to the additional effect of internal slip. The solvent slips through the teeth and at the sides of the gear wheels (line c). This effect decreases with increasing viscosity.

Pulsation There are many origins for a pulsating feed. Some of these result from the construction principle of the pump; switching between conveying pistons; compressibility of solvent; essential difference in pressure between two alternate working pistons, causing short-time inverse direction of the feed; hysteresis of pressure relief valves that are installed to ensure a minimum or maximum pressure and avoiding back flow; degassing in valves; cavitation; hot spots; etc.

Every kind of pulsation is critical for μRE and should be avoided or reduced to a minimum. An important first step in avoiding it is knowledge about the pulsation sources in μRE set-ups. If pulsating pumps cannot be avoided, pulsation dampers must be installed. The efficiency of the pulsation damper depends on the exact dimensioning of volume and pressure load relative to flow and expected pressure spikes [1].

The compressibility of the solvent has a significant influence on the effectiveness of the pump. Simple pumps work with a single piston. This results in alternating periods of feed during pushing and no feeding at all during reload of the piston. In case of two pistons or more, they must work in a synchronized, staggered way to result in minimal pulsation. If this overlapping of the streams is not good, the feed of the first piston might result in the early or delayed start of the emptying period for the second piston. A mismatch of the desired feed and pressure pulse occurs. Also, defective contaminated pump valves can lead to a pulsation in the system. They should be checked regularly, because they are intolerant of particles and deposits.

Flow Constancy and Precision   Constant flow is a very important feature of μRE, because μRE structures are very small. Volumes, back-mixing, and the chance for axial compensation of concentration deviations are very limited. The requirements on flow precision is very high. It is not enough to dose a precise volume exactly over a time period like 1 h, but it is at least as important that the instantaneous feed stream is constant and smooth. The origin for feed streams that are not constant can be defective valves in piston pumps, leakage of piston and syringe pumps, and insufficient pressure stability of the pumps when viscosity and back-pressure of the system increase. The effect of leakage can become very dramatic in μRE because feed streams and leakage volumes might be of the same order of magnitude: It is impossible to identify leakage of mL or mL per hour in most cases.

Connection of the Pump to the μRE System   Pumps for μRE systems have to feed small and precise streams to narrow structures. Like the μRE system itself, they are sensitive to particles and deposit formation. Therefore, sufficient filtering of the feed stream to the pump with sufficient filter area is necessary. The filtering system should be designed according to the vendor's specifications. The tube or capillary system ought to be designed properly to avoid cavitation, pulsation, and overpressure [2]. To assure unrestricted flow, the tube diameter on the suction side of the pump must be designed so it is larger than the diameter on the pressure side of a pump to avoid suction problems such as cavitation effects.

8.2.3 Piston Pump

Syringe Pump   Syringe pumps are one of the oldest and most widely used metering devices for chemical and medical experimental set-ups in laboratory scale. They can realize very constant and smooth feeds for one piston stroke, from beginning of delivery to the end, depending on the degree of fine stepping or a fine pitched screw-like drive. Syringes of different sizes are readily available. A limitation of this metering principle is either the insufficient volume of one syringe, insufficient pressure resistance, or greater investment in the set-up with two high-pressure syringes in parallel push-pull mode. For pulsation-free delivery the latter can make it necessary to realize fast switching and equalization of pressure at the cross-over point of the pistons. These pumps are available in different materials over a wide range of volume flow and pressure. Leakages might become a problem because they are hard to identify without a volumetric flow control. Some companies sell syringes with a special design to detect the volume behind the piston for leakage or to flush it with a sealing liquid to avoid air and humidity contact.

Syringe pumps are very suitable for μRE. The volume to be metered is completely inside the syringe and can be pressed out in a homogeneous flow into the process during the experiment's entire time frame. They have a volumetric dosing characteristic.

Disadvantages may be the low working pressure—up to 10, 20, or 40 bar—and the comparatively high cost of very-high-pressure pumps. Complete pulsation-free metering of large volumes requires more than one syringe per stream in process. For multiple-syringe configurations, it is important that switching from one syringe to the next happen after pressure equilibration. Otherwise, the direction of fluid can change for a moment until pressure is equilibrated again. To accomplish this, it is necessary to provide an individual drive for each syringe. Few manufacturers can offer or even realize this option.

Plunger Pump   Plunger pumps displace volume elements through the movement of a plunger in a chamber. To fulfill this task, it is necessary that when the plunger is moving out of the chamber to fill it the exit of the chamber is closed. During delivery the entrance must be closed to avoid back flow to the reservoir. Flow-controlled valves do this blocking and open automatically. But these valves are sensitive to gas, deposits, and disperse particles. The form of the active part in the valve (i.e., ball, needle, turning body, etc.) defines this sensitivity. A few manufacturers are selling pumps with mechanically controlled valves that are forced to open and close. These pumps are less sensitive and even able to handle gas-loaded streams or deliver fluids out of the vacuum area.

This kind of pump is very popular in laboratory practice, but delivers pulsating feeds due to alternative feeding, stop flow, and soaking of a new volume. In addition, the compressibility of the fluid has an influence on the filling, compression, and resulting pulsation. This pulsation might be eliminated or reduced with pulsation damper and pump internal compressibility compensation. Another way to reduce the pulsation is the use of two, three, five, or more heads with identical volumes that are driven staggered in series to deliver their volumes one after the other with overlapping flow.

A special approach to multipiston pumps is the use of two pistons with different volumes. One large piston (the main one) is responsible for the feed most of the time. When it refills in a fast soaking move, the smaller one (the assistant piston) maintains the original delivery rate. This principle is used in most modern HPLC pumps and can deliver liquid streams with minimal pulsation over a wide range of flow and pressure.

Problems with piston pumps such as syringe pumps might arise from unidentified leakage, valve failure, and when the free surface of the plunger coming out of the chamber comes in contact with air, humidity resulting in a thin layer of the material that disturbs the piston. Fluid handling near the boiling point can result in degassing and bubble formation when the pressure is dropping fast. This affects conventional piston pumps and piston driven membrane (diaphragm) pumps. Pumps with a cavitation control mode can be slowed down in their soaking speed to a proper limit, depending on fluid viscosity.

Rotary Piston Pump Rotary piston pumps combine the linear in and out movements of a piston with the opening and closing of the inlet and outlet due to the tournament of the piston. A soaking gap in the piston is oriented toward the entrance hole of the chamber. After filling is completed, the piston is turned and the gap is oriented toward the exit hole for release. In this way, unstable and unsteady working valves can be avoided. Pulsation is still a problem with this type of pump if unused as single head device. It has to be compensated for with a pulse damper or synchronization of multiple pumps. The feed characteristic is highly precise and reproducible. Sealing and leakage problems are the same as with syringe and plunger pumps.

Single-Head, Multiple-Head The strongest pulsation can be observed in systems with a single piston. The reason is flow breakdown while the piston is refilling. Systems with two pistons give pulsations when switching from one piston to the other. The greater the number of pistons installed in a pump working in series, the lower the pulsation, especially if the number of pistons is two or odd. In all piston systems with sinusoidal drive, the pulsation increases when the working capacity (filling or delivery level) of the pump is significantly reduced to less than 100%.

8.2.4 Diaphragm Pump

Similar to piston pumps, diaphragm pumps displace volume elements. A diaphragm moves into and out of the chamber similar to piston pumps. This way, the volume of the chamber is increased or reduced. Flow controlled valves define the direction of flow. Feed rate is unsteady and not constant. It can form sharp pulse functions in the case of magnetically driven diaphragm pumps and half sinus functions in eccentrically driven diaphragm pumps. The pulsation can be reduced by the use of pulsation dampers in which a diaphragm separates the fluid from the gas. Diaphragm pumps, especially hydraulically driven diaphragm pumps, are superior to plunger pumps because of the avoidance of leakage at the plunger and the simple mechanical pressure limitation in the hydraulic system.

8.2.5 Rotating Displacement Pumps

Gear-Type Pump Gear-type pumps utilize the rotation of two gear wheel elements for moving a fluid from the entrance side of a sealed chamber, where the cog/gear wheel elements open, to its exit side, where the same elements again close. The feed and pressure characteristic depend on the size and contact frequency of the cog/gear wheel elements, the clearance between the cog wheels and chamber walls, and last but not least, the viscosity of the fluid. It is possible to realize nearly pulse-free feeds when the specific segment volume is small.

Microannular gear pumps are most suited to μRE [3]. They use two wheels running one inside the other. They exhibit nearly no pulsation and are feasible for a wide range of volume flows and viscosities but are very sensitive to particle contamination. The pricing is similar to that for HPLC pumps.

Progressive Cavity Pumps   Progressive cavity pumps consist of a screwlike helical shaft and a stator. With rotation of the screw in the stator, the eccentric movement forms cavities in the pumps that move forward, toward the exit of the stator. The form of and delivery from the cavities result in a nearly pulse-free flow, when a pump-specific minimum flow is reached. These pumps can even handle viscose fluids up to several Pas.

Rotary Pumps and Centrifugal Pumps   Rotary pumps and centrifugal pumps are less compatible with μRE and its comparatively small feed rates in laboratory and pilot plant scale. Feeds would have to be reduced by reducing the pumps' drive below economical limits or by running the fluid in a circuit. The latter is associated with a large set-up and involves a more or less undesirable amount of energy converted into heat that has to be handled.

8.2.6   Flexible Tube Pump

Fluids can be transported via flexible tubes that are pressed from the outside with a block. For flow control, the block is moved in one axial direction of the tube. The pump set-up contains a rotating wheel with a number of blocks. The tube is applied to the pump so that with the rotation of the wheel, always one block presses the tube. These pumps are common in medical and biological research, where only low pressure is acceptable and a variety of flexible tubings that are compatible with the fluids can be found.

8.2.7   Wetted Parts, Choice of Material

Materials for Wetted Parts   Most manufacturers of μRE-compatible pumps sell their apparatus with miscellaneous materials like glass, noble minerals like diamante or sapphire, ceramics, different metals like steel, stainless steel, hastelloy®, titanium, and polymeric materials such as PEEK, PTFE, FFKM, and EPDM. A survey of these materials' main applications is given in Table 8.2.

TABLE 8.2   Characteristics of Different Pump Types for μRE

Pump Type	Material	Sealing	Maximum Pressure	Pulsation
Syringe pump	G, C, M	G, P	Medium	Low
Plunger pump	C, M	P	High	High
Diaphragm pump	P, M	M	High	High
Flexible tube pump	P	—	Low	Medium
Gear pump	M, C, P	M, P	Medium	Low
Progressive cavity pump	M, P	P	Medium	Medium
HPLC pump	G, M	P	High	Low

Note: G, glass, minerals; C, ceramics; M, metal, stainless steel, etc.; P, polymer, rubber.

Corrosion Corrosion resistance is a question of pump construction and wetted parts. For very aggressive substances, only few materials withstand corrosion attack. Corrosion in pumps has to be considered in combination with the following microstructure device. If the material of the upstream set-up is unstable to corrosion, the particulate corrosion products will influence or even block the pump or overall structure. For pumps as for microstructures, it has to be recognized that even a small-scale corrosion attack can destroy structure and reduce performance. A good survey of corrosion-resistant materials appears in [4].

A survey on the corrosion attack of noble metals and types of nickel based alloys (like hastelloy®) may be found in [5]. The diagrams there show that corrosion attacks of 100 mm per year occur under quite normal conditions and do not negatively impact macroscopic structures at all. Attempts to characterize the corrosion of microstructures and delivery systems in academia are in progress.

8.3 RANGE OF SUITABILITY

8.3.1 Volume

The volume of a pump describes the active volume to create the feed. Depending on the dimensions of the entire experimental set-up and the critical volume of μ-mixing elements, it might be necessary to adapt the pump volume. There are two extreme cases involving pumping volume: one large volume to feed the entire duration of an experiment in a single stroke, and the use of multiple small strokes with high frequency. A typical set-up combines both types to the point where they yield the greatest benefit.

Pulsating can even be reduced to some extent by damping the pressure pulses in flexible tubing following the pump. Even stainless steel tubes of small diameter and sufficient length act in this way like a pulsation damper.

8.3.2 Temperature

For an industry-related experimental set-up, a wide range of working temperatures is needed. (This is a huge difference between research at universities and industrial research.) University experiments are often designed with respect to easy-to-handle conditions (i.e., normal pressure and room temperature or only slightly above) and the properties of the easily available equipment. In industrial research, this freedom of operation is rarely encountered. Research often has to focus on the temperatures of profitable chemical systems. Temperatures between $-100°C$ and $+300°C$ are often requested. Most standard pumps without optional equipment and extra cost for high- or low-temperature resistance are tolerant in the range of 20 to 80/100°C and in an extended range of 0 to 150°C. The extended range also increases the risk of corrosion-related problems that remain less severe at 20 to 80°C.

8.3.3 Pressure

The typical pressure in an industry set-up is between 1 and 10/25 bar and extended in between 1 to 100 bar. Not all pump technologies can reach these upper limits. Some of them operate with such low-pressure, that they cannot cope with the minimum pressure needed to realize pulse-less flow.

8.3.4 Viscosity

Viscosity is the most difficult topic when discussing pump properties. Not every experimental set-up works with water-like viscosities below 50 mPa. Many pumps are declared operational up to 1,000 mPas and higher under boundaries that are not realistic for the set-up (e.g., pressure, temperature, chemical system). Polymer research has shown that the high viscosity of "liquids" and melts in the range of 1,000 mPas and greater can be handled easily. In addition salt solutions with mass contents of 20 to 50% reaching 50 to 100 mPas and higher can be processed. In both cases self-sealing pumps such as gear pumps and progressive cavity pumps show performance advantages over piston and syringe pumps.

8.3.5 Fluid

Depending on the chemical system, a reduced number of solvents is feasible. The compatibility of the underlying pump design principle, the wetted parts of a pump and the requirement of the chemical system might collide. Thus, it is important to choose the correct combination of pump type, size, material selection, conveying element, and sealing.

It is difficult to convey melts in general and especially in microstructures. In μRE set-ups a great number of problems remain unsolved concerning heating, temperature control and avoiding cold spots that block the flow in the system. It is difficult but possible to use normal low-viscous equipment for metering melts, if heating and isolation are designed well. Even special equipment like gear and progressive cavity pumps have to be designed, installed and utilized correctly.

If the fluid contains a gas phase or is conveyed near the boiling point, special care has to be taken for the selection of the pumping system. To eliminate small amounts of gas, degassers such as those used in HPLC systems can be applied. In most cases when gas-containing liquids are pumped, it is easier to increase pressure or to decrease temperature to get rid of the gas phase than to look for a degassing system or gas-tolerant pump type.

Every fluid stream contains particulate ingredients of different size and concentration. Pumps with flow-driven valves at feed and outlet are very sensitive to these impurities even on a μm scale. Gear-type pumps can handle such particles but suffer the abrasive effect of those particles. Syringe and piston pumps with their sensitive sealing suffer even more from the destructive effects of particulate impurities.

8.3.6 Pump Volume and Axial Dispersion

At the beginning of an experiment the start-up procedures should consider the internal pump volume and characteristics with respect to axial dispersion. The fluid volume inside the pump can easily reach the volume of the remaining set-up. For example, volumes and residence times for rinsing, switching between concentrations or from/to solvents, and similar procedures should be dimensioned large enough to ensure that residence time in the feeding device (pump, filters, etc.) is considered. This is even more important when reactive systems call for a stepwise start-up and inverse shut-down. Otherwise, stationary conditions cannot be guaranteed. An empirical formula suggests that stable conditions may be reached after running three to five times the volume of the total set-up.

8.3.7 Compressibility Tolerance

The linearity of the correlation between feed and pressure depends on the compressibility of the fluid. The less compressible a fluid is, the more linear the pressure characteristic of the pump. For example, mercury is one of the fluids with the lowest compressibility. Much more relevant is the compressibility of water, mineral oil, or pentane. Their compressibility shows a volume reduction from normal pressure to 1,000 bar of 4, 5, and 10% (DV/V) (see Fig. 8.2 and discussion of the compressibility of fluids in [6]). Added salt can decrease the compressibility of a fluid noticeably. When the compressibility of the fluid is relevant, it is important to use a pump type that can compensate for this effect to avoid pulsation.

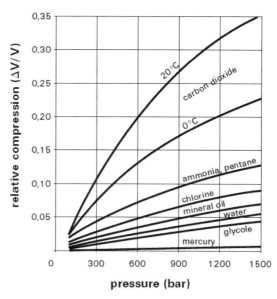

FIGURE 8.2   Compressibility of fluids (http://www.lewa.com).

8.3.8   Pressure Maintenance at the Pressure Side of the Pump

For a smooth performance, a stable and high pressure level subsequent to the pump is recommended. Some manufacturers of HPLC pumps even recommend pressures up to 40 bar for a pulsation-free flow. To generate this pressure, one could choose spring-loaded pressurizers, but the hysteresis of these valves destroys the smoothness of the flow generated by the pump. A better way to fix the pressure level is to use trim valves and adjust the desired pressure level corresponding to the actual flow manually or with a fast electromechanical control loop.

BIBLIOGRAPHY

1. http://www.pulseguard.com/index.htm (Pulse Guard Inc., 295 Sloop Point Loop Road, Hampstead NC, 28443 USA).
2. http://www.lewa.com/uploads/media/D10-311en_Installationinfo_ecodos_14.pdf (LEWA GmbH, Ulmer Str. 10, 71229 Leonberg, Germany).
3. http://www.hnp-mikrosysteme.de/hnp-english/default.html (HNP Mikrosysteme GmbH, Juri-Gagarin-Ring 4, D-19370 Parchim, Germany).
4. http://www.hnp-mikrosysteme.de/pdf/medienbestaendigkeit_hnpm.pdf. (HNP Mikrosysteme GmbH, Juri-Gagarin-Ring 4, D-19370 Parchim, Germany).
5. http://www.haynesintl.com (Haynes Int. Inc., 1020 W. Park Avenue, P.O. Box 9013, Kokomo, Indiana, USA).
6. http://www.lewa.com (LEWA GmbH, Ulmer Str. 10, 71229 Leonberg, Germany).

CHAPTER 9

MICROMACHINED SENSORS FOR MICROREACTORS

JAN DZIUBAN

9.1 INTRODUCTION

Microreactors are microscaled, integrated chemical or biochemical devices fabricated by means of microengineering methods that have been developed as a promising "micro" alternative to macroreactors, used for years in the chemical industry. Microreactors are characterized by the micrometer-size geometry of their fluidic parts (channels, chambers, mixing areas, etc.) and extremely low dead volumes of maintained fluids and/or gases in the micro-pico-nano liter range. The sensing of physical and chemical materials measurable in microreactors is a difficult task because microreactor and sensor systems should be compatible.

In the last few three decades (1975–2005), a huge world market of microengineered miniature sensors of many physical parameters (pressure, flow, temperature, etc.) and valued in the billions of dollars has been established [1–3]. Such sensors are made by micromachining techniques, with the procedures adapted from the microelectronic industry. Small dimensions, well-recognized technology, and an easy wide range of applications make microengineered sensors ideal for application in microreactor technique. Unfortunately, available custom sensors or transducers cannot be easily used in microreactor techniques although they are produced on a mass scale. Because conventionally packaged sensors are large and introduce large dead volume in a microreactor's fluidic circuit, their resistance against corrosion is poor. Packaged sensors, if applied, are localized outside of microreactors, off-channel. There are only a few reports on the on-chip integration of sensors and microreactors (what is commonly referred to as "on-microreactor"). Nevertheless, the on-microreactor integration of sensors is the only natural way of solving dead-volume problems and allowing precise in situ measurements of many important parameters.

Microchemical Engineering in Practice. Edited by Thomas R. Dietrich
Copyright © 2009 John Wiley & Sons, Inc.

This chapter discusses certain aspects of the design, fabrication, and use of micro-engineered silicon–glass sensors specially devised for microreactors and meeting the parameters of on-microreactor integration. The material presented here should not be viewed as encyclopedic. This is only a first step toward the scientific and engineering-realted discussion of the subject, based on the author's experimental experience.

Among the many parameters that have to be sensed in chemical microreactors, pressure, temperature, heat transport, flow rate, and mass transport are the most important. Electrical resistance, conductance, and impedance, pH, and the electrochemical potential of fluidic samples play an important but secondary role. Spectrometric measurements of light absorption, transmission, and generation are rarely used in chemical microreactors, although they are very popular in bio- and life-science-oriented devices. There are many types of microreactors made of varying materials [4]. "True" chemical microreactors, working in a harsh environment, are made of metals, silicon, silicon–glass, and glass. Biomedical devices are most often made of plastics (PDMS, PMMA, Teflon®, SU8) and glass, rarely from quartz and

TABLE 9.1 Measurable Parameters and Microsensors for Microreactor Technique

Parameter	μM	Type of Microsensor	Characterization
Pressure	+	Piezoresistive	Silicon, micromachined, membrane
Pressure	+	Capacitive	Silicon, micromachined, membrane
Pressure	±	Piezoresistive	Thick-film, LTTC, or ceramic
Pressure	+	Optical	Micromachined, fiber optic
Temperature	+	Thermocouple	Monolithic
Temperature	+	Thermistor	Thick-film or discrete
Temperature	+	p–n junction	Monolithic on-chip or discrete
Temperature	+	Thermoresistive	Thin-film Ni/NiCr, Pt
Heat transport	+	Thermal, constant flow of medium	Thin-film Ni/NiCr, Pt on membrane
Heat transport	+	Thermoresistive differential	Set of temperature sensors
Mass transport	+		Thin-film Ni, NiCr, Pt on dielectric membrane
Flow	?	Optical	Thermal lens/gratings
Flow	+	Hot-wire	Thin-film Ni/NiCr, Pt on membrane
Conductance/ impedance	+	Electrical	Thin-film, Pt, TiWAu, Ni, NiCr
Electrochemical potential	+	Electrical	Thin-film electrodes or FET-IG
Light absorption/ transmition	+	Spectrometric	Fiberoptic or integrated optic
Fluorescent light	+	Fluorometric	Fiberoptic or integrated optic

Note: μM, possible integration "on-microreactor"; +; ±; ?.

silicon [5–7]. The wide range of materials and many dedicated applications of micro-reactors make a description of sensor solutions difficult. The characterization of sensors for microreactors is additionally complicated by the fact that many of sensors and sensor systems have been self-made by scientists and never used under real technical conditions (Table 9.1).

As mentioned, sensors for microreactors are used as off-channel devices. Pressure and flow sensors are localized usually at the inlets of microreactors. Temperature as well as the electrical parameters of fluids (conductance, impedance, etc.) are measured at the inlets or outlets; optical transmission parameters (absorbance, trans-mission, and fluorescence) are measured by sensors localized at the outlets of micro-reactors, in close contact with fluids flowing through. In situ on-channel measurements of chosen parameters inside microreactors, which leads to the on-microreactor integration of sensors, is at a preliminary stage of development.

On-microreactor integration requires a reliable methodology for the leak-proof fixation of sensors onto the body of a microreactor. Sensors can be epoxy-bonded (glued with epoxy resins), bonded by adhesive (with thin thermofoils), bonded at low temperature with the use of spin-on glasses, and last but not least, bonded by fusion or anodically. A more extensive discussion of the bonding methods that will be taken into account here exceeds the scope of this book. Fortunately, literature on the subject is available [8].

9.2 PRESSURE SENSORS

The acceptable maximal pressure drop inside a microreactor is usually below 600 kPa. The most popular piezoresistive monolithic micromachined pressure sensors meet this limit. Sensor chips are very small (several mm^3), mass-produced, and standardized, so such sensors seem to be a very attractive option, perhaps the best one, for on-chip integration with silicon and glass microreactors.

9.2.1 Piezoresistive Silicon Pressure Sensors

The piezoresistive micromachined silicon pressure sensor chip (Fig. 9.1) consists of a thin (~20 μm thick), square (1 × 1 mm^2) membrane, anisotropically wet-etched in a monocrystalline silicon wafer. The membrane, surrounded by an ~400-μm-thick silicon frame, deflects under the influence of pressure. Micrometer deflection involves high mechanical tensile stresses near the edges of the membrane where four microminiature monolithic piezoresistors ($R = 5\,k\Omega$) configured in the balanced (for zero pressure) Wheatstone's bridge are made [9]. Under stress, the resistance of a pair of piezoresistors parallel to the edges increases, and the resistance of two other piezoresistors—perpendicular to the edges—decreases. The bridge supplied by constant DC voltage or current becomes unbalanced; at its output, an electrical voltage signal proportional to the pressure is generated. At a single silicon wafer, hundreds of pressure sensor dies are fabricated by the step-and-repeat method. The processed silicon wafer is anodically bonded (at a temperature around 400°C, under a high

(a)

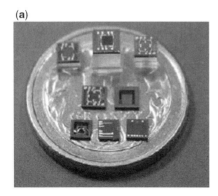

(b)

FIGURE 9.1 Piezoresistive pressure sensor chip. (a) Die on 10 grosz Polish coin; top row: die-on-glass support; middle and bottom rows, front and back-side view of die as sewed. (b) Array of membranes.

polarizing voltage of $\sim$1,000 V) to the 1- to 2-mm-thick Pyrex-like glass wafer; sealed sandwiches are sewed into particular silicon–glass chips. The bond is formed by siloxane irreversible chemical bonds between silicon and borosilicate glass. The seal is helium-tested, leak-proof, mechanically stable, and very strong. The surface energy of the seal is $\gamma > 3.5\,\mathrm{J/m^2}$. This is comparable to the fracture force of bulk silicon. The process is done at a reasonably high temperature. Borosilicate glass thermally matches silicon, so after cooling down, the bonded sandwich does not fracture. Details may be found in [8].

As a result of this procedure, described in many sources [10, 11], the silicon die with a delicate thin membrane localized onto a thick glass support (post) is manufactured. The support eliminates the mismatch of the thermal expansion coefficients of silicon and the package. In classical constructions, the sensor die is packaged into a plastic, ceramic, or metallic case; electrical connections between the thin-film contacts of the sensor die and leads of the package are wire-bonded.

Piezoresistive pressure sensors are stable and sensitive, with high (typically, more than 100 mV for a 5-V DC supply) full-scale-output signal (FSO) (Table 9.2). Off-set voltage and temperature-induced drifts of parameters can be easily compensated in coworking external electronic circuits. The maximal temperature of the work is below

TABLE 9.2 Chosen Parameters of Piezoresistive Monolithic Silicon Pressure Sensors, Chip-on-Glass Form, Most Typical Values

Pressure Range P (kPa)	Over Pressure	Full-Scale Output FSO (mV)	Sensitivity S [mV/ 100 kPa/V]	Hysteresis (%)	Linearity (%)	Zero Offset Voltage U_0 [mV/V]	Temperature Coefficient of Sensitivity TCS (%/°C)
100	3 P	100–150	>20	0.1	0.1	±5	−0.21
200	2 P	100	20	0.1–0.2	0.1	±5	−0.21
600	1.5 P	<100	<20	0.5	0.2–0.4	±5	−0.21

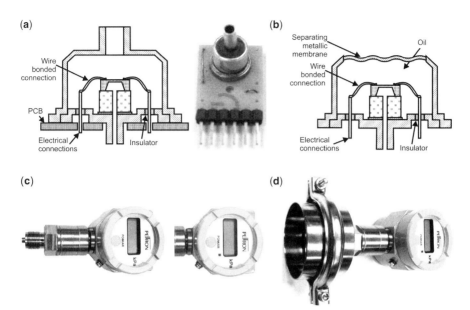

FIGURE 9.2 Packaged piezoresistive pressure sensors. (a) Cross section and view of a simply packaged sensor (courtesy ITE, Warsaw, Poland). (b) Cross section of a package with separating metal membrane. (c) Industrial pressure sensor with separating membrane (courtesy Peltron, Ltd. Warsaw, Poland). For comparison, pressure sensor die (small, black square, right-down liquid crystal display) is shown in the scale. (d) Industrial pressure sensor with Ti-separating unit for use in highly aggressive environment (courtesy Peltron Ltd. and Aplisens, Ltd., Warsaw, Poland).

$120°C$ (special SOI sensors can work at a higher temperature up to $300°C$ [12]). The minimal temperature of the work depends mainly on the method of packaging; typically, it equals $-40°C$. Sensors working in a clean, noncorrosive atmosphere are packaged in a simple package [Fig. 9.2(a) and (b)]. Pressure sensors for industrial applications and within the chemical industry, where work occurs in a corrosive, aggressive atmosphere, are packaged in a metal corrosion-resistant package, with a separating SS316 diaphragm [Fig. 9.2(c)] or they are equipped with a separation unit with a titanium membrane [Fig. 9.2(d)]. The proportion of dimensions of the pressure sensor die before packaging and of the packaged pressure sensor with a separation membrane can be 1 : 1,000. The dead volume of an unpackaged chip and a packaged sensor, as well as the delivery price, increases proportionally to this factor.

It is clear that the packaged piezoresistive pressure sensors in their existing form cannot be integrated on-microreactor. Pressure sensors for on-microreactor should be corrosion-resistant and small, preferably as small as an unpackaged sensor chip. The packaging of dies is essential; an unpackaged sensor cannot be protected against a harsh environment and electrically connected. The best solution to this contradiction is the localization of sensor die(s) directly on the microreactor body and the use of special methods of packaging.

9.2.2   Sensors on Silicon–Glass Microreactors

Silicon–glass microreactors are fabricated in a kind of sandwich of silicon and glass wafers bonded anodically [8, 13–15]. All components of the microreactor (fluidic channels, mixers, reactors, etc.) are—the most often—deeply wet- or dry-etched in a (100)-oriented silicon wafer. The processed silicon wafer is anodically bonded (from one side or either side) to Pyrex-like glass 1- to 2-mm-thick wafer(s). Two types of Pyrex-like glass substrates are used commonly: Corning 7740 from Corning in the United States (commercial name: Pyrex) or Borofloat 3.3 from Schott in Germany [16, 17].

As mentioned previously, Pyrex-like borosilicate glass (Corning 7740 or Borofloat 3.3) thermally matches monocrystalline silicon [8, 16, 17]. Borosilicate glass and silicon covered with thin-film silicon dioxide or nitride are resistant against corrosive chemical agents.

Anodic bonding/sealing of silicon and glass is helium leak-proof, chemically stable, and corrosion-resistant [18–20]. The bond strength is comparable to the fracture force of bulk silicon or glass. Sealing is time-stable and does not degrade for several years. These features make anodic bonding ideal for the direct on-microreactor integration of silicon pressure sensor dies and other microsensors made of silicon and glass. Three solutions are possible: placement of a reference silicon sensor chip directly onto the microreactor, use of a reference die-on-glass sensor chip fixed to orifice silicon onto the microreactor, or placement of an absolute sensor chip inside the fluidic channel (not preferable because of the associated poor corrosion resistance) (Fig. 9.3).

In the first solution, the silicon pressure die is directly anodically bonded to the glass surface of the microreactor, which must be made from Pyrex glass or formed as a sandwich of silicon–glass wafers. In this case, thin-film Pt or Ni/NiCr electrical

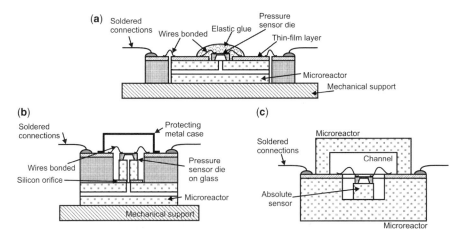

FIGURE 9.3   Three methods of on-microreactor integration of pressure sensor dies: (a) chip bonded to microreactor; (b) die-on-glass bonded to silicon orifice, which is bonded to microreactor; (c) die fixed inside channel.

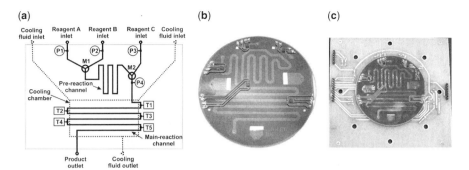

FIGURE 9.4 Silicon pressure sensor dies anodically bonded directly on-microreactor: (a) schema of microreactor (P_{1-5} = pressure sensors and T_{1-5} = temperature sensors), (b) silicon–glass microreactor with set of pressure and thin-film on-chip connecting paths, and (c) the microreactor before soldering of connections to PCB plate [23].

paths are made directly onto the top glass surface of the microreactor, electrical connections between the die and thin-film layer are wire-bonded, and the chip is encapsulated with an elastic silicone drop. An external connection between the microreactor and coworking PCB is made by wire bonding (as shown in Fig. 9.4) or soldered.

Sensor dies on the glass posts cannot be directly bonded to glass; they are bonded to small silicon orifices that are anodically bonded to the microreactor (Fig. 9.5). The height of die-on-glass (it depends on the type of die) is equal to 2 to 3 mm. Long wire-bonded connections bonded directly between the electrical contacts of the sensor die and thin-film paths made off-plane on the microreactor would tend to brake under vibration and/or mechanical shock. The epoxy PCB plate with Cu/Au electrical paths is fixed to the microreactor, and it is done (fixing) in this way that the pressure sensor die protrudes from the plate, eliminating the off-plane positioning of electrical connections between the sensor chip and making wire-bonding possible. Electrical plugs are localized at the edge side of the microreactor in both solutions.

The microreactor with assembled sensors and PCB is positioned inside a metallic case and, next, fixed fluidic connections (manufactured by Up Church in the United States) are screwed in. The method described here for the on-microreactor integration of pressure sensors is characterized by extremely low dead volumes. The dead volumes are about 0.9 and 2.2 µL for the first and second solutions, respectively (for a typical 1-mm-thick glass substrate forming the top layer of silicon–glass microreactors with machined via-holes ID 0.8 mm connecting the fluidic channels of the microreactor and the sensor, assuming a 1×1 mm^2 membrane etched in 0.4-mm-thick silicon wafer and 2-mm-high glass post bonded to a 0.4-mm-thick silicon orifice with a 0.5×0.5 mm^2 via-hole).

Custom piezoresistive pressure sensor dies for the maximal pressure range of $P_{max} = 50$ to 600 kPa are easily obtained. Technologically, the application of die-on-glass sensor chips is easier. They have been tested by the manufacturer, and the influence of the anodic bonding of die-on-glass to the silicon orifice on the

(a)

(b)

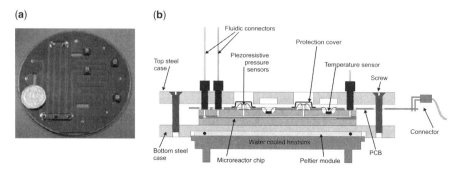

(c)

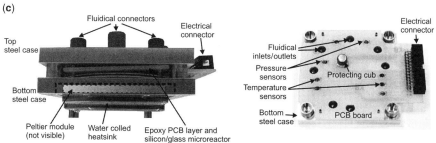

FIGURE 9.5 3-in. glass–silicon–glass microreactor with integrated on-microreactor die-on-glass chip of pressure sensors: (a) top view of the microreactor body with four sensor dies (the fifth under the coin), (b) cross section of package and PCB board and details of packaging, and (c) packaged device with Peltier's cooler and heat exchanger fixed at backside [23, 24].

metrological parameters of pressure sensors is—for properly chosen procedures [8]—negligible. Typical overpressure is 2 to 3 for $P_{max} < 200$ kPa and 1.5 for higher P_{max}. Lower P_{max} can be measured by so-called bossed-membrane sensors [21] or fusion-bonded sensors, characterized by larger dead volume and smaller FSO [22]. Pressure higher than 600 kPa cannot be measured by micromachined piezoresistive silicon pressure sensors because of "bulk effect induced" errors [9]. Fortunately, such a high pressure drop is very rarely experienced in silicon–glass or glass microreactor. The structure of the thin membrane can serve as a safety valve; the membrane mechanically destructs whenever pressure crosses the limit.

9.2.3 Sensors on Foturan® Glass Microreactors

Microreactors made of photostructured Foturan® glass [25, 26] are becoming more and more popular. The integration of monolithic pressure sensors to Foturan glass microreactors by anodic bonding procedures is difficult. Obviously, the anodic bonding/sealing of silicon and Foturan glass 1 to 2 mm thick can be easily done under standard procedure (450°C, 1,200 V), but after cooling down to an ambient temperature (~20°C), the bonded silicon–Foturan sandwich crashes [Fig. 9.6(a)] [27] because the thermal coefficients of expansion of Foturan glass $\alpha_F = 8.6$ ppm/K

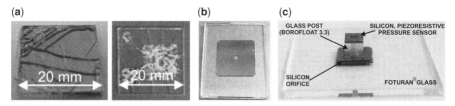

FIGURE 9.6 Bonding of silicon pressure sensors to Foturan microreactors [27]. (a) Non-optimal procedure (450°C and 1,200 V), sandwich crashes after cooling down to 20°C. (b) Optimized process (200°C and 1,000 V), $10 \times 10 \, mm^2$ silicon piece shows excellent sealing to glass, photo taken through glass sample. (c) Chip-on-glass pressure sensor anodically bonded to Foturan microreactor.

and silicon $\alpha_{Si} = 3.2 \, ppm/K$ differ significantly. Moreover, the washing and activation procedures commonly used for Pyrex-like glasses (Piranha, RCA1) [8] cannot be applied to Foturan glass because at the bonded Si–glass interface several brown voids appear and significant worsening of the quality (strength) of the seal is observed.

New recently discovered procedures [24, 27] utilize the washing of Foturan glass in a solution of $H_2O_2 : H_2O$ at 80°C for 30 min and low-temperature bonding at 200°C and 1,200 V for 5 min, followed by post bond annealing at 600 V for 20 to 30 min [Fig. 9.6(b)]. The surface of the processed Foturan glass must be mechanically polished before bonding to optical quality, but the mechanical polishing of the bondable surfaces of microreactors is technologically inconvenient because the via-holes connecting internal and external fluidic circuits may become chocked by the polishing medium. In any case, the bonding of silicon under such conditions is successful and achieves the quality needed for leak-proof assembling of chip-on-glass monolithic pressure sensors directly on Foturan microreactors [Fig. 9.6(c)]. The example of a Foturan microreactor with an on-chip pressure sensor die is addressed in other sections of this chapter in the discussion of the integration of capacitive pressure sensors (later Fig. 9.14).

9.2.4 Platform of Piezoresistive Pressure Sensors

The main principle of the new concept of integration of pressure sensors on-microreactor is the use of an intermediate sensor platform [28] [Fig. 9.7(a)] made of a 1-mm-thick Pyrex-like substrate with anodically bonded die-on-glass sensor chips at silicon orifices [Fig. 9.7(b)]. The substrate is epoxy-glued to the PCB board. The height of chips and thickness of PCB board are properly adjusted, so the top surface of the chips and surface of the PCB board are in-plane [Fig. 9.7(c)]. The electrical connections between the chips and PCB board are wire-bonded. Pressure sensor dies and delicate connecting wires are mechanically protected by metal cups [Fig. 9.7(d)]. The electrical plug is localized at the edge of the PCB board.

The packaging of the Foturan microreactor proceeds as follows: A PTFE (Teflon®) flat plate with Viton® gum O-ring orifices is placed on the microreactor; next, the sensor platform is placed on this sandwich and all parts are placed inside the metal

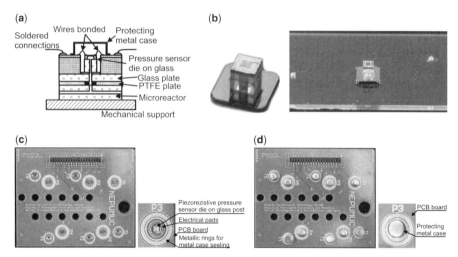

FIGURE 9.7 Platform of pressure sensors: (a) cross-sectional schematic view, (b) pressure sensor die-on-glass anodically bonded to glass plate, (c) top view of platform before encapsulation (left) and chip of pressure sensor inside via-hole in PCB board before wiring (right), and (d) platform ready to work.

package [Fig. 9.8(a)]. Two pieces of the package are screwed in, to fix all parts and seal the connections between the Foturan microreactor and sensor platform. Then fluidic connectors manufactured at Up-Church in the United States are screwed in [Fig. 9.8(b)]. The dead volume of the single sensor equals 2.2 μL. A platform with failed sensor(s) can be easily replaced, although a particular pressure sensor chip cannot. The metrological parameters of packaged pressure sensors are nonlinearity ∼0.5%, hysteresis ∼0.02%, $FSO_{200\,kPa} = 120\,mV$ [Fig. 9.8(c)].

9.2.5 Corrosion-Resistant Discrete Piezoresistive Sensor

The innovative construction of a discrete, extremely small, corrosion-resistant piezoresistive pressure sensor (its patent is still pending) is shown in Fig. 9.9. The micromechanical structure of the sensor consists of a piezoresistive pressure sensor die, a round-shaped, 0.4-mm-thick central silicon support with a via-hole,

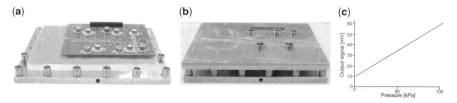

FIGURE 9.8 Microreactor with platform of pressure sensors. (a) Microreactor onto microreactors before final packaging. (b) Packaged device. (c) Example of output signal curve: voltage signal vs. pressure as measured for P3 sensor ($P_{max} = 200\,kPa$).

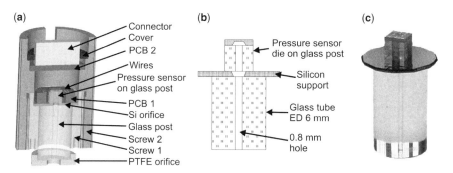

(a)
- Connector
- Cover
- PCB 2
- Wires
- Pressure sensor on glass post
- PCB 1
- Si orifice
- Glass post
- Screw 2
- Screw 1
- PTFE orifice

(b)
- Pressure sensor die on glass post
- Silicon support
- Glass tube ED 6 mm
- 0.8 mm hole

(c)

FIGURE 9.9 Corrosion-resistant discrete pressure sensor: (a) cross section of the sensor, (b) internal structure shown schematically, and (c) true image of assembled all-anodic-bonded internal structure.

micromachined in a double-side polished (100)-oriented silicon wafer and 8-mm-long Borofloat 3.3 glass tube (ED 5.9 mm/ID 0.8 mm) optically polished at both ends. Parts are anodically bonded to the central silicon support, as shown in Fig. 9.2(c). After bonding, the internal structure of the sensor is positioned inside the SS316 cartridge (screw 1) and epoxy-glued, PCB 1 part is fixed inside screw 1 and wire-bonded connections are made. Following this, screw 1 is screwed with a SS316 screw 2, and soldered wire connections are made between PCB 1 and PCB 2. PCB 2 is fixed inside screw 2, and the package is encapsulated with a cover at its top. Next, the PTFE cover with an O-ring is placed onto a glass tube protruding from the package.

The sensor is screwed to the top part of the metallic package of the microreactor, similar to the method of fixation of Up-Church fluidic connections (Fig. 9.10). The maximal working pressure of the sensor is 600 kPa; the output characteristic and other parameters are the same as those cited in Table 9.2.

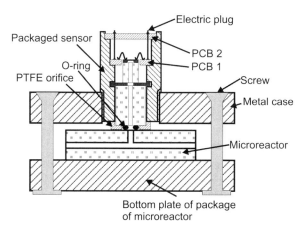

- Electric plug
- Packaged sensor
- PCB 2
- PCB 1
- O-ring
- PTFE orifice
- Screw
- Metal case
- Microreactor
- Bottom plate of package of microreactor

FIGURE 9.10 Discrete pressure sensor on-microreactor.

The most positive features of this solution are the complete corrosion resistance of the sensor against almost all cold and hot chemical agents (excluding a hot concentrated alkalic solution used for the etching of silicon) and easy replacement of only broken down sensors. The negative is larger in comparison to the previously described solutions: the dead volume of the sensor (4, 9 μL for a 8-mm-long glass tube).

9.2.6 Capacitive Silicon Pressure Sensor on Microreactor

A micromachined silicon capacitive pressure sensor consists of two flat electrodes (at least one is micromachined in monocrystalline silicon) forming a capacitor. The pressure of fluid deflects significantly one or two electrodes, changing the distance gap between them and modulating the capacitance of the sensor. The output signal is nonlinear (nonlinearity can be easily compensated for by coworking electronic). Signal value depends on the geometry of the sensor; thermal errors are negligible.

Many types of silicon capacitive sensors have been described in the literature [30–32]; some of them are available on a custom basis. The packaging of sensors depends on application. Existing packaged sensors cannot be in any way integrated on-microreactor. The main advantages of this type of pressure sensor are simple construction and thermal stability. Negatives are large planar dimensions that increase the dead volume of the chip (typical sensor is about 10×10 mm^2). The dead volume of the sensor for a 20-μm-thick, 8×8 mm^2 membrane (moving electrode) etched in a 0.4-mm-thick silicon wafer is approx. 24 μL, much higher than the dead volume of an integrated on-microreactor chip-on-glass piezoresistive pressure sensor. Fortunately, capacitive micromachined silicon sensors can be easily fabricated as fluidic flow-through devices (Fig. 9.11). This eliminates or significantly decreases dead volumes. Properly designed internal geometry eliminates the pressure drop caused by fluid flow.

9.2.7 All-Silicon Sensor

The flow-through sensor chip, designed as a device for integration onto silicon–glass or Foturan microreactors by anodic bonding/sealing is made of two micromachined

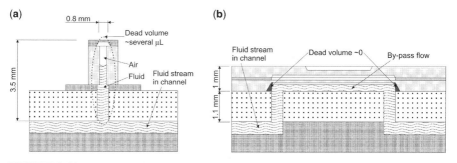

FIGURE 9.11 Dead volume problem: (a) situation for piezoresistive pressure sensor chip $2 \times 2 \times 2.3$ mm^3, where dead volume equals several microliters, and (b) situation for flow-through capacitive pressure sensor chip $10 \times 10 \times 0.8$ mm^3, where dead volume is equal to zero.

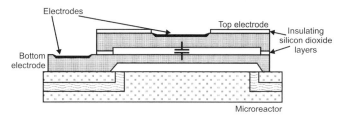

FIGURE 9.12   Capacitive micromachined pressure sensor: cross-sectional schematic view.

silicon parts assuming the role of electrodes of the capacitor (Fig. 9.12). Chips are fusion-bonded [33, 34] through a thin layer of thermally grown silicon dioxide 1 μm thick possessing excellent dielectric properties. After this, electrical thin-film Cr/Ni/Au or Ti/W/Au contacts are made at each of the electrodes. Next, the chip is positioned on the glass of the microreactor and anodically bonded, according to the accepted rules [8]. The maximal temperature of the construction sensor described here is limited by the degradation of the insulation layer, thin-film contacts, and microreactors. Such sensors work properly at 200°C.

9.2.8   Sensor with Foil

Another construction of a flow-through capacitive pressure sensor is two micromachined silicon parts (electrodes) with thin-film metallic contacts, sealed by adhesive low-temperature bonding (Fig. 9.13) [8, 35]. First, the bottom and top electrodes are double-sided KOH wet-etched anisotropically in (100)-silicon, metallic Ti/W/Au contacting area are then deposited and patterned. The bottom electrode is anodically bonded to the microreactor body, distancing Kapton® (available through DuPont in Canada) 10-μm-thick foil placed on the bottom electrode. The top silicon electrode is placed on the foil and the sandwich is bonded at 200°C for 20 to 30 min. Sealing parts have to be pressed to each other during the process. The maximal temperature of the work of the sensor is below 120°C; it is limited by the material parameters of Kapton® foil. The sensor exhibits a high nonlinear output signal (Fig. 9.14), which can be digitally improved easily. The on-microreactor integration of capacitive and piezoresistive pressure sensors onto a Foturan glass microreactor as well as the method of electrical contact to the capacitive sensor chip are shown in Fig. 9.14 (c) and (d).

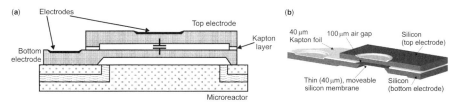

FIGURE 9.13   Capacitive pressure sensor with Kapton® foil: (a) cross section and (b) visualization.

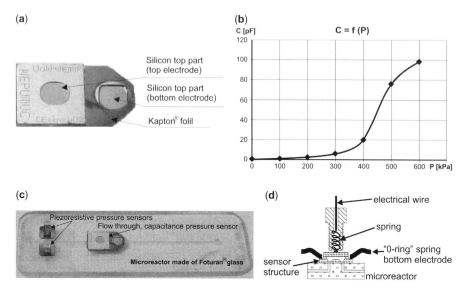

(a)

Silicon top part (top electrode)

Silicon top part (bottom electrode)

Kapton® folil

(b)

C [pF]

C = f (P)

P [kPa]

(c)

Piezoresistive pressure sensors

Flow through, capacitance pressure sensor

Microreactor made of Foturan® glass

(d)

electrical wire

spring

"0-ring" spring bottom electrode

sensor structure

microreactor

FIGURE 9.14 On-microreactor integration: (a) the chip, (b) output characteristic, (c) Foturan® demonstrator with piezoresistive and capacitive pressure sensor on-chip, and (d) electrical connecting.

9.2.9 Sensors with Optical Fibers

Pressure sensors with optical fibers utilize the effect of the modulation of light intensity or interference phenomenon. In the sensors, pressure deflects a thin silicon membrane (Fig. 9.15) [36]. Light is guided by glass fibers attached to micromachined sensor chips or by an integrated waveguide made directly on-chips [38]. The highest advantages of such sensors are the simplicity of the construction of the sensor chip and easy integration with microreactors. Negatives are the use of sophisticated external measurement equipment and the somewhat "strange" methodology of measurements.

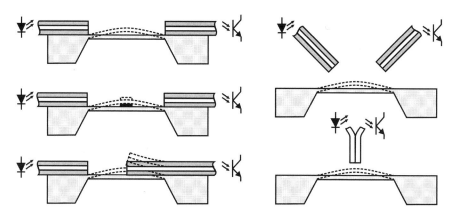

FIGURE 9.15 Optical pressure sensors with optical fibers and moving membranes.

9.2.10 Sensors with Moving Silicon Membrane

Two examples of optical sensors with glass fiber and a bossed silicon membrane—for possible on-microreactor integration—are shown in Fig. 9.16(a). The first sensor consists of a deeply wet-micromachined bossed membrane, which is anodically bonded through a thin layer of borosilicate glass [8] to the backside of a silicon holder. A glass tube made of Borofloat 3.3 tube, polished at either end, is anodically bonded to the front side of the holder. Multimode communication glass of optical fiber (ID $125/62.5\ \mu m$) is positioned inside the glass tube and fixed with UV-cured epoxy optical glue. Fluid pressure moves up and down the bossed membrane, reflecting light guided from the external 635-nm laser source. The reflection of light is proportional to pressure. The fabrication process includes the deep wet micromachining of the silicon membrane and silicon holder in KOH, thermal oxidation of silicon parts, magnetron sputtering of Borofloat 3.3 glass $0.8\ \mu m$ thick on the backside of the silicon holder, machining of the glass tube with polished front surfaces, anodic bonding of silicon parts (at 200°C, 80 V) followed by anodic bonding of the glass holder (at 400°C, 1,200 V). After this, the assembled silicon membrane–silicon holder–glass tube is anodically bonded to the glass of the microreactor. In this process, the silicon membrane plays the role of a negative electrode. Next, the glass fiber is assembled.

Proper controlling of geometrical behavior during patterning and etching of silicon parts allows one to obtain self-alignment of the optical fiber inside the silicon holder and maintain a precise distance between the front surface of the silicon membrane and end of the fiber. A second sensor consists of a V-grooved bossed membrane and optical glass fiber (whose type is mentioned above) positioned inside the V-groove [36]. The movable polished flat end of the fiber is distanced $30\ \mu m$ from a $p–n$ diode fabricated on the bulk frame surrounding the membrane [Fig. 16(b)].

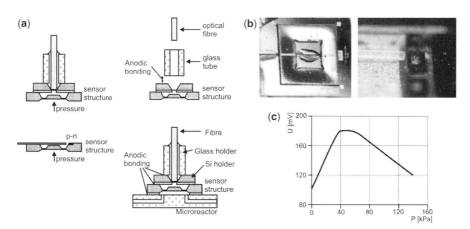

FIGURE 9.16 Pressure sensor with optical fiber and silicon membrane: (a) construction and on-microreactor integration shown schematically, (b) view of sensor chip [36] and detail showing glass fiber near $p–n$ diode, and (c) typical output signal characteristic [36].

Light guided by the fiber from an external laser source illuminates the $p-n$ diode working in the open-circuit voltaic mode. The pressure of fluid deflects the membrane up and down. The fiber moves and modulates the intensity of illumination of the $p-n$ junction; the output voltage signal follows the modulation of light. The output characteristic registers "plateau" because during montage the end of the fiber is positioned downward of the top surface of a $p-n$ junction [Fig. 9.16(c)].

Zero-pressure off-set voltage is large, but stable and positive. The fabrication process of the chip consists of two main steps: double-sided KOH deep wet micromachining of the bossed membrane and V-groove, and fabrication of the diffused $p-n$ diode with Al or Ti/W/Au thin-film contacts (one contact to the p-doped region, second to a bulk n-type silicon). The manufactured chip is anodically bonded to the glass surface of the microreactor (350°C, 1,000 V for Borofloat glass and 200°C, 1,000 V for Foturan glass). After bonding, the fiber is epoxy-glued inside the V-groove and electrical connections are made. The sensitivity of the sensor is high; it can reach 240 mV/100 kPa (for a 6×6 mm^2, 20-μm-thick, membrane with 3×3 mm^2 boss). The maximal pressure range depends on the planar dimensions and thickness of the membrane. For example, the blow-up pressure of a 6×6 mm^2, 50-μm-thick membrane is about 300 kPa. The maximal temperature of the work must be below 80°C. The sensor is corrosion-resistant and can be fabricated as a flow-through device with almost zero dead volume.

9.2.11 Sensor with Moving Thin-Film Dielectric/Metal Membrane

The optical pressure sensor with fiber, suitable for on-microreactor integration, consists of a micromachined silicon chip bonded anodically at either surface to Borofloat 3.3 glass tubes (holders) [37] (Fig. 9.17). The end of the fiber fixed inside the upper holder is self-positioned inside the cavity precisely wet-etched in silicon. The end of the fiber is positioned 20 to 30 μm over the thin-film 0.4-μm-thick multilayered TiWAu 100×100 μm^2 diaphragm formed at the SiO$_2$/Si$_3$N$_4$ dielectric membrane. The sandwich of membranes deflects under pressure and modulates the reflection of light guided by the fiber from the solid-state 635-nm laser. Reflected light is measured by an external measurement set-up. The technological process of the

FIGURE 9.17 Optical pressure sensor with optical fiber and moving thin-film diaphragm: (a) cross section, (b) view of sensors chips in two versions [8, 39], and (c) typical output characteristic.

sensor starts with the backside deep KOH anisotropic etching of a (100)-silicon substrate (covered at either side with thermally grown SiO_2 and CVD deposited Si_3N), forming a cavity ending with a 10-μm-thick silicon membrane. The front side of the processed wafer is covered with TiWAu layers by magnetron sputtering. Following this, the TiWAu metal membrane is patterned photographically at its front side and a small via-hole is DRIE-etched through a silicon membrane. After this, glass holders are bonded from either side of the silicon part; glass fiber is positioned in the upper holder and UV-epoxy-glue-fixed. The maximal pressure of work depends on the thickness of the acting membranes. For 400-nm-thick metal membranes, it is about 200 kPa [Fig. 9.17(c)], and blow-up pressure is approx. 250 kPa. The sensor has to be integrated on-microreactor before the fiber is fixed.

9.3 TEMPERATURE SENSORS

Local measurement of the temperature inside microreactors is very important because it creates the opportunity for in situ control of the course of chemical reactions and improved safety by the detection of hot points. This is especially important in microreactors for hazardous, exothermic reactions including nitration reactions. Temperature measurement on-channel is a difficult task. Two configurations of sensor localization may be considered: inside the channel [Fig. 9.18(a)] or as close as possible to the fluidic stream, inside cavities, or blind holes fabricated in the microreactor body [Fig. 9.18(b)]. Construction of the sensor depends on the expected temperature range and type of microreactor. Biochemical and life-science-oriented microreactors work close to an ambient temperature ($<41°C$) or in the range of $+40$ to $+100°C$; industrial chemical fluidic microreactors work in the range of $-30°C$ to $120°C$ (microreactors for a nitration process work in the range of -10 to $35°C$). Several principles of the work and construction of sensors covering the -30 to $+120°C$ temperature range should be taken into consideration: thin-film thermoresistors, the $p-n$ junction, and thermistors.

9.3.1 Thin-Film Thermosensors

Thermoresistive thin-film sensors are made of thin-film thermosensitive resistive materials deposited onto an insulator. Resistivity of material depends on temperature, so the resistance of the resistor is a function of temperature. The most popular are

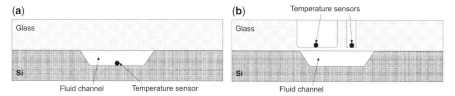

FIGURE 9.18 Positions of temperature sensors: (a) sensor inside channel and (b) sensors inside etched cavity or hole.

thin-film layers of nickel, nickel-chromium, and platinum, whose temperature coefficients of resistivity (TCR) fall in the range of 0.3 to 0.5%/K (2,000–5,000 ppm/K) [38]. Layers are deposited by magnetron sputtering and photolithographically patterned. The technologically acceptable thickness of these layers is 0.5 to 1 μm, square resistance approx. 1 Ω/□. Thin-film thermoresistive layers can be deposited onto glass, quartz, and silicon covered with dielectric layers (silicon dioxide or nitride). Adhesion is a critical parameter as well as chemical passivity. Electrical contacts are localized out of the negative influence of chemicals. Patterned contacts are usually covered with a thick layer of vacuum-evaporated gold (1,000–3,000 nm thick) for better wire-bonding or soldering of external electrical lead connections. The technically acceptable resistance of thermoresistors should be near 100 Ω. The resistive path of thermosensitive Ni, NiCr, or Pt resistors is usually 20 to 50 μm wide, so for square resistance to equal 1 Ω/□, a 100-Ω resistor has to be 2,000 to 5,000 μm long, much greater than the typical characteristic dimensions of the internal parts of microreactors (channels, mixers, etc.). This is why thin-film resistors are patterned in the form of meanders.

Stable and highly sensitive thermoresistive thin-film sensors on glass or thermal silicon oxide can be fabricated of magnetron sputtering in a vacuum of very thin (∼50-nm-thick) Ni Cr (50–50%) covered with 1-μm-thick pure Ni, annealed in N_2 at 420°C for 3.5 h followed by stabilization in 1% O_2 in N_2 at 400°C for 1 h [38, 39]. TCR equals 5,500 ppm. The etching of the bilayer can be done in a $HNO_3 : H_2O/1 : 1$ mixture at 50°C through an AZ 1350, resin mask (manufactured by Shipley in the United States). Pt resistors made of layers thicker than 1 μm should not be applied because they tend to texture. Pt layers are deposited by magnetron sputtering in a vacuum from target and patterned by the use of a lift-off technique.

Platinum resistors show excellent chemical stability, but the chemical resistivity of thermoresistive thin film is limited by a side underetching effect, caused by aggressive agents, delaminating thin-film layers, and breaking continuity of resistors. Thermoresistive sensors are often covered with protective dielectric Si_3N_4 (50–100 nm) or SiO_2 (200–300 nm) layers to improve chemical resistance. These layers are magnetron-sputtered or CVD-deposited. They protect sensors to a satisfactory degree, not influencing the quality of measurements. Highly sensitive, precise sensors are localized on a thin dielectric "bridge" micromachined inside the fluidic channel (only for silicon-based devices). Usually, such thermoresistor are made of thin-film NiCr/Ni thermoresistors that are bondable to Pyrex glass. It makes the hermetic sealing of microreactors easier by an anodic bonding technique.

Recently, new thin-film thermosensitive temperature sensors made of polycrystalline silicon doped with vanadium, zirconium, or molybdenum have been fabricated. Magnetron-sputtered poly-Si : Mo doped resistors show reasonable but negative TCR (∼1,000 ppm) and high square resistivity (18–20 Ω/□). Thermoresistors made of this material can be significantly smaller (Fig. 9.19); a resistance of 100 Ω is obtained for the resistive path 25 μm × 625 μm and square resistivity equals 20 Ω/□ [40]. Contacting paths are made of Ni/Au.

Thermoresistors and other thermosensors are always placed at thinned parts of glass or at a thin membrane etched in silicon (Fig. 9.20). This introduces several

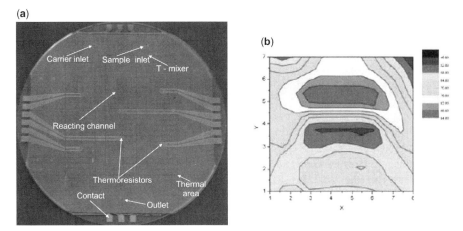

FIGURE 9.19   On-microreactor temperature measurements. (a) 3-in. DNA/PCR silicon–glass microreactor with poly-Si : Mo doped thermoresistors deposited onto thinned silicon, thin-film on-chip connecting paths are made of Ni, and gold contacts area are localized at the edge of the microreactor [40]. (b) Distribution of temperature as measured by sensors.

technological problems caused by deep etching of glass. A typical popular etch mask is patterned thin-film magnetron-sputtered Cr. Much better is the masking of glass with adhesive foil, the so-called blue tape used widely in microelectronics. Blue-tape masking is sufficient for the selective deep (1-mm-depth) long-lasting etching of all types of glass in a concentrated HF : water mixture.

Thermoresistive sensors are always used in a Wheatstone's bridge configuration. A single-resistor electrical configuration of a sensor is rarely used; most often, a pair of on-microreactor temperature-dependent resistors is configured in the bridge together with two external temperature-stable resistors. The linearity of sensors in a wide range of temperatures ($-40–200°C$) is good, and resolution and accuracy can reach $0.1°C$, depending on the particular technical solution.

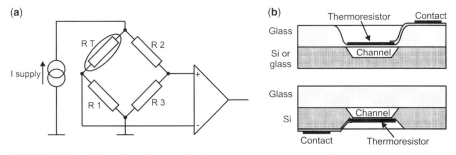

FIGURE 9.20   Thin-film thermoresistive temperature sensors: (a) electronic circuit and (b) localization of sensor at thinned glass or silicon membrane.

9.3.2 *p–n* Junction-Based Sensors

The forwardly polarized *p–n* junction shows $2\,\text{mV}/°\text{C}$ temperature-dependent voltage drift (TCU). This phenomenon is widely used in measurements of temperature. The simplest temperature sensor consists of a *p–n* junction or is supplied by stabilized current [Fig. 9.21(a)]. *p–n* junction(s) can be fabricated directly in monolithic form on the surface of the fluidic channel micromachined in silicon, so the integration on-microreactor is easy but limited to silicon-based microreactors [Fig. 9.21(b)]. Such thermometers are used rarely.

Much better is the use of thermometers in the form of a packaged integrated circuit (IC) produced on a mass scale and widely used in this technique. These thermometers are calibrated by the manufacturer, work as a voltage reference source, control the current source of digital devices, are scaled in degrees Celsius and Fahrenheit. For example, an REF-02 integrated circuit (Burr-Brown products of Texas Instruments [41]) has a special connection, which potentially increases linearly with a temperature of $+2.1\,\text{mV}/°\text{C}$ and works at -55 to $125°\text{C}$, with the accuracy of its measurements being 0.5%. Devices packaged in SMD package are small and fixed easily to the thinned glass part of the microreactor (thinning improves heat transport between the fluid in the channel and the thermometer). The integrated circuit LM 355 [42] works like Zener's diode, whose stabilization voltage V_s is linearly proportional to the absolute temperature with coefficient $+10\,\text{mV}/\text{K}$ [Fig. 3.21(c)].

For a $25°\text{C}$ output signal, V_s equals 2.982 V. The LM 355 circuit is calibrated by the manufacturer; the accuracy of its measurements is $\pm 1°\text{C}$. Integrated circuit AD 590 (manufactured by Analog Devices [42]) is a source of current whose value

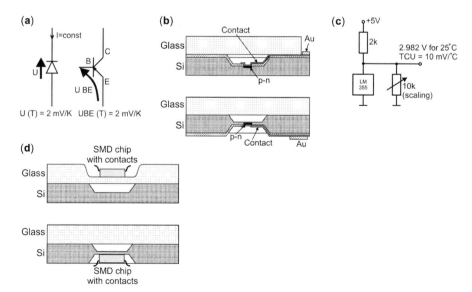

FIGURE 9.21 *P–n* temperature sensor: (a) diode/transistor as temperature sensors, (b) monolithic on-silicon-on-microreactor integration with inside and outside fluidic channel, (c) on-microreactor integration of discrete devices, and (d) temperature meter with LM 355.

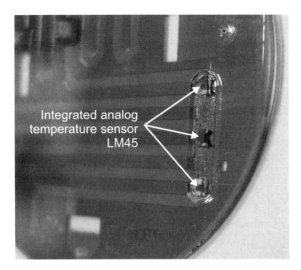

FIGURE 9.22   On-microreactor of LM 45 sensor (SMD version of LM 355 in wire-mount package) fixed to silicon–glass microreactor at its thinned glass area, as shown in Fig. 9.21(b).

is equal to the absolute temperature ($\pm 0.5\,\mu$A). At 25°C (298.2 K) the current is equal to 298.2 μA. Its temperature range is -55 to $+150$°C and total error less than 1°C. Digital output I^{2}C precise sensors AD7418 (Analog Devices [42]) work in the temperature range of -40 to $+120$°C with their total error below 1°C.

There are several integrated temperature meters on the market; go to http://www.infineon.com, http://www.national.com, and http://www.maxim-ic.com to learn about several options. Temperature meters are available as wire-mount or surface-mount devices (SMD). Small SMD meters are placed directly onto the thinned part of the microreactor, as shown in Fig. 9.21(d). An example of the technical solution of the on-microreactor integration of an SMD-packaged LM 45 thermometer is shown in Fig. 9.22. Three thermometers are epoxy-glued with a thermoconductive glue onto a thinned region of glass. Experiments show that a 200-μm-thick glass layer does not significantly influence measurements of the temperature of fluid flowing under the sensor, at least in the temperature range of 10 to 90°C.

9.3.3   Thermistors

Thermistors are semiconductor devices with a large and negative (NTC) or positive (PTC) temperature coefficient of resistance (TCR). They are produced on a mass scale in different forms. The typical resistance of thermistors is equal to several thousand ohms (typically, 10 kΩ), their TCR is about $-4\%/$°C, and temperature range is -50 to $+300$°C. They are stable, cheap, and accurate (total error <0.1–0.2°C). Varying configurations of thermistors coworking with electronic circuits are known [including Wheatstone's bridges as in Fig. 9.20(a)]. They are ideal temperature sensors for microreactors.

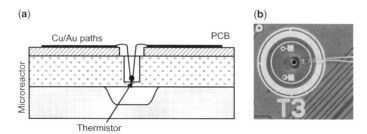

FIGURE 9.23 Glass pills like thermistor G 540: (a) principle of on-chip integration and (b) the sensor at the PCB circuit. Blind hole made in glass can be observed.

For example, a glass bead encapsulated thermistor G 540 (manufactured by Epcos in Germany) is 0.8 mm wide and 1.4 mm long; its nickel-plated leads $\Phi = 0.15$ mm are 6.5 mm long [43]. This fast sensor (has a thermal cooling time constant of approx. 0.3 s) consumes 18 mW and can be ordered in eight resistance classes from 5 to 1,400 kΩ, for a -55 to $+200°$C temperature range. The integration of G 540 sensors on-microreactor is simple, as shown in Fig. 9.23. Another small glass bead NTC temperature sensor suitable for on-microreactor measurement is S14A10325 (manufactured by Sensors Science in the United States): $\Phi = 0.35$ mm, 10 kΩ, working up to 300°C [44]. Sensors can be localized near important parts of microreactors and in mixers and reactors at inlets and outlets. An array of G-540 sensors can distinguish heat transfer along fluidic channels as well as the formation of hot points and over heated mixer areas. Replacing any broken-down sensor is easy.

9.4 FLOW AND MASS FLOW SENSORS

There are many configurations and technical realizations of flow microsensors based on heat transport in gases [45–48]. They are available on a custom basis and widely used (e.g., sensor AMV 3000 V fabricated by Honeywell in the United States [46]) (Fig. 9.24). Such sensors can be adapted for fluids, but for many reasons, in their existing form cannot be used directly in microreactors.

The simplest sensor consists of a resistive heater R_1 that is made of thermoresistive material. Heater R_1 (supplied with current I_g) is part of a balanced (for zero flow) Wheatstone's bridge [Fig. 9.25(a)] supplied with constant current I_z. The flowing medium cools down the heater, its resistance becomes a function of flow, the Wheatstone's bridge grows unbalanced, and its output voltage signal U_s is proportional to the flow. In more sophisticated constructions, two thermoelements are made: the heater and thermoresistor. Heater R_1 is supplied from a separate source with constant current I_g and kept in a constant temperature. The flowing medium transports heat from heater R_1 to thermoresistor R_2, a part of the Wheatstone's bridge, supplied with constant current I_z [Fig. 9.25(b)]. The output signal of the bridge is a function of flow. To eliminate the potential time-related instabilities of R_1 sensors, the heater and thermoresistor are sometimes doubled.

(a)   (b)

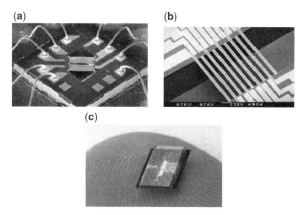

(c)

FIGURE 9.24 Examples of construction of flow sensors. (a) Internal view of gas flow sensor AVM3000 V of Honeywell (USA) [46]. (b) Internal view of gas flow sensor of 3T (the Netherlands) [47]. (c) Scale factor: 3T chip on a finger.

Under constant flow of the medium, flow sensors become a mass transport sensor. Two-element configuration is widely used in gas microchromatography as a mass detector of eluents separated by a column (so-called thermal coefficient microdetector TCD).

Thermoelements of flow and mass microsensors are realized as thin-film devices, most often made of Pt or Ni/NiCr just like thin-film thermosensors. To minimize parasitic heat flow caused by the high thermal conductance of bulk silicon, thermo-elements are patterned onto thermally isolated thin dielectric (S_3N_4 on SiO_2) perfo-rated membranes (bridges) suspended over the fluidic channel (Fig. 9.26). The fabrication process proceeds, in short, as follows: A (100)-oriented silicon wafer is covered with an SiO_2/Si_3N_4 dielectric layer. The layer is photographically patterned to form a perforation and window for etching of the fluidic channel. The thermoresistive thin-film metal layer is deposited onto the substrate and pat-terned. The channel is etched in TMAH (trimethyl-ammonium-hydroxide) in 80°C. This etchant is passive against dielectric and metal layers but etches anisotro-pically silicon. After etching, a borosilicate glass cup with a shallow isotropically

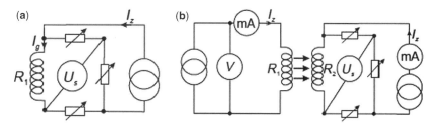

FIGURE 9.25 Three configurations of electrical circuits: (a) heater/thermoresistor as single element, (b) heater and thermoresistor separated.

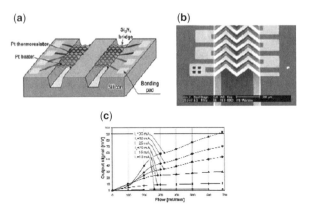

FIGURE 9.26 Flow sensors with thermoelements on a membrane [35]: (a) typical construction (before sealing to glass cover), (b) Pt meandered heater and thermoresistors at perforated SiO_2/S_3N_4 1000-nm-thick membrane, and (c) output characteristic.

etched channel is anodically bonded and sewed to open access windows to the contacts. At a single chip, two sensors are fabricated; one of them is used for referencing measurements.

Flow/mass microsensors are linear and precise. Flow range depends on the type of sensor and is typically from a single μL/min to hundreds of thousands mL/min. The output signal depends on the thermal conductivity of the measured medium, so this type of sensor must be scaled. On-off transient time is about 60 ms.

The unique construction of a flow sensor with microspirals made of Pt $\Phi = 12$ μm wire is shown in Fig. 9.27 [35, 38]. Two spirals, 800 μm long and 300 μm wide (the first spiral works as a heater, the second as a thermoresistor), are positioned inside a

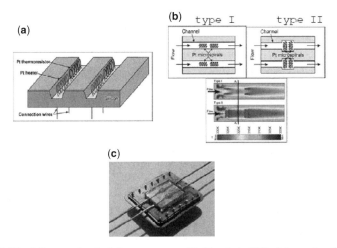

FIGURE 9.27 Micromechanical flow sensors with Pt spirals [35]: (a) construction, (b) two configurations and heat transports, and (c) sensor in metal case.

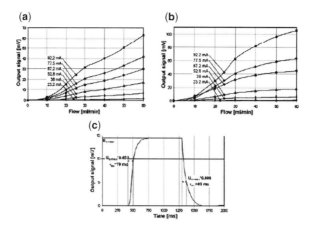

FIGURE 9.28 Flow characteristics of sensor from Fig. 9.27: (a) signal vs. flow for type I, (b) the same curve for type II, and (c) on-off curve version I.

fluidic channel etched in silicon substrate and covered with Borofloat 3.3 glass sealed by anodic bonding. This sensor can be used for corrosive fluids flow measurement because the maximal temperature of the heater can be lower than $100°C$ and Pt wires are chemically inert. In a gas environment, the sensor exhibits high sensitivity and an extremely fast on-off reaction (Fig. 9.28).

9.5 CONDUCTOMETRIC/AMPEROMETRIC SENSORS

Conductometric/amperometric microsensors (C/A), included in a group of electro-chemical detectors, can be integrated on-a-chip with microreactors or used as off-channel devices [49]. The C/A sensor consists of planar electrodes immersed in a fluid stream and deposited onto the inner surface of the fluidic channel. There are

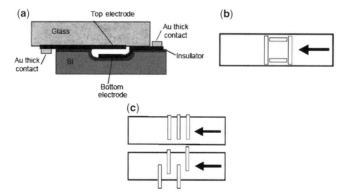

FIGURE 9.29 Configuration of electrodes of conductometric and amperometric microsensors: (a) top-down, (b) Wheatstone's bridge, and (c) three in-line and four double-in-line.

many configurations of the electrodes of C/A detectors, the most popular being the in-line positioning of three electrodes and Wheatstone's bridge (Fig. 9.29). Working electrodes have to be chemically passive; the electrochemical potential of the material has to agree with chemical standards and be stable in time. The usability of a particular kind of electrode material is limited by the technological procedures for the fabrication of microreactors. On-microreactor sensors are deposited and patterned before final sealing of the microreactor.

Sealing must be done at a temperature that will not destroy patterned layers, so only low- and medium-temperature procedures are acceptable. Thin-film Pt, Pd, Ti/W/Au, CrNi/Ni, and CrNi/Ni/Au tolerate the low-temperature sealing of glass and quartz, and anodic bonding of silicon and glass below 450°C; poly-Si can handle higher temperature up to 900°C but is rarely used. Mo thin-film layers survive high-temperature annealing at 1,100°C but tend to form dendrites and cannot be applied. Ti/W/Au layers do not tolerate washing in hydroxide peroxide, a part of the standard activation procedure before anodic bonding [8]. The thickness of electrodes is maintained below 1 μm to ensure the tight and leak-proof sealing of parts of microreactors or sensors. C/A sensors for microreactors sealed by high-temperature bonding (over 650°C, microreactors which are made of Foturan, borosilicate glass, or quartz) are designed as off-channel devices, because the high-temperature sealing of parts always influences the quality of electrodes.

The dead volume of a micromachined conductometric sensor with a channel 200×150 μm^2 and three in-line electrodes 250×25 μm^2 distanced 50 μm from each other is about 5.25 nL. The dead volume of standard glass conductometric sensors equals to approx. 1 mL^3. This clearly shows a scale factor and the advantages of integrated sensors (Fig. 9.30).

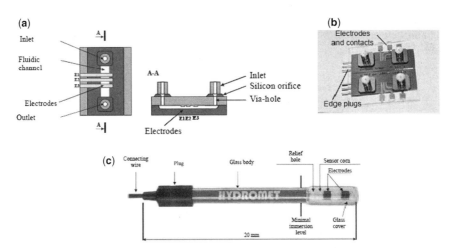

FIGURE 9.30 Conductometric sensor: (a) schematic view of the off-channel conductometric sensor device; (b) its technical realization [50], with channel cross section of 0.8 × 0.3 mm^2 and dead volume of 5.25 nL; and (c) for comparison, a 20-cm-long, 1-mL deadvolume glass conductometer commonly used in chemical practice (HYDROMET, Poland).

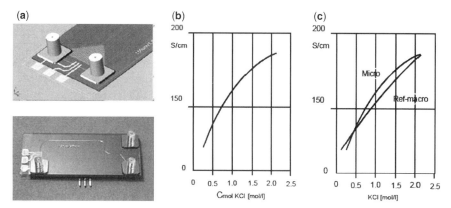

FIGURE 9.31 On-microreactor conductometric sensor: (a) technical realization of Y-mixer and reacting channel with three electrodes sensor at its end, (b) typical output characteristic for $f = 5$ kHz and $U = 1$ V supply, and (c) comparison of characteristics of microsensor and reference conductometer.

The fabrication procedures for the sensor starts with the selective isotropic etching of the fluidic channel in a silicon substrate in NH solution (65% HF plus 40% HNO_3) through 1.5-μm-thick thermal oxide. After residual postetching oxide is removed, the silicon is covered again with a new chemically passive thermal silicon dioxide 1 μm thick. Next, thin-film metal layers (CrNi/Au or Ti/W/Au) are

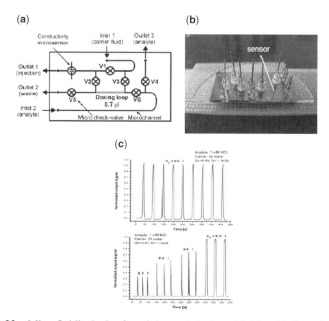

FIGURE 9.32 Microfluidic device for mixing and dosing of fluids with five microvalves and conductometric sensor on-chip [51]: (a) schema of fluidic circuit, (b) true image of the device with inlet/outlet connections, and (c) output signals for repetitive dosing of 0.1 to 0.6 of μL samples.

magnetron-sputtered and patterned photolithographically on a Borofloat 3.3 glass cover. The silicon is subjected to Piranha washing and activated in 30% hydrogen peroxide; the glass is degreased in hot acetone and ETOH and activated by boiling in DI water for 30 min. The parts are anodically bonded at 400°C for 1,200 V. After this, the electrical connections are soldered and the sensor is packaged. Fluidic connections can be made by anodic bonding of glass tubes on silicon orifices, gluing, or crewing.

This fabrication method can be used for manufacturing off-channel devices as well as for the on-microreactor integration of conductometric sensors (Fig. 9.31). Output characteristics are comparable to any mandated standards; they depend on the configuration of electrodes and frequency and amplitude of polarization. The sensor is fast and repeatable. The dynamic repetitive work of a microconductometer is illustrated in Fig. 9.32 [51]. The microfluidic mixer-doser with six on-chip valves, with a built-in conductometric sensor, has been used for dosing portions of 1-mM KCl into a DI water stream. Accurate $\pm 1\%$ dosing has been noted for doses of 0.6 μL; the minimal recognized doses are 0.1 μL. The repetition time of measurements, below 50 s, depends on the fluidic behavior in the microfluidic device because the response time of the sensor is below 1 s.

Microconductometric sensors are an interesting alternative to widely used macro-conductometric sensors. Their extremely low dead volume and fast response allow the use of such sensors in real-time monitoring of reactions in microreactors. They can also be constructed as completely corrosion-proof devices, although electrocorrosion always shortens their life-span.

9.6 OPTICAL PHOTOMETRIC AND FLUOROMETRIC SENSORS

Photometric and fluorometric sensors (Fig. 9.33) are very attractive from the point of view of microreactor technique. Spectrophotometric sensors measure the spectrum of transmission or absorption of light; fixed-wavelength sensors use laser light or light

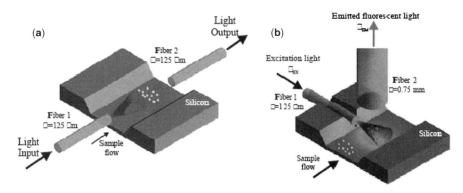

FIGURE 9.33 Optical fiber sensors for microreactors: (a) photometric sensor where light is absorbed in sample and (b) fluorometric sensors where light is emitted from sample.

filtered by narrow bandpass optical filters. Fluorometric sensors measure the fluorescence of a fluorophor added to a sample subjected to excitation light λ_{ex}. The intensity of fluorescent light λ_{em} is the direct measure of the concentration of investigated molecules.

Photometric and fluorometric sensors have been used for years in analytical chemistry, but their use in microchemistry has been limited by the lack of reliable microsensors. The miniaturization of such sensors is an interesting task; the creation of different discrete microsensors with glass fibers or optics-based integration has been reported recently [52–56], including the use of a photonic-structure fiber [57]. In the next sections of this chapter, we discuss several new microsensors suitable for the microreactor technique: photometric integrated sensors working in the visible (VIS) and near infrared (NIR) light spectrum and fluorometric microsensors, including the most recent fluorometric VIS microsensors and NIR microsensors designed for DNA/real-time PCR microreactors.

9.6.1 VIS Spectrophotometric Sensor

This sensor (Fig. 9.34) [58] is composed of a silicon chip ($25 \times 25 \times 0.36$ mm^3) with a C-shaped fluidic channel, two V-grooved channels, two standard multimode fibers $\phi = 125/62.5$, and a 1-mm-thick Borofloat 3.3 glass cover sealed to the silicon chip. Fibers are positioned inside the V-grooves at opposite corners of the fluidic channels, thus forming an optical cell several milimeters long. For a 10-mm-long, 150-μm-deep, and 700-μm-wide fluidic channel, the volume of the optical cell is about

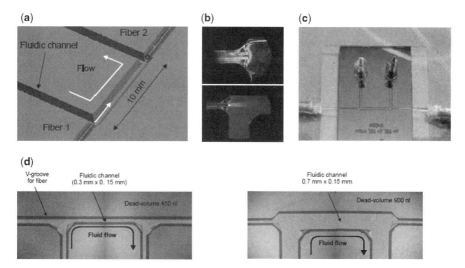

FIGURE 9.34 Spectrophotometric microsensor [58]. (a) Principle of work. (b) From top: assembled fiber, true image of working sensor. (c) Assembled sensor chip before encapsulation. (d) Close-up view of fluidic channel and V-grooves; at left dead volume of 450 nL, at right 900 nL.

900 nL. Light guided by input fiber 1 is absorbed by a fluidic sample flowing through the C channel; transmitted light is collected by fiber 2.

The fabrication process includes three main steps: the formation of channels, bonding of the glass cover with machined fluidic inlet/outlet holes, and assembling of optical fibers. The fluidic channel and V-grooves (150 μm deep) are wet-etched anisotropically (in 10 M KOH at 80°C) in a (100)-oriented silicon wafer through a 0.6-μm-thick thermal SiO_2 mask layer. After etching, the SiO_2 mask is removed and a new 0.3-μm-thick layer of thermal silicon dioxide is formed. The silicon wafer is then divided into chips. Two in-out holes ($\phi = 0.9$ mm) are mechanically drilled in the glass cover. The silicon and glass are washed and bonded at 450°C and 1.5 kV. Next, the optical fibers are positioned in the V-grooves and immobilized by a small drop of UV-cured optical glue.

Tests conducted with varying types of standard spectrometers used in analytical chemistry—in which a microsensor replaced the optical "macro-cuvette"—show good agreement of these microsensors and standards [58]. For example, the spectra of a water solution of potassium permanganate ($KMnO_4$), measured in the flow mode (100–150 μL/min) with the use of a microsensor, are in good agreement with corresponding spectra obtained for PC-2000 (manufactured by Ocean Optics in the United States) and DA (Zeiss in Germany) spectrometers equipped with classical cuvettes [Fig. 9.35(a) and (b)].

The fixed-wavelength mode tests, done for a light wavelength of 650 nm, also indicate the analytical usefulness of this microsensor. Measurements of the spectra of absorption for solutions with different phosphate concentrations (0.2–3.0 μg P/mL) for a flow rate of 100 μL/min show a detection limit of 0.2 μg P/mL and linear response of the microsensor up to 1.6 μg P/mL [Fig. 9.35(c)]. The microsensor has been calibrated with 1 mg P/mL KH_2PO_4 in a DI water test solution.

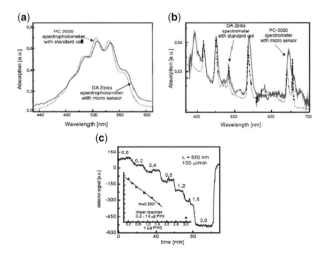

FIGURE 9.35   Characterization and work of microsensor [58]: (a) spectra of solution of $KMnO_4$ in DI water, (b) spectra of solution of $Ho(ClO_4)_3$ in DI water measured conventionally and with microsensor, and (c) detection of phosphates in water.

The positive features of the photometric VIS microsensor described here are relatively simple construction, small dead volume, valuable spectrometric and fixed-wavelength measurements, and finally, possible on-microreactor integration. The negatives are an application limited to noncorrosive medias and a low-temperature range, below 80°C. This makes the sensor ideal for biomedical applications, but its use in industrial microreactors is doubtful although possible.

9.6.2 NIR Spectrophotometric, Corrosion-Resistant Sensor

Sensors for a harsh environment must be corrosion-resistant over the long term and ideally leak-proof. Low dead volume and—at least—reasonable parameters of measurement are obvious requirements.

The corrosion-resistant microspectrometric device shown in Fig. 9.36 [59] meets, almost ideally the features of the perfect NIR spectrometric sensor for corrosive media: The fluidic stream is completely separated from optical fibers by a separation wall made of silicon, transparent for infrareds; the anodic bonding/sealing of parts ensures the leak-proofness of the sensor up to 600 kPa; and special NIR glass

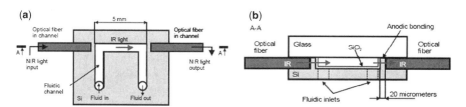

FIGURE 9.36 Corrosion-resistant NIR spetrophotometric microsensor: (a) lay-out and (b) cross-section.

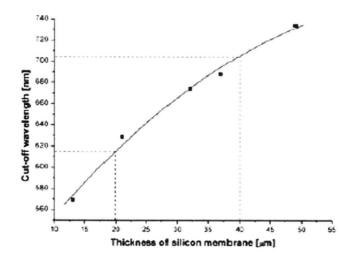

FIGURE 9.37 Cut-off wavelength of silicon membrane of different thickness.

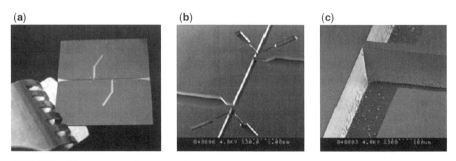

FIGURE 9.38 Details of construction: (a) micromachined silicon substrate, (b) view of fluidic channel and grooves for fibers (SEM image, 30X magnification), and (c) silicon separation wall (SEM image, 300× magnification).

fibers from Ocean Optics (United States based), manufacture with 100 μm of corn, ensures the ability of the sensors to work in the 700- to 4,000-nm wavelength range. As mentioned above, the principle behind the work of the NIR sensor is the thin layer of silicon that remains transparent for infrareds (Fig. 9.37). The cutoff wavelength of a 20-μm-thick layer is approx. 620 nm. Longer wavelengths are transmitted almost perfectly. Thus, a thin separation wall is sufficient to ensure the ideal separation of fluid and fibers up to 600 kPa.

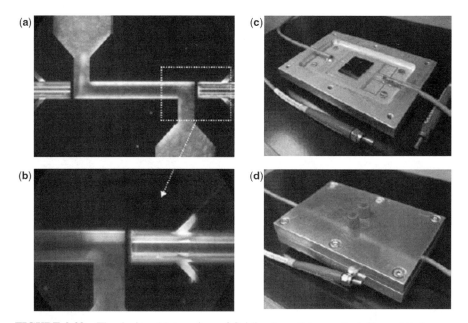

FIGURE 9.39 The device: (a) top view of fluidic channel two optical fibers, (b) close-up showing precise fixation of the fiber, (c) sensor chip inside package, and (d) packaged sensor. Up-Church fluidic connections and standard optical plugs of fibers are shown here.

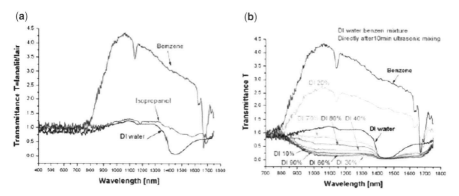

FIGURE 9.40 Spectra transmittance taken with NIR microsensor: (a) identification of hydrocarbons and (b) characterization of water, benzene mixtures.

The fabrication process for the sensor starts with frontside deep DRIE etching in a silicon substrate 140-μm-deep/130-μm-wide fluidic channel and grooves for positioning of fibers, through a Cr thin-film mask (Fig. 9.38). Next, a second backside deep DRIE process opens via-holes of fluidic connections at the ends of the fluidic channel. Following this, micromachined silicon substrate is thermally oxidized to the thickness of 0.3 μm. The glass Borofloat 3.3 cover and silicon substrate are washed and anodically bonded at 450°C and 1.5 kV. Fibers are positioned inside etched grooves and fixed by gluing (Fig. 9.39). The chip with fibers is placed onto PCB inside the metal package, fibers and PCB are fixed by gluing, top part of the package is screwed, and Up-Church fluidic connections are assembled. The packaged sensor is shock-resistant, corrosion-resistant, and leak-proof. Concentrated hot nitric/sulphuric acids and their mixtures do not damage or influence the work of the sensor. Tests done with the AQ-6315B spectrometer (manufactured by ANDO in Japan) confirm the good work of the device (Fig. 9.40).

The advantageous features of the NIR microsensor are chemical corrosion resistance, ideal leak-proofness, relatively simple technology and easy on-microreactor integration, and wide spectrum of wavelengths. The negative feature is the limitation of the spectrum to the NIR/IR range.

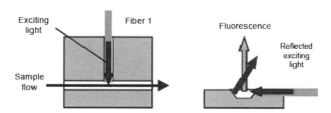

FIGURE 9.41 T-type sensor: the principle of work.

9.6.3 T-Type Fluorometric Microsensor

The principle behind this sensor lies in the T-type configuration of the sample channel and fiber 1 (Fig. 9.41) [39, 58, 60]. A (111) crystallographic wall-forming fluidic channel is angled 54.47° vs. glass fiber 1 that is aligned in-plane to the (100)-oriented surface of the silicon chip. Exciting light guided by fiber 1 is reflected at the (111)-wall and does not blind highly sensitive detectors sensing fluorescence. Thus, the exciting light beam is attenuated so sufficiently that—at least in typical chemical measurements—the optical filters usually applied in spectrofluorometry are not used.

9.6.4 VIS Sensor with Excitation Light Guided by Glass Fiber

Version 1 of the microsensor consists of a silicon chip with a wet-etched sample channel, glass cover, and two fibers: fiber 1, a standard telecommunications multimode glass fiber, with a diameter of $\phi = 125\,\mu m$ and core size of $62.5\,\mu m$, and fiber 2, $\phi = 0.75\,mm$, made of (polymethyl methacrylate PMMA) (Fig. 9.42). Fiber 1 is positioned in the V-groove etched perpendicular to a sample channel. Its dimensions are carefully designed to allow the precise alignment of fiber 1 and the sample channel. Fiber 2 is positioned perpendicular to the top surface of the sensor inside a holder. This fiber collects the fluorescence signal. Its external diameter is much larger than the dimensions of the area in which fluorescence takes place. Therefore, all the fluorescent light emitted by the sample is guided to an external photodetector. The types of fibers have to be adjusted to the spectrum of exciting light and fluorescence. The basic construction of version 2 of the microsensor is similar; fiber 2 and its holder are replaced by a high-efficiency detecting diode (BPW 21).

The fabrication process begins with deep wet anisotropic KOH etching of the sample channels and V-groove (for positioning fiber 1) in a (100)-silicon wafer

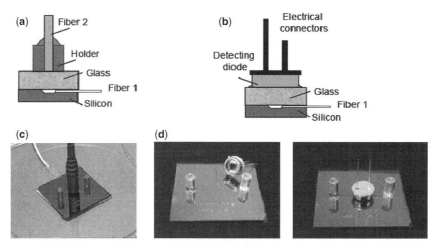

FIGURE 9.42 VIS fluorometric microsensor: (a) version 1 with two fibers, (b) version 2 with on-chip detector, (c) finished $33 \times 35\,mm^2$ chip of version 1, and (d) version 2 before and after placement of the detecting diode.

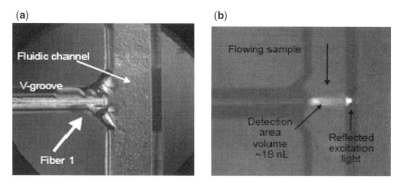

FIGURE 9.43 VIS spectrofluorometric microsensor [58]: (a) close-up picture of manufactured device and (b) true image of working sensor.

through a thermal oxide mask. After etching, residual oxide is removed and the wafer is thermally oxidized to a thickness of 0.5 μm. A Borofloat 3.3 cover with via-holes is prepared; then the silicon wafer and glass are washed and anodically bonded (450°C, 1,500 V). The fluidic connections and glass holder (only for version 1) are placed and fixed at the top surface of the sensor. Next, fibers are placed in their proper places and glued. The fixation of fiber 1 inside the V-groove proceeds as follows: The fiber is positioned in the groove under the microscope, a drop of fluidic epoxy UV-curable

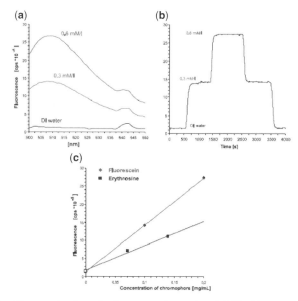

FIGURE 9.44 Characterization of microsensor in version 1 [58]: (a) spectra of fluorescence for different concentration of fluorescein, (b) time-dependent curves of luminescence of small portion of sample with varying concentration of fluorescein, and (c) output signal vs. concentration of fluorescein and erythrosine.

glue is placed inside, capillarian force drives in the glue precisely to the end of the fiber [Fig. 9.43(a)], and the glue is hardened with a UV lamp.

The dead volume of detection equals 18 nL [Fig. 9.43(b)] and corresponds to the volume of sample inside the fluidic channel illuminated by fiber $\phi = 125/$ 62.5 μm. Tests done for fluorescein ($\lambda_{ex} \sim 490$ nm, $\lambda_{em} \sim 510$) added to DI water show the reasonable sensitivity (a detection limit of 5 μM) and fast reaction of the sensor (Fig. 9.44). The sensor can be easily adapted for on-microreactor integration, especially version 2, but its application is limited to noncorrosive and low-temperature media because of the degradation of epoxy glue.

9.6.5 T-Type VIS/NIR Spectrofluorometric Sensor with Direct Illumination

The construction of this sensor is based on the principle of a T-type chip as presented in Fig. 9.45. In this sensor, however, the excitation light directly illuminates the fluidic channel and excites the fluorescence of a chromophor near the end of the on-chip integrated SU8 waveguide [Fig. 9.45(a) and (b)] [61]. Fluorescence light is collected by a SU8 waveguide and transmitted to the glass fiber, which guides light to the external spectrometer. The first steps of the fabrication process are similar to those described earlier, but after the channels are sealed by anodic bonding, a small drop of fluidic SU8 resin is dropped inside the V-groove channel. Capillarian forces

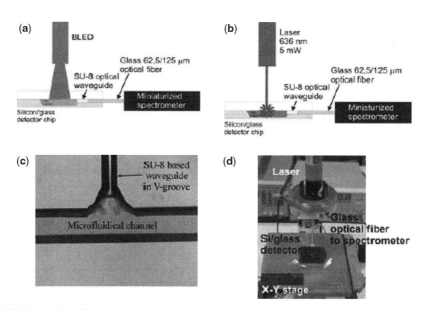

FIGURE 9.45   VIS spectrofluorometric microsensor with SU8 waveguide: (a) version A with BLED (Bright-LED) excitation used for fluorescein detection; (b) version B with laser light source for red-line chromofores; (c) close-up of the SU8 waveguide, microlens, and detection area; and (d) true image of version B with added small aperture over the detection area.

(a)

(b)

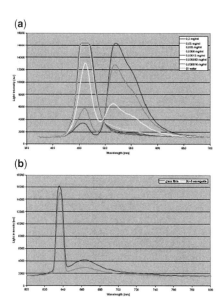

FIGURE 9.46 Characterization of microsensor with SU8 waveguide: (a) tests with fluoroscein and (b) identification of 100-ng sample of DNA of Salmonellosis bacteria. No optical filters; spectra are taken with Optel microspectrometer (manufactured in Poland).

pull the SU8 so it flows along the V-groove. The flow stops precisely at the widened end of the V-groove [Fig. 9.45(c)] where, a SU8 microlens is self-formed. After this, glass fiber is introduced inside the V-groove, carefully pushed to the moment when SU8 wets the glass, optically coupling the SU8 waveguide and fiber. Next, SU8 is solidified by baking at 90°C. The maximal length of the SU8 waveguide is 10 to 20 mm; glass fiber length may be up to 0.5 m.

The sensitivity of a sensor working in a spectrometric mode depends mainly on the type of spectrometer used. The detectability of fluorescein in DI water is 130 ng/mL for the sensor in version A and a LIGA-based miniature spectrometer with diode array detector [62] [Fig. 9.46(a)]. The detection of about 100 ng/mL of DNA of Salmonellosis bacteria marked with Cy-5 fluorophor is clearly seen in the curves of Fig. 9.46(b), where on the left the 635-nm peak represents exciting laser light and on the right the 672-nm peak represents the fluorescence of the DNA sample.

9.6.6 T-Type NIR Sensor with Excitation Light Guided by Glass Fiber and CCD Camera Read-Out

In this sensor, the exciting light of the laser is guided by the glass fiber, illuminates the sample, and is reflected by the (111)-wall as shown in Fig. 9.41. Fluorescence light passes through long-pass and/or bandpass optical filter(s) and is collected by a highly sensitive CCD camera (Fig. 9.47). This configuration increases the sensitivity of the T-sensor to the limit acceptable in DNA tests. The high sensitivity of

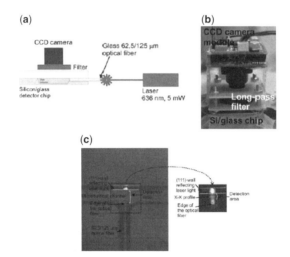

FIGURE 9.47 NIR fluorometric microsensor with the CCD camera module [63]: (a) optical configuration, (b) 3-cm-high measurement set-up, and (c) close-up of detection area (18 pl). True image as recorded by the camera module; no additional optic was used.

the sensor is shown in this example: samples of DNA marked with red-line fluorophor TOPRO-3 excited by 635-nm laser light, flourishing at about 670 nm. One 650-nm long-pass filter or in-line 650-nm long-pass and 660-nm bandpass filters of THORLABS are used. The applied detector is a YK-2015 CCD color camera manufactured by Sony (with a sensitivity of 0, 4 lux) or more sensitive YK-3043 CCD black-white camera also available from Sony (with a sensitivity of 0, 2 lux). The detection limit of a sensor equipped with the YK 2015 CCD camera is ~ 10 ng/mL of DNA for a concentration of 0.3 μg/mL of TOPRO-3. The more sensitive YK3043 CCD camera detects ~ 2 ng/mL of DNA for 0.3 μg/mL of TOPRO-3 in a sample (Fig. 9.48). The low-frequency (~ 1 KHz) amplitude modulation of laser light (60%

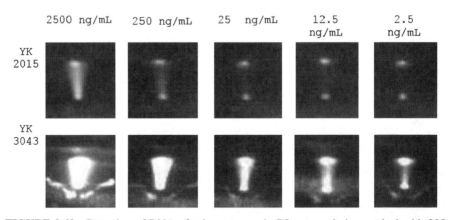

FIGURE 9.48 Detection of DNA of salmon sperm in DI water solution marked with 385 ng/mL of TOPRO-3. Numbers in the columns indicate the concentration of DNA in ng/mL. True images: top line, YK-2015 color camera; bottom line, YK-3043 black-white camera.

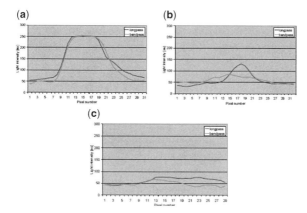

FIGURE 9.49 X-X profiles of fluorescence of samples with varying concentration of DNA (images in from Fig. 9.48). YK 2015 CCD camera: (a) 2500 ng/mL, (b) 25 ng/mL, and (c) 12.5 ng/mL.

deep) is strongly recommended. CCD camera images are processed digitally to evaluate measurable data, which are presented in numerical or graphical form (Fig. 9.49).

This highly sensitive sensor is very attractive for the microreactor technique, although its use is limited to red-line fluorophores. Its detection limit falls in the range of several attograms of the DNA sample (10 ng/mL detected in a volume of approx. 20 pL) and is comparable to the detectability of widely used macroinstruments. The sensor is cheap, small, and can be easily adapted for on-microreactor integration. The solution presented here is a key component of the read-out optical unit of the portable DNA real-time PCR medical device [63].

9.7 CLOSING REMARKS

Increasing demand on low-cost miniature chemical reactors is the driving force behind the development of instruments capable of controlling reactions on a microscale. Measuring a set of physical and chemical parameters becomes the key factor; commonly used technical solutions cannot be applied because of scale-factor problems. The design and fabrication of microengineered sensors of physical and chemical parameters dedicated especially to the microreactor technique have become an important task. The mainstream strategies of development are:

- Integration of on-channel microsensors on-microreactor, and/or
- Use of off-channel microsensors and instruments developed according to the rules of microengineering.

According to this author's best knowledge, micromechanical silicon–glass sensors seem to be the best option for both strategies. Glass and silicon are technologically recognized; these materials are chemically inert and can be covered with thin-film chemically passive dielectric layers (SiO_2, Si_3N_4, $SiON_x$). On-channel sensors can be integrated on-microreactor, although the spectrum of microreactors is limited to silicon–glass or glass devices. Off-channel sensors are able to cowork with almost all types of microreactors, including microreactors working with extremely corrosive medias and biomedical devices.

Some of the solutions described here have been designed especially for use in nitration microreactors [64]: pressure sensors, a sensor platform, an NIR corrosion-resistant sensor. Others have been designed as useful sensing tools for integrated analytical devices [65]: thin-film temperature sensors, flow sensors, T-type VIS photometric sensors, conductometric sensors. Still others play the role of detectors in biomedical microreactors [66]: T-type NIR fluorometric sensors. All of these sensors, from a technological viewpoint, are relatively simple; they can be manufactured by small enterprises or spin-off laboratories specializing in silicon–glass microengineering. Wet isotropic and anisotropic etching, DRIE, anodic bonding, 3-μm rule photolithographic patterning, set of fiberoptic-related assembling techniques, mechanical machining of glass and silicon–glass, and wire-bonding techniques are all typical manufacturing methods. The most difficult tasks are coming up with a corrosion-resistant design, final packaging, and the electrical or fluidic connecting of the sensors.

The space constraints of this chapter do not allow extended discussion of the microsensors that might be potentially suitable for microreactors, although a continuously growing number of publications on the subject is enabling such a dialogue [67–72]. Many similar technical concepts discussed here can be found in other books and papers. For example, the integration of pressure sensor chips inside channels etched in glass substrates is explored in [73], a nano-gap on-chip capacitor for the detection of DNA in [74], an IR detector with light transmission through a silicon membrane in [75], a thin-film chemiresistor concept in [76], on-chip remote control of flow inside a channeled microreactor by the use of a laser light beam in [77], an on-chip photometric sensor with glass optical fibers in [78], another flow-sensor principle with a differential pressure-sensitive unit and interferometric system in [79], photometric sensors with C-channel shape and SU8 glass fiber in [80], a silicon aperture improving the sensitivity of UV absorption in an on-chip detector in [81], the application of a spectrometric sensor for the control of chemical reactions in a microreactor in [82–85], on- and off-channel fluorometric sensors in [86–88], an integrated atomic emission flame spectrometer, and plasma detector for integrated gas chromatography in [89, 90]. In addition, conductivity and impedance sensors are discussed frequently (examples appear in [91, 92]); the new SU8 UV-LIGA Mach-Zender interferometer for on-channel measurements of fluids is described in [93, 94]; the utilization of the SNOM microscope as an on-chip universal tool is explored in [95]. New ideas on the fabrication of microreactors and coworking on-channel sensors through the use of LTCC technique are presented in [96–98], in which on-channel temperature, photometric, and conductometric sensors were successfully fabricated and tested.

Hiroyuki Fujita, from the University of Tokyo, made the following comment at the MicroTAS 2006, Tokyo, 5–9 November: "Scaling down biochemical systems does not mean mere miniaturization of macroscopic counterparts but requires revolutionary change in fabrication and also will bring about revolutionary change in chemistry." This is the most accurate description of the potential behind micro-reactor technique.

ACKNOWLEDGMENTS

The author would like to extend his thanks to members of his scientific team on the Faculty of Microsystem Electronic and Photonics at the Wroclaw University of Technology and his colleagues from the Institute of Electronic Technology in Warsaw: Anna Gorecka-Drzazga, Pawel Knapkiewicz, Rafal Walczak, Sylwester Bargiel, Jan Koszur, Bogdan Latecki, and Pawel Kowalski.

The author also wishes to thank the Universite Franche-Comte and FEMTO ST of CNRS in Besancon, France, for his warm reception as a visiting professor during the stage when this chapter was finalized.

BIBLIOGRAPHY AND OTHER SOURCES

1. MEMSentry and MICRONEWS. *Yole Development Custom Analysis of MEMS.* http://www.yole.fr.

2. Salomon, P. (2005). A Nexus market analysis for MEMS and Microsystems III 2005–2009. *MST News*, 5: 33–34.

3. Salomon, P. (2006). Micro-sensors world wide markets and economic impact. In *Proceedings of 20th Eurosensors Conference*, Sept. 17–20, Goeteborg, pp. 30–31.

4. Finitshenko, Y., and van der Berg, A. (1998). Review: Silicon microtechnology and micro-structures in separation technique. *J. Chromatogr.*, A819: 3–12.

5. Manz, A., and Becker, H. (1999). *Micro-System Technology in Chemistry and Life Sciences.* Berlin: Springer Verlag.

6. Erhfeld, W., Hessel, V., and Loewe, H. (2000). *Microreactors.* Weinheim: Wiley-VCH.

7. Nguyen, N.-T., and Wereley, S. T. (2002). *Fundamentals and Applications of Microfluidics.* London: Artech House.

8. Dziuban, J. (2006). *Boding in Microsystem Technology.* Berlin: Springer.

9. Bao, M. (2000). *Micromechanical Transducers, Pressure Sensor, Accelerometers and Gyroscopes.* New York: Elsevier.

10. Maluf, N., and Wiliams, K. (2004). *An Introduction to Microelectromechanical Systems Engineering.* London: Artech House.

11. Bryzek, J., Petersen, K., Mallon, J., and Pourahmedi, F. (1991). *Silicon Sensors and Microstructures.* Fremont, CA: NovaSensor.

12. Hase, J., Bessho, M., and Ipposhi, A. (1994). SOI type pressure sensor for high tempera-ture pressure measurement. *SAE Transact.*, 103(3): 992–996.

13. Manz, A., Graber, N., and Widmer, H. M. (1999). Miniaturized total-analysis sytems—a novel concept of for chemical sensing. *Sens. & Act.*, B1: 244–248.

14. Harrison, D. J., Glavina, P. G., and Manz, A. (1993). Toward miniaturized electrophoresis and chemical analysis systems on silicon: an alternative to chemical sensors. *Sens. & Act.*, B10: 107–116.

15. Dziuban, J., Górecka-Drzazga, A., Nieradko, L., and Malecki, K. (1999). Silicon-glass micromachined chromatographic microcolumn. *J. Capillary Electrophor. Microchip Technol.*, 6(1–2): 37–41.

16. Corning Web site. http://www.corning.com.

17. Borofloat Web site. http://www.schott.com/borofloat.

18. Pommerantz, W. D. (1969). Field assisted glass metal sealing. *J. Appl. Phys.*, 40(10): 3346–3349.

19. Cosma, P., and Puers, R. (1995). Characterization of the electrostatic bonding of silicon and Pyrex glasses. *J. Micromech. Microeng.*, 5: 98–102.

20. Choi, W. B., Ju, B. K., Lee, Y. H., Jeong, S. J., Lee, N. Y., Sung, M. Y., and Oh, M. H. (1999). Glass-to-glass bonding for vacuum packaging of field emission display in an ultra-high-vacuum chamber using silicon thin film. *J. Electrochem. Soc.*, 146(1): 400–404.

21. Lucas-Nova Sensors 1055. Product information. http://www.novasensor.com.

22. Petersen, K., Brown, J., Veurmellen, T., Bart, P., Mallon, J., and Bryzek, J. (1990). Ultrastable high temperature sensors using fusion bonding process. *Sens. & Act.*, A21–23: 96–101.

23. Knapkiewicz, P., Walczak, R., and Dziuban, J. (2006). On integration of silicon/glass micromachined sensors to microfluidical devices—toward intelligent microreactor. *Proc. Eurosensors XX.*, 1: 40–41.

24. Knapkiewicz, P. (2007). Intelligent microreactor for nitration reactions. Ph.D. thesis. Wroclaw University of Technology, Faculty of Microsystem Electronics and Photonics. Wroclaw, 15 October 2008.

25. Mikroglas Technik (2003). *Foturan: a Photostructurable Glass.* http://www.mikroglas.com.

26. Freitag, A., and Dietrich, T. (2000). Glass as a material for microreaction technology. *Proc. IMRET*, 4: 48–54.

27. Knapkiewicz, P., Walczak, R., Dziuban, J., and Dietrich, T. R. (2006). Integration of silicon and silicon/glass sensors structures to Foturan glass. *Proceedings of Conference on Optical and Electronical Sensors*, 415–418.

28. Knapkiewicz, P., Walczak, R., Dziuban, J., Latecki, B., Koszur, J., and Dietrich, T. R. (2007). Pressure/temperature sensors platform for microreactors working with highly corrosive medias. Submitted to 2007 ELTE Conference.

29. Knapkiewicz, P., Walczak, R., Dziuban, J., Latecki, B., and Koszur, J. (2007). Packaged pressure sensor, patent application P-382967 from 20 07 2007.

30. Puers, R., van den Bossche, E., and Sansen, W. (1990). A capacitive pressure sensor with low impedance and active suppression of parasitic effect. *Sens. & Act.*, A21–23: 108–114R.

31. Puers, R., and Blasquez, G. (1994). Low cost high-sensitivity integrated pressure and temperature sensor. *Sens. & Act.*, A41–42: 338–401.

32. Dziuban, J., and Walczak, R. (2001). A silicon capacitive pressure sensor micromachined by EMSi method. *Proc. MME*, Sept. 16–18, Cork, Ireland, 123–127.

33. Klaasen, E., Petersen, K., Noworolski, J. M., Logan, J., Maluf, N. I., Brown, J., Storment, C., Culley, W., and Kovacs, G. T. A. (1995). Silicon fusion bonding and deep reactive ion etching: A new technology for microinstrumentation. *Tech. Dig. Eurosensors IX*, June 25–29, Stockholm, Sweden, 556–559.

34. Tong, Q. Y., and Goesele, Y. (1999). *Semiconductor Wafer Bonding.* New York: Wiley.

35. Dziuban, J. A., Mróz, J., Szczygielska, M., Malachowski, M., Górecka-Drzazga, A., Walczak, R., Bula, W., Zalewski, D., Nieradko, L., Lysko, J., Koszur, J., and Kowalski, P. (2004). Portable gas chromatograph with integrated components. *Sens. & Act.*, A115: 318–330.

36. Dziuban, J., Gorecka-Drzazga, A., and Lipowicz, U. (1992). Silicon optical pressure sensor. *Sens. & Act.*, A32: 628–631.

37. Górecka-Drzazga, A., Dziuban, J., and Bargiel, S. (2000). Micromechanical pressure sensor with optical fiber. *Proceedings of 6th Optoelectronics and Electronics Sensors*, COE 2000, Gliwice, June 13–16, v.2, 144–149.

38. Nieradko, L., and Malecki, K. (2000). Silicon components for gas chromatograph. Presented at Shaping the Future, international forum for graduates and young researchers at EXPO 2000. http://www.shaping-the-future.de/pages/abstracts/abstract_203.htm.

39. Dziuban, J., Górecka-Drzazga, A., Nieradko, L., Malecki, K., Mróz, J., and Szczygielska, M. (2001). Silicon components for gas chromatograph. *SPIE Proc.*, 4516: 249–257.

40. Bak, T. (2004). Technology of flow-through PCR reactor. *Proceedings of 6th Optoelectronics and Electronics Sensors*, COE 2004, Wroclaw, June 27–30, v.1, 587–588.

41. Texas Instruments Web site. Listing of Burr-Brown products. http://www.focus.ti.com/lit.

42. Analog Devices Web site. http://www.analog.com.

43. Epcos Web site. http://www.epcos.de.

44. Sensors Science Web site. http://www.sensorsci.com/letter.htm.

45. MTI Web site. http://www.chrompack.com.

46. Honeywell Web site. http://www.honeywell.com/sensing.

47. 3T B. V. Web site. http://www.3t.nl.

48. Popescu, D. S., Dunare, C., Lerch, P., Renaud, P., Modreanu, M., and Dascalu, D. (1998). Integrated flow sensor for fluids with dynamic velocity, temperature and conductivity profiles. *Proc. CAS*, 98(2): 561–564.

49. Huikko, K., Kostiainen, R., and Kotiaho, T. (2003). Introduction to micro-analytical systems: Bioanalytical and pharmaceutical applications. *Eur. J. Pharmaceut. Sci.*, 20: 149–171.

50. Górecka-Drzazga, A., Dziuban, J., and Bargiel, S. (2002). Integrated fluidic detectors. *Proc. COE*, 2: 215–220.

51. Bargiel, S., Walczak, R., Knapkiewicz, P., Górecka-Drzazga, A., Olszewski, T., and Dziuban, J. (2005). Micromachined silicon-glass dosing device with built-in conductivity detector. *Proceedings of 20th Eurosensors Conference,* Sept. 17–20, Goeteborg, Sweden, 292–293.

52. de Graaf, G., and Wolffenbutel, R. F. (2006). Fabrication of micromachined spectrometer system for evanescent wave infrared spectroscopy. *Proceedings of 20th Eurosensors Conference*, Sept. 17–20, Goeteborg, Sweden, 292–293.

53. Meulebroeck, W., Ottevaere, H., Scheir, K., Clicq, D., Desmet, G., and Thienpoint, H. (2006). A novel optical detection system for chromatographic applications. *Proc. SPIE*, 6188: 1111–1112.

54. Goel, S., McMullin, J. N., Qiao, H., and Awadelkarim, O. O. (2003). Optical detection systems for biochips using plastic fiber optics. *Rev. Sci. Instrum.*, 79(9): 4145–4149.

55. Camou, S., Tixier-Mita, A., Fujita, H., and Fujii, T. (2004). Integration of microoptics in bio-micro-electro-mechanical systems towards micro-total-analysis systems. *Jap. J. Appl. Phys.*, 43(8B): 5697–5705.

56. Keel, J. S., Puiu-Poenar, D., and Yobas, L. (2006). Integrated polydimethylsiloxane waveguide. In *Proceedings of Micro TAS 2000*, Enschede, The Netherlands, May 14–18, pp. 732–736.

57. Rindorf, L., Høiby, P. E., Jensen, J. B., Pedersen, L. H., Bang, O., and Geschke, O. (2006). Towards biochips using microstructured optical fiber sensors. *Analyt. Bioanalyt. Chem.*, 385(8): 1370–1375.

58. Bargiel, S., Gorecka-Drzazga, A., Dziuban, J. A., Prokaryn, P., Chudy, M., Dybko, A., and Brzozka, Z. (2004). Nanoliter detectors for flow systems. *Sens. & Act.*, A115: 245–251.

59. Bargiel, S., Dziuban, J., Walczak, R., Knapkiewicz, P., and Gorecka-Drzazga, A. (2006). Integrated optical microsensor for NIR spectroscopy of highly corrosive analytes in micro-reactors, *Proc. Eurosensors XX*, 2: 42–43.

60. Bargiel, S., Dziuban, J. A., Walczak, R., Knapkiewicz, P., Górecka-Drzazga, A. (2006). Corrosion resistant integrated optical microsensor for NIR spectroscopy and its application in microreactors. In *Proceedings of Micro TAS 2006*, Tokyo, Japan, November 5–9, pp. 741–743.

61. Dziuban, J., Walczak, R., Bargiel, S., Koszur, J., Kowalski, P., and Latecki, B. (2006). Optical platform for portable DNA/PCR system. Annual report of OPTOLABCARD PR6 Project.

62. Boeringher-Ingelheim, http://www.boeringher-ingelheim.de.

63. Schulze, J. L. M., Verdoy, D., Olabarria, G., Berganzo, J., Elizalde, J., and Ruano-López, J. M. (2006). Micro SU-8 chamber for PCR and fluorescent real-time detection of Salmonella DPNA. In *Proceedings of Micro TAS 2000*, Enschede, The Netherlands, May 14–18, pp. 1243–1245.

64. Loebecke, S., et al. (2005). New eco-efficient industrial process using microstructured unit components. Documentation of NEPUMUC Project of 6th Framework Programme financed by EC.

65. Gorecka-Drzazga, A., Dziuban, J., and Bargiel, S. (2002). Integrated fluidic detectors. *Proc. COE*, 2: 215–220.

66. Ruano-Lopez, J. M., et al. (2005). Mass produced optical diagnostic labcards based on micro and nano SU8 layers. Documentation of OPTOLABCARD Project of 6th Framework Programme financed by EC.

67. *Proceedings of International Conferences on Microreaction Technology (IMRET)*, 1990–2006.

68. *Proceedings of Micro Total Analysis (μTAS) Conferences*, 1994–2006.

69. *Proceedings of Eurosensors Conferences*, 1986–2006.

70. Lee, G.-B., Lin, C.-H., Chang, G.-L. (2003). Micro flow cytometers with buried SU-8/SOG optical waveguides. *Sens. & Act.*, A103: 165–170.

71. Saliterman, S. S. (2006). *BioMEMS and Medical Microdevices*. Bellingham, WA: SPIE Press.

72. Lee, A. P., Lee, L. J., and Ferrai, M. (2006). *Biological and Biomedical Nanotechnology*. New York: Springer+Business Media.

73. Becker, T., Muelberger, S., Bosch-von Braunmuehl, C., Mueller, G., Meckes, A., and Benecke, W. (2001). Microreactors and microfluidic systems: An innovative approach to gas sensing using tin oxide-based gas sensors. *Sens. & Act.*, B77: pp. 48–54.

74. Kang, H. K., Seo, J., Di Carlo, D., Choi, Y.-K., and Lee, L. P. (2003). Planar nanogap capacitor arrays on quartz for optical and dielectric bioassays. In *Proceedings of Micro TAS 2003*, Squaw Valley, CA, USA, October 5–9, pp. 697–700.

75. Jackmann, R. J., Floyd, T. M., Schmidt, M. A., and Jensen, K. F. (2000). Development of methods for on-line chemical detection with liquid-phase microchemical reactors using conventional and unconventional techniques. In *Proceedings of Micro TAS 2000*, Enschede, The Netherlands, May 14–18, pp. 155–158.

76. Lu, C.-J., Tian, W.-C., Steinecker, W. H., Guyon, A., Agah, M., Oborny, M. C. Sacks, R. D., and Wise, K. D. (2003). Functionally integrated MEMS micro gas chromatograph subsystem. In *Proceedings of Micro TAS 2003*, Squaw Valley, CA, USA, October 5–9, pp. 411–415.

77. Kraus, T., de Mas, N., Guenther, A., Schmidt, M. A., and Jensen, K. F. (2003). Integration of a flow regime sensor into a three dimensional multichannel microreactor. In *Proceedings of Micro TAS 2003*, Squaw Valley, CA, USA, October 5–9, pp. 809–812.

78. Mensinger, H., Richter, T., Hessel, V., Doepper, J., and Erhfeld, W. (1994). Microreactor with integrated static mixer and analysis system. In *Proceedings of Micro TAS 1994*, Enschede, The Netherlands, November 21–22, pp. 237–243.

79. Yang, Z., Matsumoto, S., Tsaur, J., Ichikawa, N., and Maeda, T. (2005). Bi-directional optical flow sensor fro online microfluidic monitoring. In *Proceedings Micro TAS 2005*, Boston, MS, USA, October 9–13, pp. 961–963.

80. Leeds, A. R., van Keuren, R. R., Durst, M. E., Scneider, T. W., Currie, J. F., and Pananjape, M. (2004). Integration of microfluidic and microoptical elements using a single photolithographic step. *Sens. & Act.*, A115: 571–580.

81. Nishimoto, T., Fujiama, Y., Abe, H., Kanai, M., Nakanishi, H., and Arai, A. (2000). Microfabricated CE chips with optical slit for UV absorption detection. In *Proceedings of Micro TAS 2000*, Enschede, The Netherlands, May 14–18, pp. 395–398.

82. Quiram, D. J., Hsing, I.-M., Franz, A. J., Jensen, K. F., and Schmidt, M. A. (2000). Design issues for membrane based gas phase microchemical systems. *Chem. Eng. Sci.*, 55: 3065–3075.

83. Keoschkerjan, R., Richte, M., Boskovic, D., Scnuerer, F., and Loebbecke, S. (2004). Novel multifunctional unit for chemical engineering. *Chem. Eng. J.*, 101: 469–475.

84. McCready, T. (2000). Fabrication techniques and materials commonly used for the production of microreactors and micro total analysis system. *Trends Analyt. Chem.*, 19(6): 396–401.

85. Svasek, P., Svasek, D. E., Lendl, Vellekoop, M. (2004). Fabrication of miniaturized fluidic devices using SU-8 based litography and low temperature bonding. *Sens. & Act.*, A115: 591–599.

86. Lin, C.-H., Lee, G.-B., Chen, S.-H., and Chang, G.-L. (2003). Micro capillary electrophoresis chips integrated with buried SU-8/SOG optical waveguides for bio-analytical applications. *Sens. & Act.*, A107: 125–131.

87. Sant, J. H., and Gale, B. K. (2003). An integrated optical detector for microfabricated electrical field flow fractionation system. In *Proceedings of Micro TAS 2003*, Squaw Valley, CA, USA, October 5–9, pp. 1259–1262.

88. Friis, P., Hoppe, K., Leistiko, O., Huebner, J., Kutte, J., Wolff, A., and Telleman, P. (1999). Integrated optics for bio/chemical systems. *Proc. Eurosensors XIII*: 565–567.

89. Zimmermann, S., Wischhusen, S., and Mueller, J. (2000). A μTAS—atomic emission flame spectrometer (AES). In *Proceedings of Micro TAS 2000*, Enschede, The Netherlands, May 14–18, pp. 135–137.

90. Eijkel, J. C. T., Stoeri, H., and Manz, A. (2000). An atmospheric pressure plasma on a chip emission detectors applied as a molecular in gas chromatography. In *Proceedings of Micro TAS 2000*, pp. 591–594.

91. Schasfoort, R., Guijt-van Duijn, Schlautmann, S., Frank, H., Biliet, H., van Dedem, G., and van der Berg, A. (2000). Miniaturized microcapillary electrophoresis system with integrated conductivity detector. In *Proceedings of Micro TAS 2000*, pp. 391–394.

92. Fuller, C. K., Hamilton, J., Ackler, H., Krulevitch, P., Boser, B., Eldrege, A., Becker, Yang, J., and Gasconye, P. (2000). Microfabricated multi-frequency particle impedance characterization system. In *Proceedings of Micro TAS 2000*, pp. 265–268.

93. Shew, B. Y., Kuo, C. H., Huang, Y. C., and Tsai, Y. H. (2005). UV-LIGA interferometer bio-sensor based on the SU8 optical waveguide. *Sens. & Act.*, A120: 383–389.

94. Gorecki, C. (2000). Optical waveguides and silicon-based micromachined architectures. In P. Rai-Choudhury (ed.), *MEMS and MOEMS Technology and Applications*. Bellingham, WA: SPIE Press.

95. Kurihara, K., Iwasaki, Y. C., Niwa, O., Tobita, T., and Ito, S. (2002). Fiber-optic microdevices for surface resonance applied to microfluidic devices. *Proc. Micro TAS*, 1: 269–271.

96. Golonka, J. L., Roguszczak, H., Zawada, T., Radojewski, J., Grabowska, I., Chudy, M., Dybko, A., Brzozka, Z., and Stadnik, D. (2005). LTCC based microfluidic system with optical detection. *Sens. & Act.*, B111–112: 396–402.

97. Golonka, J. L., Zawada, T., Radojewski, J., Roguszczak, H., and Stefanow, M. (2006). LTCC microfluidic system. *Int. J. Appl. Ceram. Technol.*, 3(2): 150–156.

98. Wasilek, P., Golonka, L., Gorecka-Drzazga, A., Roguszczak, H., and Zawada, T. (2003). LTCC liquid conductivity detector. In *Proceedings of 26th International Spring Seminar on Electronics Technology*, High Tatras, Slovakia, pp. 202–206.

CHAPTER 10

AUTOMATING MICROPROCESS SYSTEMS

THOMAS MÜLLER-HEINZERLING

10.1 AUTOMATION: WHY?

The last years have seen a strong increase in the interest in microstructured process systems. This is mirrored by a growing number of conferences and publications on the subject.

However, most microprocess systems are operated today with zero or little automation. This is partly due to a lack of apt sensors and actuators, but it also indicates the state of this field that is still in an early phase of development.

Of course, the first question has to be: Why should I automate my microprocess system set-up at all? Running it manually is cheaper, easier, and has worked for me until now—why should I change? Some answers may be derived from the general field of process automation, where similar questions were asked decades ago and are still being asked today.

In general, the following factors often justify automating a process:

1. *Reproducibility of Process Control* Eliminating differences in the quality of process control across shifts, times of year, or load situations normally improves product quality and reduces power consumption.

2. *Optimization of Process Working Point* Conducting the process closer to an optimum working point may increase product yield or quality, and reduce waste.

3. *Automated Reactions to Process Upsets* Fast and reliable reactions to process upsets may reduce the risks of product deterioration or loss, and prevent the activation of safety shut-downs. In hard-to-control processes, the operation close to an optimum is often only possible through automation.

4. *Reduction of Operator Load* Freeing operating personnel from monotonous, repetitive, or even dangerous tasks enables them to control larger parts of a plant and to concentrate on superordinate objectives.

5. *Longer Operating Times* In many cases, only automation makes it viable to keep processes running around the clock for weeks or years without the need for continuous human supervision.

6. *Automatic Recording and Archiving of Process and Operator Behavior* Through automation, the automatic recording of measured values and of process alarms, operator acknowledgments, and interventions is achieved as a "side effect." Long-term archiving enables later analysis as a basis for further improvements.

7. *Integration/Coordination with Other Equipment* The interplay of different pieces of equipment can be enabled or facilitated through automatic data exchange.

Let us check to what extent these arguments can be applied in the case of microprocess systems:

1. *Reproducibility of Process Control* Yes—this is definitely desirable for microprocesses as well!

2. *Optimization of Process Working Point* Again, yes—it is one of the well-known advantages of microstructured systems that tighter process control is made possible through their high surface-to-volume ratio and intense mixing. To make full use of these possibilities, however, the externally controllable variables have to be controlled tightly as well!

3. *Automated Reactions to Process Upsets* Although this may be less crucial compared to conventional processes in view of the small hold-up volumes involved in microprocess systems, this feature is still desirable in order to prevent damage to the equipment or spoiling an experiment/production run.

4. *Reduction of Operator Load* Today, the typical microprocess is run in a laboratory, so the term "operator" should probably be replaced by "laboratory worker," but reducing those individuals' workload is no less valuable. In addition, the first production applications for microprocesses have started to show up already, and especially when numbering-up concepts start to gain ground, this aspect will be as important as in a conventional process.

5. *Longer Operation Times* Another definite yes—whether in laboratory experiments (with an automated parameter screening) to get faster results or in production, where continuous operation is always a prerequisite.

6. *Automatic Recording and Archiving of Process and Operator Behavior* Again, yes—in laboratory experiments, these data (together with the analysis of the end product) may actually be the most important result, more so than the produced material!

7. *Integration/Coordination with Other Equipment* Although not directly obvious, this is also required in both the laboratory and production. Each

microsystem needs to interact with its macro environment (raw material tanks, product storage facilities, heating and cooling supply, etc.). Using standard automation tools makes it easy to achieve such integration. This will become even more important in the case of production applications with numbering up and integration in an existing production environment. A third future scenario might be the use of microprocess systems for analysis purposes in "normal" processes.

To sum up: Practically all aspects of automation are useful for microprocess systems, too.

Of course, the effort to achieve automation has to occur in proportion to the resulting benefits, so the question is how to get a microprocess system automated with adequate effort. Can we just use standard process control systems from normal process automation for microprocess systems as well? Or is the development of dedicated automation solutions just for microsystems required and feasible?

A second question is whether there are specific requirements for the automation of microprocess systems not encountered in "normal" process automation. Some special requirements are indeed caused by the specific properties of microprocess systems:

- Due to the small volumes, some process values will have very fast change rates that are not easily covered by standard automation solutions → some dedicated, fast control may be needed.
- Automation needs sensors and actuators, but in view of the sizes of microprocess systems, commercially available process instrumentation is not really applicable. This is due to their sizes, but also to the dead volumes caused by the "macro" connections → new generations of miniaturized sensors and actuators including appropriate connection technologies will have to be developed.
- A laboratory set-up will be changed much more often than in a production process, and in the laboratory workforce there are no automation system specialists available → an easy set-up and change of the automation functions without deep automation knowledge are required.

The thesis of this chapter is that standard process control systems—with a few enhancements and in the right configuration—are indeed suited to the task of automating microprocess systems.

In the chapters that follow, some examples of emerging solutions to these specific requirements and answers to the question about automation with adequate effort will be presented.

10.2 SENSORS AND ACTUATORS FOR MICROPROCESS SYSTEMS

10.2.1 Conventional Process Instrumentation

For some microprocess systems, small conventional field devices may be used. As opposed to laboratory devices, these offer the reliability and long-term stability

FIGURE 10.1 Example of a miniaturized field device: Siemens' MASS 2100 DI 1.5.

required in process applications. An example would be the MASS 2100 DI 1.5, the smallest member of Siemens' family of Coriolis mass flow meters SITRANS F C MASSFLO. With a pipe diameter of 1.5 mm, head diameter of 129 mm, and measuring range of 50 to 65 kg/h, it is applicable in small-scale plants. The Coriolis principle has the advantage that it measures the true mass flow (not only volume flow) and that it does not require calibration. See Fig. 10.1.

10.2.2 Miniaturized Sensors

However, to make full use of the capabilities of microprocess systems, specialized new sensor designs are needed. For example, to measure the pressure in a microprocess system, using a conventional pressure transducer would be rather impractical: The dead volume of the device itself would typically be larger than the total working volume of the microprocess system.

To solve this problem, Siemens has developed a totally new in-line micropressure sensor (Fig. 10.2). It is smaller than a 1 cent coin and includes the piezo-based sensor itself and electronics.

The dead volume is practically zero because of the in-line construction. The set-up (Fig. 10.3) consists of a fluidic channel between an inlet and an outlet port that is filled by the fluid and a circular diaphragm between the ports that is deflected by the fluid pressure and acts as a separation membrane between chemistry and electronics. To convert this deflection into an electrical signal, an additional chip similar to a standard pressure sensor is mounted to the outside of the membrane such that any deflection of the separation membrane is transferred mechanically to the sensing chip through a central boss structure and converted to electrical signals

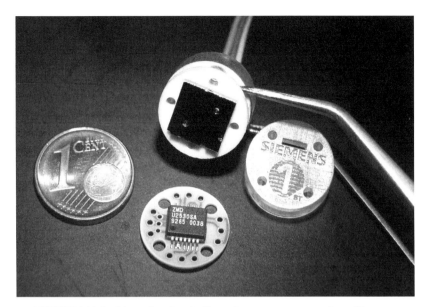

FIGURE 10.2 Example of an in-line micropressure sensor from Siemens.

by piezoresistors in a Wheatstone bridge. In its present form, it can measure up to 30 bar, but adaptation to other measuring ranges would be easily possible.

10.2.3 Future Developments

Many more miniaturized sensors and actuators have been demonstrated as prototypes and would be interesting candidates for the automated control of microprocess systems. The challenge here is the transition from a laboratory stage to industrial production and stability of the devices.

   The need also exists for a reduction in the size of the connections, combined with low dead volume and easy assembly in the field. All these implementations would help to reduce size, minimize hold-up, and increase the efficiency of the system. The development of bulk materials, of stable coatings, and of fabrication techniques for these devices so they withstand chemical attacks on the fluid contacting surfaces remains a future challenge.

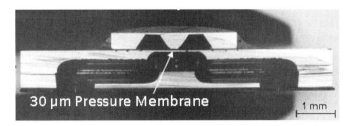

FIGURE 10.3 Internal structure of the micropressure sensor.

10.3 TYPICAL ARCHITECTURES AND FUNCTIONALITIES OF CONTROL SYSTEMS

The basic aspects of a control system are the same, independent of the size of the automated system or type of controlled process. The purpose of this chapter is to describe such basic features. This should be helpful for anyone wanting to understand how a microprocess system can be automated, or how an existing control system works. We will first have a look at:

- The fundamental automation tasks that are independent of the control system used
- Then implementation issues
- And finally some of the typical automation families available today and applicable to microprocess systems

10.3.1 Fundamental Automation Tasks

In all control systems, certain tasks have to be fulfilled, independent of which hardware structures or software tools are used for realization. We discuss these in the sections below.

Types of Logic   The logic of the control task has to be described and implemented. Examples of such control tasks are as follows:

Open-Loop Control   This refers typically to binary values: The output values of the control system are set based on certain input values.

Example: If the pump switch (input 1) is on and the emergency switch (input 2) is not pressed, the corresponding pump (output 1) must be running and the valve behind the pump (output 2) must be open. If the pump switch is off or the emergency switch is pressed, the pump must be switched off and the valve must be closed.

Closed-Loop Control   This refers typically to analog values: The system tries to move a measured process variable toward a desired value (setpoint) and keep it there by manipulating an output variable.

Example: A fluid has to be kept at 85°C by controlling the current flowing through a heating jacket around the fluid.

Interlocks   These are used to prevent certain actions from happening/certain outputs from being set if the required prerequisites are not met.

Example: The pump from the first example must be switched off (and could not be started in the first place), if a high-level alarm of the target tank occurs (to prevent overfilling) or if the protective cover over the motor and pump (supervised by a contact switch) is removed.

Calculations   These may be prerequisites for open-loop control, closed-loop control, or interlocks.

Example: The pump from the first example must be switched off additionally, if the change rate of the temperature of the pump motor is higher than some limit (indicating an overload situation). For this, the measured temperature has to be differentiated, and the result probably has to be smoothed as well.

Sequences   These serve to automate the execution of several steps one after the other where the stepping is determined by reaching certain conditions.

Example

- First, a tank is filled with a fluid.
- When the tank level has reached the desired value, the fluid is heated up to 85°C.
- After reaching the setpoint, the pump is switched on to circulate the fluid through a filter.
- After a delay of 10 min, the pump is switched off, and the setpoint is lowered to 50°C.
- After reaching the new setpoint, the drain valve is opened to transfer the fluid to the next processing stage.
- When the tank is empty, the drain valve is closed.

In terms of formal descriptions, these control tasks can be solved by different means: human interaction, discrete hardware modules (relays, specialized control electronics, etc.) or a programmable device of some sort [personal computer (PC), programmable logic controller (PLC), distributed control system (DCS), etc.].

Specification Languages   Independent of what solution is chosen, in order to avoid the ambiguities of natural languages, the description of the task should preferably be given in a formalized language. For this purpose, a well-known standard, IEC 61131-3, is useful. Although the standard languages described therein were originally developed for the system-neutral programming of programmable logic controllers (PLCs), they can also be used as solution-neutral descriptions of automation tasks.

The languages described in IEC 61131-3 are:

- *Instruction List (IL)*   Text-based simple command language.
- *Ladder Diagram (LD)*   Graphical language that depicts the logic in the form of electric current flow networks.
- *Function Block Diagram (FBD)*   Graphical language that allows the graphical interconnection of (possibly complex) function blocks.
- *Structured Text (ST)*   Text-based high-level language similar to well-known programming languages such as Pascal.
- *Sequential Function Chart (SFC)*   Superordinate graphical language to describe the sequential execution of steps that are themselves written in one of the other languages.

Typically, specific tasks will be best described by a specific language, although this is not necessarily always clear-cut. The "best" solution also depends on the taste/experience of the specifying person, and it might be influenced by the best mapping to the target system (if this is already determined).

Here are some examples of typical applications of the languages for certain use cases: Open-loop control tasks can easily be described in IL or LD—but in certain control systems, the graphical handling of function blocks is the preferred way, and then it is advantageous to describe such tasks in FBD or the system's equivalent. If there is a very complex logic involving, for example, the conditional execution of different branches, ST might be better suited.

Closed-loop control tasks will normally be best described in FBD. However, in some systems, programming is only possible in textual form, and such a task can be solved by calling predefined functions (e.g., for a closed-loop controller) from IL. Then it might be easier to use this form of specification from the start. Again, for very complex cases or if the predefined function blocks do not offer the required functionalities, ST might be used.

Interlock control tasks will naturally be described in FBD using AND and OR function blocks. However, it is also possible to use IL or LD for this, either to make mapping to text-based systems easier or because the specifying person tends to think in these terms.

It is easiest to describe calculations in FBD using math function blocks. However, in complex cases it might be better to use ST for this, for example, if complex math is needed or if constructs like IF-THEN-ELSE are needed.

Sequences will best be described in—as the name suggests—SFC, and if the target system supports this language, it is certainly a good idea to use it for such a purpose. Again, other languages could be used. However, this makes it in most cases more difficult to understand what the original intent was. Thus, in this case, using the correct specification language is especially important and helpful, both to avoid specification errors and to facilitate implementation.

Example   Let us look at one rather simple example in order to see how this could be expressed in different specification languages. Disclaimer: The following examples are only meant to give you a rough idea of these languages and are not formally correct in every detail; full details and more background information may be found in the ample literature on IEC 61131-3.

Example in IL   The example for open-loop control was previously given on page 250. The description in Instruction List would be:

> LD   Pump Switch
> AND   NOT (Emergency Switch)
> ST   Pump On
> ST   Pump Valve Open

Comments and Explanations   This language is mainly useful for the description of binary logic. It may seem rather primitive, but it is widespread among automation programmers. LD is short for LOAD; it means that some input value is read into the "register" that contains the current result of the evaluation. AND and NOT are the well-known logical operators. ST is short for SET. An output is set to the current result of the preceding evaluations; this can be repeated in order to set a second output to the same value. The switching off of the outputs need not be specified explicitly. It is an implicit consequence of the execution rules: If the cyclic evaluation of the first two lines evaluates to FALSE, the outputs are set to FALSE, too.

Example in LD The description of the same example in a Ladder Diagram would look like this:

```
|          _____                                  |
|-- ]Pump  Switch[--]Emergency  Switch[-┬-(Pump On)--------- |
|                                       └-(Pump Valve Open)-- |
|                                                            |
```

Comments and Explanations This semigraphical language is for people who think in terms of electrical flow diagrams. The vertical lines on the left and the right are the "power rails" that are horizontally connected by "contacts" [in other words, logical inputs, written in the form "(xxx)"] and "indicators" [in other words, logical outputs, written in the form "(yyy)"]. If the correct contacts are "closed" (in other words, the inputs are TRUE), the indicators come "ON" (in other words, the outputs are set to TRUE). The negation of a "contact" is indicated by a line over the contact name—as for the EmergencySwitch signal above. AND is expressed by having the contacts in sequential order; OR would be expressed by placing the contacts in parallel.

Example in FBD The description of the same example in a Function Block Diagram would look like this:

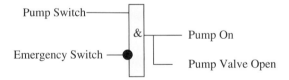

Comments and Explanations This graphical language is broadly used in control systems and can be used for both binary and analog values. There are standard functions for binary and mathematical functions, like the AND used above. The negation is here expressed by the dot. The concept lends itself also to much more complicated uses, for example, to encapsulate an entire technological function such as a closed-loop control in a single function block.

Attempts have been made to define a standard set of function blocks for such technological use cases, but most control system suppliers offer additionally a large number of their own function blocks. With some systems, the supplier-defined library of function blocks can be extended using one of the other languages to create user-defined function blocks.

Example in ST   The description of the same example in Structured Text could look something like this:

> Pump Switch AND NOT(Emergency Switch) =: Pump On
>
> Pump On =: Pump Valve Open

Comments and Explanations   This text language can be used for complex tasks; it offers constructs like IF-THEN-ELSE or WHILE-DO, etc. It is offered by most modern control systems. The "danger" of using ST lies—as for any high-level programming language—in the possibility of creating very complicated structures that are difficult to understand and maintain. If the required result can be achieved using graphical languages such as FBD or SFC, this should be preferred.

Example in SFC   The description of the same example in Sequential Function Charts could be done as well, but this would be a rather inappropriate application of SFC. It would look like this:

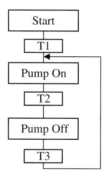

The internal structure of the transition T1 could be

> Pump Switch .EQ. TRUE AND Emergency Switch .EQ. FALSE

The internal structure of the step Pump On could be

> 1 =: Pump On =: Pump Valve Open

The internal structure of the transition T2 could be

> Pump Switch .EQ. FALSE OR Emergency Switch .EQ. TRUE

The internal structure of the step Pump Off could be

> 0 =: Pump On =: Pump Valve Open

The internal structure of the transition T3 would be like T1.

Comments and Explanations   This graphical language consist of steps (the larger boxes) and transitions (the smaller boxes). In a step, commands are set; in a transition, conditions are checked. But not all steps and not all transitions are active at the same time. After start-up, the start step is the active step. The transition(s) (there could be branches) after the active step is (are) checked; if a transition is true, the following step(s) becomes the new active step(s) (again there could be branches), and the competing transitions are not checked any longer. Instead, the transition(s) after the new active step(s) are checked. In this way, the process state is tracked/mirrored in the state of the SFC, and state-dependent commands are issued.

Implementation   The implementation of such control tasks in a given control system should be easy if the specification was done in one of the standard ways described above. Actually, utilizing modern control systems means that as part of the standard software, an editor is supplied that allows the user to describe the task in one or several of these languages (often including automatic syntax checks, version management, etc.), which is then automatically compiled to (or interpreted in) the internal language of the chosen system. Thus, especially in the case of graphical languages, programming is substituted by "configuring."

Visualization and Operation   Another important task of process control systems is the visualization of the process. Of course, it is not so much the display of the mechanical layout of the plant, but the display of live process values that is important to the user.

But beyond the sheer presentation of actual values, additional information can be displayed: trends, live execution of sequences or interlocks, characteristic curves, operating points, alarms, etc. And then, of course, it is normally required that the operator should be able to influence the running process by changing setpoints, executing switching operations, etc.

The design of the static part of the graphic displays is done today with Windows-like drawing programs. Especially in the case of a microprocess system, this is not a major task because normally only one single display is required (as opposed to large production plants which required dozens or hundreds of displays). The inclusion of live values into the displays can however present more work, depending on the control system.

Offering the possibility of entering a new setpoint or a command brings additional questions to mind: Does every operator/user have the same rights? Can such entries be tracked? If so—by time of initiation or even to the individual who entered them? Especially if the system is used in regulated industries, there is a long list of expectations from the regulating authorities regarding security against uncontrolled access, management of access rights, audit trails, etc.

To facilitate the engineering of all these types of displays, certain predefined displays are normally delivered as part of the standard software of a control system. If the logic has been implemented in an object-oriented way (typically achieved by using FBD or SFC), the link between the graphical representation and logic can be created automatically by assigning a predefined "faceplate" (a collection

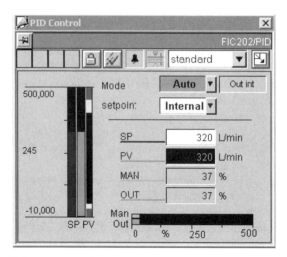

FIGURE 10.4   Example of an operator faceplate for a closed-loop controller.

of all operator-relevant data shown in a logical grouping) to each operator-relevant object in the logic. For example, in the Siemens control system SIMATIC PCS 7, the following faceplate (Fig. 10.4) is supplied automatically for each closed-loop controller. To enable access to the less often used special parameters of the same object, additional "views" of the same faceplate can be called by the operator (Fig. 10.5).

In addition, the handling of alarms is an important subject. During the configuring of the system, alarms have to be defined and classified. At run time, they have to be displayed, sorted, acknowledged, and stored. For this purpose, standard alarm displays are normally created by control systems (Fig. 10.6). Trends are also an

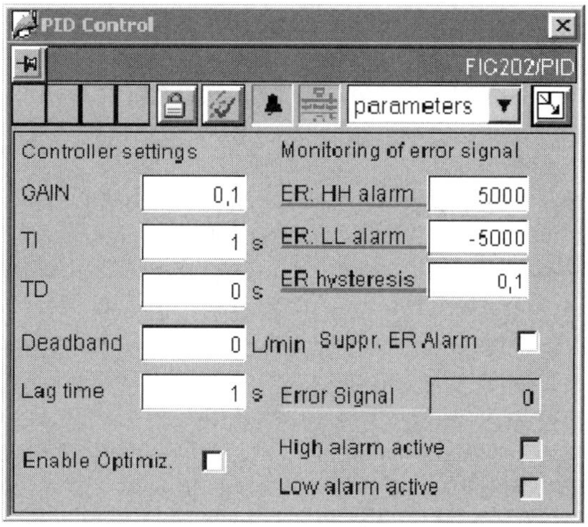

FIGURE 10.5   Additional view of the closed-loop controller faceplate: controller parameters.

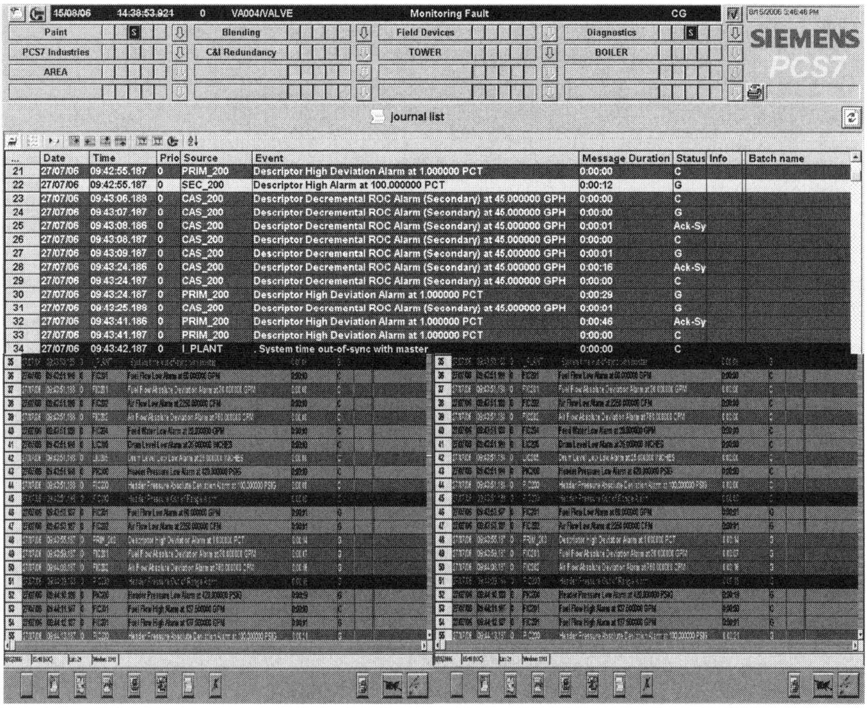

15/08/06 14:38:53.921 0 VA004/VALVE Monitoring Fault CG
Paint
Blending
Field Devices
Diagnostics
PCS7 Industries
C&I Redundancy
TOWER
BOILER
AREA
SIEMENS
PCS7
journal list
Date Time Prio Source Event Message Duration Status Info Batch name
21 27/07/06 09:42:55.187 0 PRIM_200 Descriptor High Deviation Alarm at 1.000000 PCT 0:00:00 C
22 27/07/06 09:42:55.187 0 SEC_200 Descriptor High Alarm at 100.000000 PCT 0:00:12 G
23 27/07/06 09:43:06.188 0 CAS_200 Descriptor Decremental ROC Alarm (Secondary) at 45.000000 GPH 0:00:00 C
24 27/07/06 09:43:07.187 0 CAS_200 Descriptor Decremental ROC Alarm (Secondary) at 45.000000 GPH 0:00:00 G
25 27/07/06 09:43:08.186 0 CAS_200 Descriptor Decremental ROC Alarm (Secondary) at 45.000000 GPH 0:00:01 Ack-Sy
26 27/07/06 09:43:08.187 0 CAS_200 Descriptor Decremental ROC Alarm (Secondary) at 45.000000 GPH 0:00:00 C
27 27/07/06 09:43:09.187 0 CAS_200 Descriptor Decremental ROC Alarm (Secondary) at 45.000000 GPH 0:00:01 G
28 27/07/06 09:43:24.186 0 CAS_200 Descriptor Decremental ROC Alarm (Secondary) at 45.000000 GPH 0:00:16 Ack-Sy
29 27/07/06 09:43:24.187 0 CAS_200 Descriptor Decremental ROC Alarm (Secondary) at 45.000000 GPH 0:00:00 C
30 27/07/06 09:43:24.187 0 PRIM_200 Descriptor High Deviation Alarm at 1.000000 PCT 0:00:29 G
31 27/07/06 09:43:25.188 0 CAS_200 Descriptor Decremental ROC Alarm (Secondary) at 45.000000 GPH 0:00:01 G
32 27/07/06 09:43:41.186 0 PRIM_200 Descriptor High Deviation Alarm at 1.000000 PCT 0:00:46 Ack-Sy
33 27/07/06 09:43:41.187 0 PRIM_200 Descriptor High Deviation Alarm at 1.000000 PCT 0:00:00 C
34 27/07/06 09:43:42.187 0 I_PLANT . System time out-of-sync with master 0:00:00 C

FIGURE 10.6 Alarm display.

important aspect of visualization. The development of a measured value over time is often very valuable information to judge the behavior of the process. Again, standard displays are normally offered (Fig. 10.7). Finally, internal system settings of the control system may need to be accessed and manipulated by the user (e.g., clock setting, dialog language setting, start of archiving activities, etc.).

Data Management/Archiving The final but also very important aspect of a control system's tasks relates to the management of data. These could be data

TIC
90.8
79.0
62.0
45.0
28.0
11.0
-6.0
-23.0
-40.0
08/15/06
3:52:00 PM
3:53:00 PM
3:54:00 PM
3:55:00 PM
3:56:00 PM
Trend in the foreground Red_IO/TICos.U

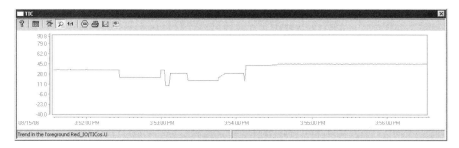

FIGURE 10.7 Trend display.

that the system has collected during its operation: for example, trends, alarms, and operator inputs. It must be possible to display, analyze, print, or export these data.

These could also be data that need to be entered in the system for the automatic execution of several consecutive experiments/production runs in the context of a parameter screening; we will call these "schedules." They could include:

- Number/name of the individual experiment
- Timeframe
- Parameter set/formula (sometimes, the term "recipe" is used in this context; however, this term is defined in the area of batch control (IEC Standard 61512-1) slightly differently; therefore, the term "formula" is used here.)

Thus, for an automated microprocess system, it should be possible to create and manage schedules and their related data including formulas. Of course, these data should be printable, exportable, and reimportable, too.

10.3.2 Implementation Issues

Although we discussed some principal features and their implementation-independent specification in Section 10.3.1, additional aspects have to be taken into account for a real implementation.

Scanning Frequency for Continuous Functions   Most of the functions in a control system (open- and closed-loop control, interlocks, calculations, etc.) have to be executed "continuously." However, using a programmable digital system implies that real execution is cyclical with a finite interval between executions.

A famous mathematical theorem, the sampling theorem, states that the effects of this finite scanning speed in further processing are negligible if the scanning frequency is notably higher than the fastest change frequency of the scanned signal.

Obviously, it is a good thing if the scanning frequency of each function can be adjusted individually. Otherwise, the fastest signal determines the required scanning frequency of the entire program.

In a conventional process automation task, most signals do not change very rapidly—temperatures and levels mostly slower than 1 s, pressures and flows mostly slower than 200 ms. In microprocesses, however, faster change rates can occur. These aspects have to be taken into account when designing an automation solution.

Execution Sequence for Continuous Functions   The sequential execution of individual functions during one scan cycle may lead to additional effects. Of course, the different parts of the whole automation task can only be executed one after the other. This execution sequence has to be chosen—either by the programmer or by the system itself. If this sequence is chosen in an infelicitous way, it could have adverse effects.

Example   If a calculation is executed after the control task that needs the result of the calculation as an input, the system will always work with values that are "one cycle old." For a slowly changing analog value this is not critical, but in binary calculations with feedback loops this may lead to unexpected results.

Execution Rules for Sequential Functions   Sequential function charts (SFC) were introduced above. They are an intuitively understandable tool to describe the sequential execution of functions ("sequential" here means from a process point of view, not from a system point of view, so the individual process steps are typically active for many system cycles).

The finite execution speed and sequential execution in real systems lead to questions here, too:

- Can the signal to switch to the next step get "lost," if it is TRUE only for a very short time?
- Can several steps of an SFC be executed during one scan cycle?

The answers to these questions are system-dependent and may depend on the details of realization.

Ergonomics   This aspect of implementation refers mostly to the visualization. There are "hardware aspects" such as:

Screen resolution (standard today is 1280×1024)

Display quality (standard today is an LCD screen; if CRT is still used, the screen refresh cycle should be at least 70 Hz)

Avoidance of reflections (do not place the screen opposite or in front of windows, etc.)

There are "software aspects" such as:

Minimum character size

Easily readable fonts

Recommended use of colors

And there are logical aspects such as:

Screen segmentation

Operator guidance in case of alarms

Standard navigation through hierarchies

Standard placement of windows

This is a wide area, and there is a lot of corresponding literature. However, it is mostly targeted at conventional process automation applications; in the case of microprocess systems, any problems in this area are usually not as crucial as in large production applications because of the smaller number of signals and displays. Using a standard system that has been designed for similar uses should help to implement a user-friendly solution.

10.3.3 Automation System Families

As mentioned above, control tasks can be solved by different means: human interaction, discrete hardware modules (such as relays, specialized control electronics, etc.), or programmable devices of some sort. In classical process automation, human labor and discrete automation components have been largely displaced today by programmable systems. These are either the so-called distributed control systems (DCS) or, for smaller or less complex cases, programmable logic controllers (PLC) with a superordinate PC-based visualization (supervisory control and data acquisition systems or SCADA).

In the laboratory environment, discrete modules and/or the PC are today the typical tools. The discrete solution has obvious disadvantages regarding data acquisition and management, but also regarding the possibility of interconnecting individual parts. Therefore, for somewhat more complicated tasks, PCs with specialized software packages are used in this environment, where the somewhat lower dependability of such a solution compared to DCS / PLCs is acceptable.

For microprocess systems, solutions from both worlds are applicable (and are being applied). Their respective advantages and disadvantages are discussed in the following sections.

Industrial Process Control Systems So-called distributed control systems (DCS) originated from the background of large and complex process plants; they have been on the market since the late 1970s. As a tool for smaller or less complex process automation tasks, programmable logic controllers (PLC), together with a superordinate PC-based visualization, supervisory control and data acquisition systems (SCADA), have gained importance since the 1990s. Originally developed for machine control in discrete production, PLC/SCADA solutions have grown to encompass more complex requirements, making it possible to apply them to process control. On the other hand, vendors of DCS systems have over the years strived to make their systems smaller and more economical, and the two worlds are merging today.

Typical Features of Such Systems

Signal acquisition occurs through standard industrial input/output cards (4–20 mA, 0–10 V) or via a field-bus.

The logic is executed in special hardware (controller or AS) to ensure the highest reliability and availability.

Visualization and data management are done by standard software packages (SCADA or OS) on a PC (respectively, for large installations, in a number of PCs, typically in a client-server architecture).

Advantages

Have all the features expected from process automation (e.g., integrated visualization and archiving, interfaces to other parts of automation environment).

Very stable operation through special, industrial-proven controller.

Input/Output of the same type as in a production environment.

Large libraries of predefined functions.

Integrated visualization and data management.

Can be easily extended/adapted to a pre-/postprocessing environment.

Easy to integrate in a production environment after scale-up or for analysis purposes.

Disadvantages

DCS are typically not suitable for very fast (<100 ms) control tasks.

DCS allow many adjustments and features, which makes handling for a laboratory application complicated. DCS are typically rather expensive.

PLC typically require extensive automation know-how to do the programming. PLC and SCADA do not offer the same integrated approach as DCS.

Examples of DCS systems include SIMATIC PCS 7 from Siemens, Experion PKS from Honeywell, 800xA from ABB. Examples of PLC systems include SIMATIC S7 from Siemens, ControlLogix from Allen-Bradley, Modicon from Schneider Electric. Examples of SCADA systems include WinCC from Siemens, InTouch from Wonderware, automationX from AutomationX GmbH.

PC-Based Control Systems   These systems come from the laboratory environment; they have been available since the 1990s.

Typical Features of Such Systems

PCs with a Windows operating system are used.

Signal acquisition occurs through dedicated PC input/output cards.

The logic, visualization, and data management are done by special software packages on a PC.

Advantages

Large libraries of predefined functions

Integrated visualization and data management

Lower cost due to the simpler hardware

Relatively easy handling

Disadvantages

Typically not suitable for very fast (<100 ms) control tasks.

Do not support all features expected from process automation.

Input/Output is not of the same type as in a production environment.

Difficult to integrate in a production environment after scale-up or for analysis purposes.

Examples are LabView from National Instruments and LabVision from HiTec Zang.

10.4 EXAMPLES OF AUTOMATED MICROPROCESS SYSTEMS

There are a number of suppliers for microprocess systems on the market, for example, CPC Systems, Ehrfeld Mikrotechnik BTS, Mikroglas, Syntics, and Syrris. In the following examples of current systems, we illustrate the two different approaches to automation described above.

10.4.1 mikroSyn from Mikroglas

This system is available as a ready-to-run system in different versions (with one or two reactors, without or with integrated IR analytics). See Fig. 10.8 and [1]. The control system used is a PLC/SCADA solution (SIMATIC S7 and WinCC from Siemens) (Fig. 10.9). The S7 controller consists of:

One S7 CPU 315-2 with integrated interface for the bus system PROFIBUS DP One CP 340 communication card for serial coupling to the syringe dosing subsystems

One digital input card for up to 32 binary signals

One digital output card for up to 32 binary signals

Two analog input cards for up to 16 thermocouples

Two analog input cards for up to 16 analog signals

One analog output card for up to four analog output signals

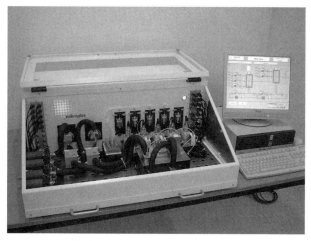

FIGURE 10.8 mikroSyn system from mikroglas for research on catalytic reactions.

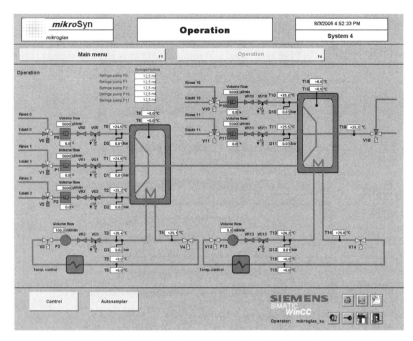

FIGURE 10.9 Automation of the mikroSyn system from Fig. 10.8 with Siemens' SIMATIC S7 and WinCC.

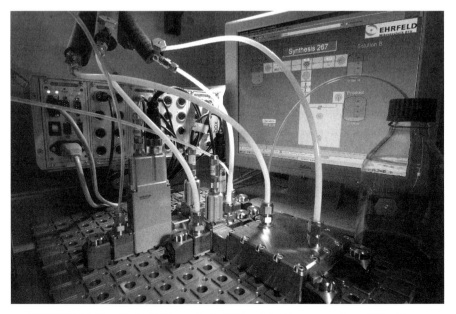

FIGURE 10.10 Ehrfeld Mikrotechnik BTS and LAB-manager from HiTec Zang.

The SCADA system is a Siemens PC Esprimo with Windows XP running the WinCC package.

In a two-reactor system, there are:

Two temperature controllers (one per reactor) including a circulation pump for each and control of the external heating/cooling system

Control of the five educt syringe pumps via a serial link, five miniature valves for educt selection and cleaning, tracking of current dosing speed, dosed volume, and remaining volume

Display of 17 measured temperatures, seven pressures, and two flows

10.4.2 Ehrfeld Mikrotechnik BTS and LAB-Manager from HiTec Zang

This is a modular microprocess system from Ehrfeld Mikrotechnik [2] combined with a PC-based laboratory control system: LAB-manager from HiTec Zang [3] (Fig. 10.10).

10.5 SUMMARY

The automation of microprocess systems is still in its infancy. Nevertheless, existing solutions for the automation of large processes and from laboratory automation can be adapted and used.

For pure research applications, manual operation of the microprocess system may still be sufficient. But in many applications, the benefits of automation are notable. The typical results:

- Reproducibility of process control
- Optimization of process working point
- Automated reactions to process upsets
- Reduction of operator load
- Longer operating times
- Automatic recording and archiving of process and operator behavior
- Integration/coordination with other equipment are all welcome improvements that will help to make use of the full potential of microprocess technology.

These aspects will grow in significance, as microprocess systems become a more standard tool and as their use for production applications begins to gain importance.

BIBLIOGRAPHY

Fa. Mikroglas Chemtech GmbH, http://www.mikroglas.com/mikrosyn.htm.

Fa. Ehrfeld Mikrotechnik BTS, http://www.ehrfeld.com/english/

Fa. HiTec Zang, http://www.hitec-zang.de/en/microreaction-systems.html

PART IV

MICROREACTION PLANTS

CHAPTER 11

STRATEGIES FOR LAB-SCALE DEVELOPMENT

DIRK KRISCHNECK

11.1 INTRODUCTION

11.1.1 General Considerations for Lab-Scale Development

One of the fundamental ideas behind micro-chemical engineering is the optimization of mass and heat transfer. This means that microchemical engineering can reduce the distance of a molecule to its reaction partner. Reaction heat can be efficiently added or removed. Microchemical engineering does not influence the kinetics of a reaction. There are indirect possibilities to influence the reaction speed.

Microchemical engineering shifts from time-dependent processes (batch) to place-dependent processes (continuous). The reactions or other unit operations are carried out under steady-state conditions. The four main target values are:

1. Time to process
2. Quality of the process
3. Quality of the product
4. Time to production start-up

The expected main advantages will be improved processing costs and product properties. This chapter focuses on lab-scale equipment use, mainly for chemical reactions. As examples, microreaction modules have been chosen, because the development of other unit operations, for example, down stream processing lags far behind their application in the field of reaction technology. However, the main principles can be used in other unit operations as well.

Microchemical Engineering in Practice. Edited by Thomas R. Dietrich
Copyright © 2009 John Wiley & Sons, Inc.

11.1.2 Microchemical Engineering as a Development Tool

Most companies divide development into two different parts: first, chemical development as done by chemists and then, in a second step, process development as carried out by chemical engineers. These two steps are completed in two different departments, one after the other, and separated from each other. In microchemical engineering, it is useful to connect them. They evolve together and will interact on a permanent basis. This means the process will be developed according to the needs of the chemical reaction. The lab-scale plant will be the development tool for both the chemist and the chemical engineer. The chemist defines the needs for the reaction. As Fig. 11.1 shows, the necessity and importance of pilot-scale experiments will decrease, since a lot of process development can be done on a lab scale. Expensive pilot-scale experiments are reduced to a minimum, because the same dimensions of microstructures are used on every scale. Therefore, microchemical engineering enables fast and cost-efficient chemical and process development and provides easy scale-up [1].

Another important paradigm will result from the microstructured devices used in the process industry: The devices and plant will be designed according to the needs of the process and not the other way around. The technology offers the possibility of shifting from multipurpose plants based on batch technology to completely automated small-scale production plants. This has three distinct advantages.

Processes are very effective, with no compromise to an existing plant or device. Less manpower is needed due to the high automation level and no campaign changes. No unused equipment stands dormant, accruing unnecessary costs. This means that all new developed processes are "intensified." The advantages of large-volume processes can now be reached although for low-volume processes. Even MRT plants can be designed for multipurpose applications in an automated way.

The two philosophies lead to a natural competition. In some cases, the implementation of microchemical engineering as a development tool causes trouble in the analytical department. If an experimental set-up works well, a large number of samples are produced in a very short period of time. In other cases, one or two weeks of

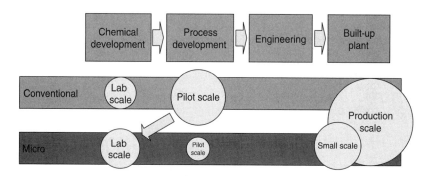

FIGURE 11.1 Principal changes in process development characteristics.

experimental runs produce so many samples that the analytic department needs months to determine all product parameters.

11.1.3 Characterization of the Heat Transfer

If someone wants to carry out an exothermal reaction in a microreactor, it is not only important to know how much heat is released. The time dependence of the heat release must also be known. Otherwise, the product can become overheated even if the heat exchanger is able to swap the overall heat. A continuous reaction calorimeter has been developed by Loebbecke et al [2]. Only very few exothermic reactions need a cooled reactor. In most cases, the reactor and heat exchanger can be set up in a serial way. Residence time of a few milliseconds is usually far too short to overheat the product. But at the pilot and production scale, it is still difficult to exchange the heat exactly at the point where it is needed. The point of maximum heat release is not fixed in the plant and strongly dependent on throughput. That means heat exchangers with more than one compartment are needed for strong exothermic processes. Special considerations for the process design inside heat exchangers are necessary. Special plant start-up and shut-down procedures need to be considered, since the point of heat release moves during these changing conditions. The same method can be used for endothermic reactions. In them, a heat exchanger provides the necessary heat. For strong endothermic reactions, an electrically powered heat exchanger can be beneficial.

11.2 CRITERIA FOR CHOOSING THE CORRECT MRT DEVICE

11.2.1 Characteristics of Liquid–Liquid Systems

Typical Mixing Principles   Microreactors can be used for homogeneous and heterogeneous liquid–liquid reactions [3]. Most types are based on the multilamination concept. This means that thin films of liquid are placed next to each other to optimize mass transport distances by passive mixing. The mixing energy is provided by the pumps. Two different concepts are commonly applied at the moment:

- Parallel multilamination
- Serial multilamination

The use of other concepts based on jet and nozzle principles are less common than these two in the area of microchemical engineering. An upcoming principle could be chaotic mixing, where mixing is provided by direction changes of the fluids. An in-depth look at the mixing principles of micro-chemical engineering may be found in Hessel et al. [4]. The two main principles cited above have a huge impact on applications, yielding special advantages and disadvantages. Micromixers using the parallel multilamination concept generate the multilamination field all at once (Fig. 11.2). The size of the structures is in the area of the size of the lamellae.

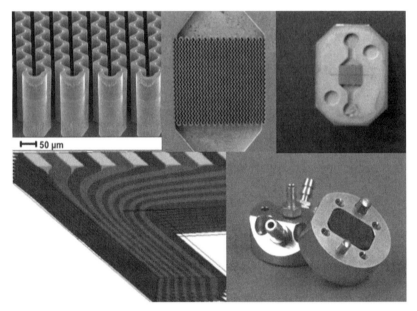

FIGURE 11.2 Parallel interdigital mixing (from Imm Mainz, 2006 with permission [3]).

The interfaces are all the same age. Since the structures are fine, typically between 10 and 250 μm, the process media need to be filtered to avoid any blocking of the microstructure. It is not possible to operate these mixers if the media contain any solids. Therefore, the progress of the reaction is equal in the lamellae field. A further improvement in mixing may be acheived by compressing the lamellae into each other. For example, the superfocus mixer uses this principle [5, 6].

The second important concept, serial multilamination, is often referred to in the literature as split-and-recombine [7, 8]. In this scenario, the lamellae field is generated step by step. Every step doubles the number of lamellae and to half its size, as shown in Fig. 11.3. The size of the structures is much larger than the final lamellae and they can usually cope with some solids in the feedlike nano particles. The disadvantage is that the interfaces are not of the same age and therefore a different reaction progress results. This can cause a loss in performance, especially in complex systems containing following reactions. But not only micromixers use the split-and-recombine principle. Conventional static mixers can be used in a similar way for some homogenous applications. One major difference is that static mixers carry out their split-and-recombine in a turbulent regime, while micromixers usually operate in a laminar regime. For two-phase applications, the turbulent field causes dispersion, whereas the laminar regime offers the possibility of processing both continuously.

Mixing Quality The mixing quality is for some reactions essential. Others are not very sensitive to it. The kinetics of a reaction determines the necessary mixing quality. The fastest reaction needs the highest mixing quality. That means microreaction

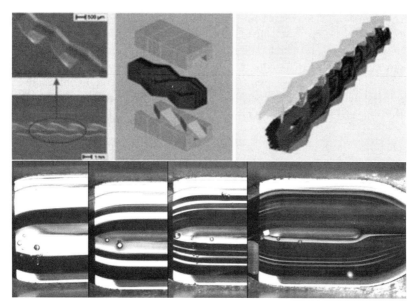

FIGURE 11.3 Serial interdigital mixing (from Imm Mainz, 2006 with permission [3]).

technology is usually very beneficial for fast reactions. The most common method to determine the mixing quality is the Villermaux–Dushman reaction [9]. In the reaction, the amount of iodine corresponds to mixing quality: More iodine means less mixing quality. Some manufacturers deliver performance data for their microstructured devices according to this reaction. Due to method differences, it is not always possible to compare values measured by different groups. To make this comparison easier, a segregation index is used to characterize micromixing phenomena and to quantify them [10, 11]. The segregation is the ratio between the actual micromixing quality and the maximum micromixing quality. The segregation index can have values between 0 and 1, where 0 is perfect micromixing and 1 total segregation. Some applications are quite tolerant of mixing performance as long as basic performance is reached. That means not every reaction requires the maximum possible mixing quality. Handling is as important as the mixing quality—sometimes even more important.

Not very much work has been done to characterize different phase ratios and higher viscosities. Tekautz et al. [12] show that the mixing quality gets worse when utilizing different phase ratios or higher viscosity. There also seems to be a trend suggesting that the mixing quality improves if the velocities of the mixed liquids in the microchannels are equal. One has to use different channel widths to achieve this when different phase ratios are used. Such a result has been obtained with the StarLaminator, in which the channel width can be changed easily due to its foil assembly. Work is ongoing in this field.

Experience has shown that the new possibilities of continuously operating lab-scale plants requires a shift in the development process. New parameters like residence times offer new possibilities for efficient reaction technological solutions worked out by the chemist and the chemical engineer. More aggressive reaction conditions for pressure and temperature are usually chosen for the microreactor. Pennemann et al. [13] review these conditions and give a good overview: Temperatures are usually $10°C$ to several $10°C$ higher than in conventional processes. Even the temperature of cryogenic processes can often be increased, since microchemical engineering provides good heat transfer. Local hot spots can be excluded.

A pressure of several bars is typically applied in these plants. The usage of overpressure offers the possibility of slightly increasing the boiling point of the substances or solvents. This has an indirect impact on kinetics. The rate of reaction can be increased. New process windows can be reached. Concentrations will be lowered a little bit to prevent possible precipitations and to ensure a stable process. Residence times in microreactors are typically milliseconds, sometimes seconds. Residence time tubes provide residence times in the area of seconds to minutes.

Further Topics for Microreactor Evaluation The mixing principle and quality are two important factors to consider in making this decision, but there are a wide range of other factors and no universal solution exists. Handling is a very important factor especially on the lab scale. The following questions will help one choose the correct device for a given application [3]:

- Do I need to open and clean the reactor?
- Do I need to see what is happening inside (transparent material)?
- Is corrosion relevant?
- Is it necessary to adjust the device to solve the problem?
- How can I transfer the results to production scale?

11.2.2 Characteristics of Liquid–Solid Applications

Different research groups have shown that microreactors are useful tools to carry out precipitation reactions. The main target is usually to produce small particles (e.g., polishing powders, pigments, or nano particles) with narrow particle-size distributions. One of the most interesting effects is to distinguish between the nucleation phase that is usually very fast (in the area of milliseconds to seconds) and particle growth. This offers the possibility of separating the nucleation step in a microreactor from particle growth. Depending on the character of the precipitation, the second step can be carried out inside or outside of the microstructure. Different researchers have shown that precipitation can be handled in microstructured devices. Werner et al. [14] have shown how to operate microreactors with precipitation systems. A special type of device was designed using a separation liquid to ensure stable operation. Kirschneck et al. [15] have demonstrated that it is possible to eliminate milling steps, since these steps are safety-relevant because of the risk of powder explosions: It has been shown that the particle size could be reduced by a factor of 10.

Liquid phase

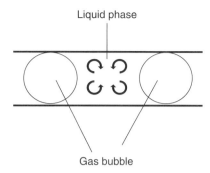

Gas bubble

FIGURE 11.4 Flow pattern generated by gas injection.

A production-scale process is under development. The very narrow particle-size distributions of nano particles can be reached by using segmented flows [16]. During the period of particle growth, it is very important to have perfect plug flow. This ensures the same residence time for all particles. All particles from the nucleation phase require the same time for growth under the same concentration conditions. This can be achieved by injecting gas bubbles into the residence time microchannel. Separation by means of the gas forms different reaction compartments. The friction of the liquid at the channel wall introduces a flow pattern (Fig. 11.4). The gas causes a circular flow inside the flow compartment and thus provides a perfect mixture within it.

11.2.3 Characteristics of Liquid–Gas Applications

Basically, we can distinguish between the two main concepts of liquid–gas microreactors by means of the contacting concept of the governing phases. This contacting concept normally determines the principal characteristics of the microreactors. The main concepts are two continuous-phase contactors without phase intermixing, meaning that both the liquid and gas phases are continuous [17], and dispersive mixing.

Two-Continuous-Phase Contactors Without Phase Intermixing The principle behind this reactor is the wetting of a surface by a liquid stream (continuous phase), governed by gravity force, which thus spreads to form an expanded thin film. In order to guarantee a very thin stable film, this liquid stream is spread to many parallel streams by means of microchannels. The liquid phase is contacted by the gas phase continuously as a co- or counterflow. The type of device used is known as a microfalling film reactor in the literature. The most important parameters are the film thickness, which can be obtained by the microchannel geometry; its material and scale; and the liquid flow rate, typically between 10 and 100 mL/h. It sets the velocity distribution in the liquid face, interfacial area, and surface-to-volume fraction, respectively, which are the dominant parameters for mass transport. Microchannel geometries are typically 35 to 1,500 μm in width, 35 to 1,500 μm in depth, and 10 to 200 mm in length. The number of channels per lab-scale device are 1 to 120. A surface-to-volume fraction of $>20,000 \, m^2/m^3$ can be reached. This type of device is typically operated with residence times between 10 and 30 s and film thickness of a few hundred μm.

Two-Phase Dispersive Mixers Generating Dispersed Phases   These devices have gas and liquid streams that merge by special feed arrangements. From them, both phases flow as coflow in the same encasing, that is, a microchannel. Due to phase instability, fragmentation of the gas jet occurs under given conditions, forming a dispersion. The most important parameters are the interfacial area, which has the same magnitude as falling film, and the velocity of the dispersive phase, which is determined by the flow rate of both phases. The first parameter is set by the gas flow rate, which determines the form and size of the gas phase (bubble flow, slug flow, or annular flow), and the microchannel geometry. Typical velocities are 0.001 to 10 m/s for the liquid phase and 0.1 to 100 m/s for the gas phase. A micro-bubble column is a typical device in this area. However, parallel interdigital liquid–liquid mixing devices can sometimes be used for liquid–gas applications. Loeb et al. [18] have shown that this approach works reasonably well. To adapt these structures, the gas channels need to be 5 to 10 times smaller; otherwise, not every channel is served with gas. A certain pressure loss is needed to reach an equal gas distribution.

Flow Patterns in Microchannels   De Mas et al. [19] have investigated the different flow patterns of liquid–gas flows in microchannels. They have used acetonitril for the liquid phase and nitrogen for the gas phase. Figure 11.5 summarizes the results. Bubbly, slug, churn, and annular flow patterns were found. The velocity of the gas phase is shown on the x-axis marked with j_G, and the corresponding velocity of the liquid phase on the y-axis marked with j_L. In general, annular flow was determined to

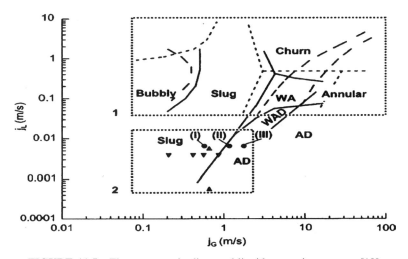

FIGURE 11.5   Flow patterns in dispersed liquid–gas microreactors [19].

be the preferred flow pattern for liquid–gas processes, since it has the largest interfacial area. In the annular flow pattern, the gas forms a continuous cylinder in the middle of the channel and the liquid moves continuously at the surface of the microchannel walls.

All liquid–gas microsystems show significantly higher specific interfacial areas, a factor 10 to 100 higher, than in conventional equipment. Whereas in a conventional type of equipment interfacial areas of 50 to 500 m^2/m^3 are reached, microstructured devices reach several 1,000 to several 10,000 m^2/m^3.

11.2.4 Characteristics of Gas-Phase Reactions in Presence of Solid Catalysts

Reactor Principles and Basic Function Most gas-phase reactions are carried out in the presence of a solid catalyst phase. This is different from experiments in liquid–liquid systems where the common type of experiments has a batch continuous experimental set-up. Keil [20], Kiwi-Minsker and Renken [21], and Hessel et al. [17] provide a good overview of research activities in this field. Microreactors in gas-phase applications can be divided into two classes. The first are the micropacked bed reactors and the second group are the catalytic wall reactors using microchannels. The micropacked bed reactor typically uses catalyst particles in the range of 35 to 75 μm. Since packed beds cause high-pressures losses and difficulty in maintaining a well-defined laminar flow, they are used far less often than the second type. Usually, the microchannel reactor type is prepared with a wall coating and works under laminar flow conditions. The residence time and residence time distribution strongly influence the product yield and selectivity. The contact times are usually very short, from a few ms up to a few 100 ms. Only half of the commercially available catalyst types work in a microstructured device. For the other half of the commercially catalysts, the contact times are too short. The development of special catalysts is necessary.

Preparation of Catalyst Coatings Microchannels are coated with thin layers (1 to 50 μm) prepared mainly by the following methods: a wash coat, sputtering, sol-gel-technology, or chemical vapor decomposition. The most common technique is the wash-coat preparation; Zapf et al. [23] describe it. A suspension of metal oxide carrier is wiped inside the microchannels and dried at room temperature. The wash coat is calcinated at 500 to 600°C where the binder is burned out. In a second step, the catalyst is brought into the microchannels by means of an impregnation method. Common characterization methods are surface imaging, determination of the porosity, cross-sectional profiles, and SEM analysis.

11.3 APPLICATIONS FOR MRT PLANTS

11.3.1 General Remarks on Principal Strategies

The following sections describe the methodology for developing plants containing microstructured devices. They often insights into various development strategies and suggest how to adapt a plant for the specific needs of each reaction. The question is how to adjust a plant for maximum process performance. The aim is to generate a maximum of knowledge given very different operating conditions. The following sections present possible strategies for different applications and offer new solutions for insufficiently solved problems in the process industry. Reviews providing an overview of different applications may be found in the literature [17, 22, 24, 25]. Economic considerations have been described by Roberge et al. [26].

11.3.2 Mass Transfer Enhancement

Mass transfer enhancement is one of the principal ideas in microchemical engineering. As already shown, short and defined distances for the molecules to arrive at the "point of reaction" are wanted. This is of fundamental advantage if the process is mass transport-limited. For example, the ethanol extraction of phosphatidylcholine from highly viscous raw lecithine (with a viscosity of 0.3 to 1 Pa s) can be strongly increased by reducing the mass transfer distances [27]. The increase in heat transfer is responsible for short reaction times. Whereas a reaction in a flask or batch vessel usually takes several hours, the same reaction can be carried out here in seconds to minutes. Pennemann et al. [13] compare the reaction times and conversions of different reactions. The comparison in Table 11.1. shows that a significant change in the magnitude of reaction times may be observed.

This illustrates that microreaction technology delivers a large amout of results in a very short period of time. More experiments can be carried out at the same time, as long as the analytic determination of results does not become a bottle neck.

11.3.3 Heat Transfer Enhancement

One of the principal advantages is the effective heat transfer of exothermic reactions. Fischer–Tropsch synthesis experiences a fundamental problem in the control of generated heat [28]. This lack of heat transfer causes hot spots, where long-chain waxes are formed. These waxes block the catalyst and limit the up-time of the plant. Yoshida et al. have shown that the effective heat transfer of a micromixer can narrow the molecular mass distributions of exothermic polymerizations, while the distributions

TABLE 11.1 Comparison of Different Reaction Times and Conversion According to Pennemann et al. [13]

	Microreactor		Batch Reactor	
	Reaction Time	Conversion	Reaction Time	Conversion
Suzuki coupling	6 s	68%	8 h	60%
Kumada–Corriu reaction	10 min	60%	25 h	70%
Michael addition	20 min	100%	24 h	89%

of nonexothermic polymerizations are not affected in the same set-up [29]. Microreaction technology plants are however a good tool to investigate temperature effects and thermodynamic properties.

11.3.4 Processing of Explosive or Unstable Substances

Janicke et al. [30] explored the controlled oxidation of hydrogen with oxygen (up to 50% by volume) from an explosive mixture of gases in a microstructured reactor/heat exchanger device. The hydrogen was completely converted to water without explosions by maintaining a heterogeneously catalyzed reaction using a Pt/Al_2O_3 catalyst. This showed new possibilities for operating a explosive reaction under controlled and safe conditions.

Microinnova has developed a concept for the production of an unstable intermediate exactly at the point of use [31]. The concept "chemical production anywhere" offers the possibility of producing unstable intermediates shortly before they are needed. The system consists of a completely automated small-scale production plant usually situated at the customer's production site; it can be operated remotely. Concepts such as this offer a wide range of new business opportunities. Since the methods will cause changes in the business, modifications are expected to be slow. This presents an excellent opportunity to choose MRT plants as a production tool for safety-sensitive reactions.

11.3.5 Test of New Chemistry: New Process Windows

Microchemical engineering offers a wide range of possibilities for investigating reactions under transient conditions [17]. This can be temperature or concentration cycles. System response to a shift in hydrodynamic conditions or a sensitivity analysis to define impurities can be carried out very easily. This is mainly done for gas-phase reactions at the present time, but there is no reason not to do things like this in other regimes. Such an approach would offer the possibility of generating fundamental knowledge about a certain reaction and its mechanistic behavior in a limited period of time. New intermediates may be discovered, observing reaction system responses to conditions in an extremely short timeframe. Solvation steps can be determined using continuous FTIR measurement.

11.3.6 Test of New Chemisty: Reaching New Process Regimes and Extreme Conditions

Jähnisch et al. have shown that the direct fluorination of toluene using elemental fluorine can be carried out in liquid–gas microreactors such as a microbubble column or falling film microreactor [32]. On a technical scale, reactions like this are carried out in several steps (via the Schiemann process) since exothermal heat and explosive regimes could not be controlled with conventional equipment. Even on the lab scale, performance in terms of selectivity using the direct route was poor. In the falling film microreactor, 76% conversion to the monofluorinated ortho and para product could be achieved, similar to the results with the Schiemann multi-step process.

Most batch processes are operated at a boiling temperature. Since microreaction set-ups are closed, it is very easy to operate the plant under pressure. This opens a new window of operation due to the higher boiling points needed to speed up reactions. It offers the possibility of optimizing a specific reaction, meaning that using higher temperatures has an indirect influence on process kinetics unlike the common process to speed up the reaction [33].

11.3.7 Ways and Alternative Ways of Providing Energy for Chemical Reactions

Thermal energy is usually provided to carry out chemical reactions. Schwalbe et al. [34] show that the way thermal energy is provided to a reaction can have a significant impact on selectivity. Figure 11.6 describes the impact of the potential energy curve on a reaction system with a side reaction. The potential energy can be correlated to different temperatures, since microtechnology can provide narrow temperature distributions compared to batch processes. This offers the possibility of avoiding side reactions with higher activation energy levels.

Stankiewicz et al. discuss other methods for providing energy to chemical reactions [35]. Since microreaction systems are continuously operated, this energy can be provided in compact flowthrough devices. They explore the process intensification potential of different types of energy and identify five different types of alternative energy and its possible impacts:

- High gravity field (impact on reaction time and on liquid side mass transfer)
- Electric field (impact on interfacial area and on heat transfer)

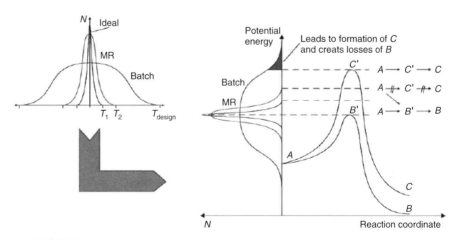

FIGURE 11.6 Impact of the activation energy for side reactions on quality [34].

- Microwaves (impact on reaction time and on distillation time)
- Light (impact on reaction selectivity and yield)
- Ultrasound (impact on reaction time, liquid–gas mass transfer, and liquid–solid mass transfer)

Stankiewicz et al. show several examples width these kinds of energy. The enhancement factors for different impacts are estimated to be between 5 and 1,250. Examples are a spinning disk reactor with 10 times higher local mass and heat transfer coefficients, and microwave-enhanced reactions where specific heat can be applied to chemical reactions.

11.3.8 High Throughput Screening

De Bellefon et al. carried out kinetic investigations of asymmetric hydrogenations (liquid–gas reactions) to discover new chiral catalysts [36]. Batch catalyst tests were compared with tests carried out in a microreaction system. During the same time period, 14 tests were carried out in batch and 214 in a microreaction system, with 11 tests estimated to be significant in batch and 170 in the microreaction system. The maximum number of tests possible per day is one magnitude higher. The reaction volume was lowered from 10 to $0.2\,cm^3$. In addition, to these results with the microreaction system offer the possibility of operating the plant in a completely automated manner.

Microchemical engineering plants offer a lot of possibilities for automation. Plants can be equipped with parameter variation procedures and online analytics. This presents the possibility of high throughput screening. These processes can be optimized without the presence of lab technicians during experiments. Advancing one step further even seems to be within reach: A special bond of the desired molecule is monitored, for example, by IR spectroscopy and an algorithm optimizes (maximizes) the signal for the bond. This means that processes can be developed by means of self-optimizing plants.

11.4 FURTHER ASPECTS OF LAB-SCALE DEVELOPMENT

11.4.1 Lab-Scale Method as Production Tool

Of the chemicals produced in Europe, 60% are needed in amounts of less than 10 tons per year. Small-scale demand is increasing as products become more and more customer-specific and time-to-market is of growing importance. Microchemical engineering offers the possibility of runnig a continuous plant in a lab environment, for example, in a walkable fume cupboard as shown in Fig. 11.7. The advantages compared to those of conventional production methods fall in two areas. First, the lab-scale method can be used directly for production. Scaling factors up to 50 cause minor changes in the plant, which can be done in several hours. This becomes even more important if the process requires special conditions such as

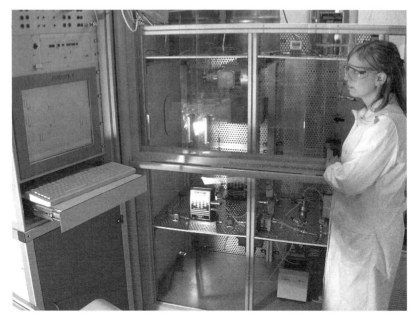

FIGURE 11.7 Small-scale production plant at Microinnova Engineering GmbH.

high pressure or low temperature. The synthesis for phenyl boronic acid could be increased from a cryogenic temperature $-35°C$ to $50°C$ [37]. This example demonstrates the high energy-saving potential of intensified processes.

The second important advantage is the new chance to operate production in a dedicated plant and not a multipurpose plant. The continuous process becomes a serious competitor to multipurpose processes. A problem associated with the multi purpose plant is the fact that only a part of installed equipment is in use. Manpower is needed to run the plant and to then rebuild it for the next product. MRT plants can be designed for multipurpose applications; however, the range of difference in use is usually smaller. An other point to make is that a drop in performance needs to accepted since the equipment has to fulfill more than one process requirement specification. The drop in performance is generally small compared to the benefits resulting from the shift from batch to MRT processes. MRT plants can be equipped with a high level of automation.

Three points need to be considered if a dedicated plant is to be produced in a lab environment:

- Storage of educts and products
- Separation steps that are necessary to clean the product
- Energy supply (heat and cooling)

These three considerations are not very different from those corresponding to conventional production. Separation in microtechnology is still under development and it depends on the application, if it is already available. If separation steps are not available in micro, the plant size increases and you must go to the next level. The next level will be a dedicated plant in a pilot-scale environment or plant in a container. The first and third considerations, storage and energy supply, are just a question of infrastructure. These capabilities are not that common in a lab environment right now, but no principal obstacle exists to installing them there. Figure 11.7 shows such a plant in fume cupboard. For reasons of infrastructure, it may be useful to install such a system in a pilot plant environment, since the infrastructure needed is already presented. If possible, it might be wise to install the plant in a fume cupboard, since this could be a way of coping with ATEX regulations.

The technology has the potential to reach an area far beyond the development lab inside large companies. It offers the possibility of preparing customized products at the point of sale. For example, pharmacies could prepare creams according to the skin type of a specific person, and drug delivery systems could be adjusted for diagnostic results.

The use of microreactors in lab-scale plants has a strong impact on the safety of critical substances and processes. First, these systems are usually operated in closed systems. Contact with problematic substances can be reduced to special cases:

Loading and unloading

Unexpected events (e.g., leaking)

During normal processing, contact can be prohibited. This is especially important for toxic, self-ignitng, or explosive substances. Since microreactors have a very small reaction volume, it is even possible to carry out reactions in an explosive regime. The possible hazards of an accident can be eliminated or reduced to a minimum effect. This has a big impact in terms of obtaining government certification. It becomes easier to get any necessary permits, if it can be proven that even an accident would have little impact.

ACKNOWLEDGMENTS

The auther thanks Doris Hofer, Guenter Tekautz, Lukas Wiesegger, and Daniela Windisch of the Microinnova team for support during the preparation of this chapter.

BIBLIOGRAPHY

1. Kirschneck, D., Kober, M., and Marr, R. (2005). *Chem. Eng. Tech.*, 28(3): 314–317.
2. Antes, J., Schifferdecker, D., Krause, H., and Loebbecke, S. (2004). *Chem. Ing. Tech.*, 76(9): 1332.
3. Kirschneck, D. M., and Marr, R. (2006). *Chem. Ing. Tech.*, 78(1–2): 29–38.
4. Hessel, V., Loewe, H., and Schoenfeld, F. (2005). *Chem. Eng. Sci.*, 60: 2479–2501.
5. Hessel, V., Hardt, S., Loewe, H., and Schoenfeld, F. (2002). *AIChE J.*, 49(3): 566–577.
6. Hardt, S., and Schoenfeld, F. (2002). *AIChE J.*, 49(3): 578–584.

7. Hessel, V., Loewe, H., Hofmann, C., Schoenfeld, F., Wehle, D., et al. (2001). In *Proceedings of the 6th International Conference on Microreaction Technology*, IMRET 6, New Orleans, LA.

8. Schwesinger, N., and Frank, T. (1995). Merck Patent No. GmbH WO 96/30113.

9. Panic, S., Loebbecke, S., Tuercke, T., Antes, J., and Boskovic, D. (2004). *Chem. Eng. J.*, 101: 409–419.

10. Guichardon, P., Falk, L., and Villermaux, J. (2000). *Chem. Eng. Sci.*, 55: 4245–4253.

11. Guichardon, P., and Falk, L. (2000). *Chem. Eng. Sci.*, 55: 4233–4243.

12. Tekautz, G., and Kirschneck, D. (2006). In *Proceedings of Achema 2006*, Frankfurt.

13. Pennemann, H., Hessel, V., and Loewe, H. (2004). *Chem. Eng. Sci.*, 59: 4789–4794.

14. Werner, B., Donnet, M., Hessel, V., Hofmann, C., Jongen, N., Loewe, H., Schenk, R., and Ziogas, A. (2002). *Chem. Eng. Comm.*, 189.

15. Kirschneck, D., Tekautz, G., and Linhart, W. (2006). In *Proceedings of the 17th International Congress of Chemical and Process Engineering*, 27–31 August. CHISA, Prague, Czech Republic.

16. Khan, S. A., Guenther, A., Schmidt, M. A., and Jensen, K. F. (2004). *Langmuir*, 20: 8604–8611.

17. Hessel, V., Hardt, S., and Loewe, H. (2004). *Chemical Microprocess Engineering: Fundamentals, Modeling, and Reactions*: Weinheim: Wiley-VCH.

18. Loeb, P., Pennemann, H., and Hessel, V. (2004). *Chem. Eng. J.*, 101(1–3): 75–84.

19. De Mas, N., Guenther, A., and Schmidt, M. A. (2003). *Ind. Eng. Chem. Res.*, 42(4): 698–710.

20. Keil, J. F. (2004). *Chem. Eng. Sci.*, 59: 5473–5478.

21. Kiwi-Minsker, L., and Renken, A. (2005). *Catalysis Today*, Vol. 110, pp. 2–14.

22. Pennemann, H., Watts, P., Haswell, S., Hessel, V., and Loewe, H. (2004). *Org. Proc. Res. & Dev.*, 8(3): 422–439.

23. Zapf, R., Becker-Willinger, C., Berresheim, K., Holz, H., Gnaser, H., Hessel, V., Kolb, G., Löb, P., Pannwitt, A.-K., and Ziogas, (2003). *Trans IChemE*, A81: 721–729.

24. Jänisch, K., Hessel, V., Loewe, H., and Baerns, M. (2004). *Angew. Chemie.*, 116: 410–451.

25. Gavrilidis, A., Angeli, P., Cao, E., Yeong, K. K., and Wan, Y. S. S. (2002). *Trans. IChemE*, A80: 3–30.

26. Roberge, M. D., Ducry, L., Bieler, N., Cretton, P., and Zimmerman, B. (2005). *Chem. Eng. Technol.*, 28: 3.

27. Kirschneck, D., Schmitt, H., and Marr, R. (1999). In *Proceedings of ISEC'99*, Barcelona, Spain, 1999.

28. Wang et al. (2006). Velocys Inc. Patent U.S. 7,084,180.

29. Iwasaki, T., and Yoshida, J.-I. (2004). *Macromolecules*, 38: 1159–1163.

30. Janicke, T. M., Kestenbaum, H., Hagendorf, U., Schüth, F., Fichtner, M., and Schubert, K. (2004). *J. Catalysis*, 191: 282–293.

31. Kirschneck, D., Kober, M., Wojik, A., and Marr, R. (2004). *Chem. Ing. Tech.*, 76(9): 1328.

32. Jähnisch, K., Baerns, M., Hessel, V., Ehrfeld, W., Haverkamp, V., Löwe, H., Wille, C., and Guber, A. (2000). *J. Fluo. Chem.*, 105: 117–128.

33. Hessel, V., Löb, P., and Löwe, H. (2005). *Curr. Org. Chem.*, 9(8): 765–787.

34. Schwalbe, T., Autze, V., and Wille, G. (2002). *Chimia*, 56(11): 636–646.

35. Stankiewicz, A. (2006). *Chem. Eng. Res. Des.*, 84: 511–521.

36. De Bellefon, C., Pestre, N., Lamouille, T., Grenouillet, P., and Hessel, V. (2002). High Throughput Kinetic Investigations of Asymmetric Hydrogenations with Microdevices. *Chem. Eng. Comm.*, 345(1–2): 190–193.

37. Hessel, V., Hofmann, C., Löwe, H., Meudt, A., Scherer, S., Schönfeld, F., and Werner, B. (2004). *Chem. Proc. Res. Des.*, 8: 511–523.

CHAPTER 12

MICROREACTION SYSTEMS FOR EDUCATION

MARCEL A. LIAUW and DINA E. TREU

12.1 INTRODUCTION

The chapters in this volume well illustrate the fact that microreactors and microsystems are a powerful option for R&D and may even be used in production. With progress in this field being very fast, it has become in an industrial context highly relevant and is experiencing a strong pull from the industry. As a result, micro-reactors of all sorts have become commercially available: These days, 80% of those in use are of commercial origin, whereas five years ago, 80% were hand-made [52].

In general, these microsystems have been embraced likewise by small companies and numerous global players, which in the process have looked into drastically changing their production sites. Degussa is a good example of this development, which will be discussed in more detail below.

Obviously, at the same time, the job specifications for employees in this field have accordingly changed to a substantial degree. In-depth knowledge of and experience with microreactors and microsystems are much sought after these days, making it crucial for practicing scientists and engineers, as well as those in training, to learn about this new technology. The need for microreaction-related education has already been pointed out in the Potential and Applications of MicroReaction technology (PAMIR) study [46].

This chapter will therefore first focus on how microreactors have impacted the industry's structure—as illustrated in a case study in Section 12.2—as well as the resulting requirements in the industry for employees and students. Then, we will explore how in response:

- Academia teaches microreactors, that is, either as a subject or vehicle for illustrating generic as well as specific engineering issues (Section 12.3).

Microchemical Engineering in Practice. Edited by Thomas R. Dietrich
Copyright © 2009 John Wiley & Sons, Inc.

- Where and in what manner classes on microreactors are currently being offered (Section 12.4), while also listing opportunities for continuing education and other channels of comprehensive information on the subject.

In the final Section (12.5), the further development of the above-mentioned academic classes through governmental support programs will be discussed.

12.2   INFLUENCE OF THE INDUSTRIAL SECTOR ON MICROREACTOR EDUCATION

Although the issue of green chemistry as well as increased economical pressure are the driving forces behind implementing microtechnology, its successful use within industry heavily depends on management offering the appropriate company structures. These need to encourage employees to mentally shift gears and readily embrace paradigm-shift technology over traditional, proven but less efficient technology. At the same time, these revised corporate structures will need to bring together available skills in a new team to meet the current challenges in the shortest time possible. To do so, Degussa, like other large companies, has introduced so-called project houses, bringing together several business units and its subsidiary Creavis, with a budget of 15 million € and 50% cofunding from the corporate division. Following this scheme, delegated from the business units to a central campus, 15 project team members are investigating new technologies for a limited time of three years in cooperation with academic and research institutions [10].

One of the "project houses" that focuses on process intensification and is led by H. Hahn, ran from January 2005 to December 2007 [23]. With comprehensive internal and external cooperation, including participation in the publicly funded projects IMPULSE and DEMiS along with the unusually focused and dedicated nature of Degussa's research, the educational aspects of this project are being very carefully covered.

Other major companies have been active in the field: BASF, Bayer, Clariant, DSM, DuPont, Merck, to name a few. Although these large corporations have already strongly embraced microtechnology, the situation with Small and Medium-sized Enterprises (SMEs) in Germany, for example, is quite different: A questionnaire developed by Dechema addressing 1,500 out of 1,700 SMEs in German chemical process industries indicated that they are in need of immediate and comprehensive education and training in this field [4].

Of the approximately 100 SMEs that returned the same questionnaire, half of them replied that they did not know anything about microreaction technology and 87% noted that this technology was not yet in use. Considering that in 2004, SMEs made up 91% of Germany's chemical industry (22.5% of the employees) [50], this situation calls for the immediate and widespread dissemination of information on the subject.

It must be noted that the fundamentally different structures of large corporations and SMEs demand quite different educational programs. In addition, large corporations have many different departments involved in different aspects of implementing a new technology: They range from R&D to process design to the marketing department, all with their own, differing educational needs.

Furthermore, every company has its own culture that also influences employees and therefore the kind of knowledge sought during employee recruitment. Any lack of knowledge among employees is best dealt with by offering highly modular educational material that can be easily tailored to fit individual corporate needs.

12.3 ACADEMIC APPROACHES TO TEACHING ABOUT MICROREACTORS

Due to these complex internal industrial structures and the arrival of a radically new microtechnology, the industry's need for training has significantly broadened in range. The main challenges academia faces in meeting these complicated demands are based on the fact that for decades, chemists have learned their trade by means of batch processes: test tubes and reaction flasks. Although batch chemistry is also prevalent in the industry's R&D departments, its translation into continuously operated chemical production adds to the difficulties of scale-up and commercial implementation. However, thanks to microreactor technology, training in continuous operation (and in decent mixing/cooling, etc.) may already be introduced at a very early stage in chemistry education.

Although in a first step, teaching the fundamentals of microreactors is obviously of great importance, students should develop the corresponding sound engineering judgment that will allow them to reliably apply this technology to unknown situations. Such fundamental understanding is gained by studying the microreactors' physicochemical background.

Generic issues (such as scaling behavior; see below) are equally important and well suited to being taught in interdisciplinary classes [34], which unfortunately at this point are still scarce. Likewise, the industry is asking that the issues of green chemistry, which strongly overlap with microreaction technology, be not only taught in separate, elective classes, but also addressed in every course, and thus benefit from all educational experiences had there (see, e.g., [31]). It may be noted, however, that education in "microreaction technology" is quite well defined, which is not necessarily true for education in sustainable chemistry [49].

With globalization and economic pressure increasing steadily, the industry must be able to react with ever-increasing flexibility. Creativity and strong problem-solving capabilities ensure such ability and are best trained using a goal-based scenario. In addition, it is well understood that in the long run mainly team players will advance a company, a skill that may be trained during lab classes when students are working on group assignments.

In part, to date, the academic world has managed to serve this large variety of quests. And although microreactors are still being taught mostly as a subject of education, they are also being used more and more as tools of education, aiding in conveying certain concepts and discussing generic issues such as scale-up,

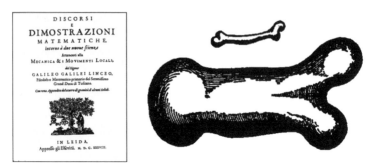

FIGURE 12.1 Cover of Galilei's work on scale-down (left) and original illustration showing the proportions of a mice bone (top right) and an elephant bone (bottom right) [16].

continuous processing, in-line analytics, or process control. As an example, a chemical engineering class on scale-up may start off with photographs of microdevices in which their small size is illustrated by the juxtaposition of a fly or an ant. Students then are asked whether this actually demonstrates the smallness of the devices, or whether the fly or ant shown may just be a giant insect, like those sometimes seen in B movies. This leads to the question of whether small animals may be scaled up to monster size. The answer ("No") is given in the very first treatment of scale-up, published by Galileo Galilei as early as 1638 [16]. Galilei noticed that corresponding bones in a mouse and an elephant were different in not only size but also proportion (Fig. 12.1). This example of a scale-up problem nicely catches students' attention [9]. They then comfortably turn to problems of scale-up from lab-scale dimensions to production-scale size, using recurring examples from microreaction technology.

12.4 ACADEMIC COURSES ON MICROREACTORS AND INTERDISCIPLINARY CLASSES

From an industrial perspective, there are two main educational sources for acquiring knowledge on innovative technologies:

- By hiring knowledgeable coworkers, particularly those recently graduated
- By educating the staff, for example, via continuing education, scientific journals, trade magazines, in-house training, conference participation, etc.

The continuous of life-long learning is outlined in (Fig. 12.2). It may be noted that in terms of scientific education, much ground can already be covered prior to high school. As a matter of fact, it has been increasingly acknowledged [39] that scientists are thought to benefit greatly from a very early introduction to the sciences, meaning in elementary school and preschool. Recent research indicates that young children (5–7) are already highly interested in scientific experiments and their background.

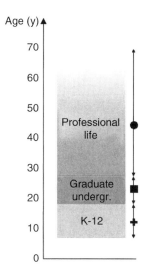

FIGURE 12.2 Insights into new technologies may be conveyed during academic education (■) or, after graduation, during life-long learning (●). Increasingly, early education in natural sciences is perceived as crucial (✚).

Asked to reproduce an experiment, even several months after a class on it, they still recalled the steps and results of the experiment as well as its underlying theories.

Along these lines, it should be noted that as small and toylike as they are, microreactors as well as microcomponents in general may be well suited to such introductory classes. The microdevices developed by T. Fujii, Tokyo, and Fuji Film serve as an example (Fig. 12.3). Note that devices in microreaction technology are miniaturized continuous devices. As such, they are different from the miniaturized batch devices used in chemistry education, as described by [5] on microchemistry and [7] on small-scale chemistry, supporting the trend to reconcile (batch) chemistry and (continuously operated) reaction engineering. A prerequisite for introducing microreaction technology is obviously that school teachers remain reliably informed on the progress in this field (see, e.g., [29]).

At a high-school level, microtechnology has proven very valuable from many different perspectives: For one thing, such devices offer the possibility of reintroducing experiments that have been discontinued for safety reasons. At the same time, they introduce students to the fact that these days many reactions can be much better monitored through microfluidic devices [27].

On a university level, introductory as well as advanced classes on, or touching on, microreactors are offered at numerous sites around the globe (Table 12.1). For more detailed information on specific aspects of some students' experiments, see [30].

Given the rapid evolution of the microreactor field, it has become necessary for educational efforts to train students with the skills to research and understand diverse new technologies on their own. State of the art in one semester becomes an aged

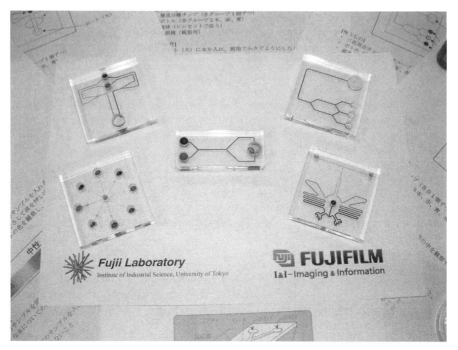

FIGURE 12.3 Microfluidic devices for education, developed by Teruo Fujii et al. (IIS, University of Tokyo) in collaboration with Fuji Photo Film. These chips are used for science festivals and comparable events [27].

technology by the time the students enter the workforce. For this reason, one newly developed course at Mississippi State University focused on providing fundamental knowledge and a coupled problem-based learning climate where students pulled from archival literature to envision and design a novel new microtechnology. This approach empowered the students with the ability to problem solve and contribute to this rapidly evolving field.

For advanced training on problem solving, a Web-based tool for remote experimentation in process control is available, as offered by J. Henry at the University of Tennessee, Chattanooga, since the early 1990s [45]. Students log on from any location to an actual experiment that they run remotely. This combines high flexibility with a real-life experiment, as opposed to pure computer simulations or recorded experiments. Remote experimentation with reactive experiments like saponification is offered via VIPRATECH in Germany [44]. Reactive mixing experiments, remotely accessible via the internet, were developed in a combined effort by the North German Universities of Oldenburg, Hamburg, Rostock and Breman, with accompanying university courses starting in summer 2009 [53, 54].

Microreaction technology has only started to be described in student textbooks. In the English-language market, B. A. Finlayson's *Introduction to Chemical Engineering Computing* includes several examples from the field of microfluidic devices [12–14], (see also the corresponding entry in Table 12.1), and the latest edition of a textbook by F. P. Incroprera et al. contains some material on μRT [24]. The textbook by J. O. Wilkes has a chapter on "microfluidics and electrokinetic flow" [51], and H. S. Fogler's book includes μRT examples [15]. More textbooks will certainly follow in due course.

In Germany, the textbook *Technische Chemie* by G. Emig and E. Klemm [11] is available; may well be the first to offer a full-fledged chapter on microreaction technology. In its Fifth Edition, Klemm has added a 20-page chapter in which reaction engineering issues such as pressure drop, diffusion limitation, or residence time distribution are discussed from a microreactor perspective. The main objective is an unprejudiced view of the strengths and weaknesses of miniaturized equipment for given problems. Another German textbook with a new μRT section has been written by M. Baerns et al. [3].

12.4.1 Other Sources of Education

One source of microreaction technology education for the staff of chemical companies may be enrollment in a corresponding master's program, for example, the pharmaceutical master's degree program at the New Jersey Institute of Technology [1]. The pursuit of an advanced degree may pose some challenges, however, especially in terms of the time and economics involved.

Regular continuing education may be preferable. Here, courses on microreaction engineering are offered by several organizations such as Dechema [47], IMM and VDI [22], or ASME [8]. A continuing education class by GDCh with the title "preparative chemistry in microreactors," organized by W. Reschetilowski, is successful in drawing many synthesis chemists to look into reaction engineering

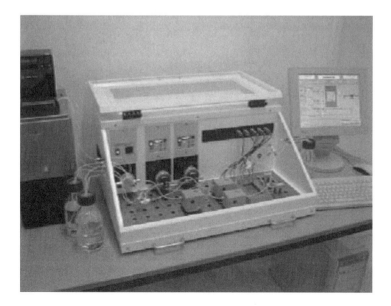

Mikrosyn edu: A dedicated microreactor system for educational purposes (mikroglas chemtec).

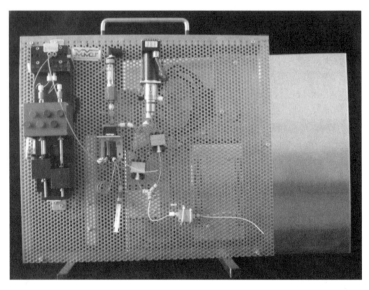

MEMR: A micro enzyme membrane reactor for hands-on laboratory classes at RWTH Aachen. Left: dosing pumps; gray blocks: reactors [37].

Microheat exchanger system with co-countercurrent (IMM) and cross-current (FZK) heat exchanger at FSU in Jena, Germany.

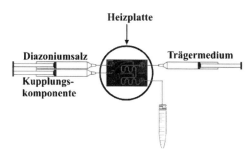

Microreactor system for azo-coupling at TU Ilmenau, Germany.

TABLE 12.1 Selected Academic Courses on Microreaction Technology

Location, Teacher(s)	Course Description	Remarks
France: INPL, Nancy, Jean-Marc Commenge	Elective course	
Germany: FH Kaiserlautern, Zweibrücken, Thomas Dietrich	Elective course, described as "microprocess engineering—chemistry in microstructures," including student experiments and a field trip	[9]
Germany: FSU Jena, Bernd Ondruschka	Elective course, lecture course with, hands-on experiments	[18, 19, 43]
Germany: North German Universities, Michael Schlüter, Frank Rößner, Udo Kragl, Hans-Ulrich Moritz	Hands-on experiments and Web-based practical hands-on training	[53, 54]
Germany: RWTH Aachen, Marcel Liauw	Elective course, lecture course	[32]
	Core practical course, hands-on experiments	[2, 37]
Germany: TU Chemnitz, Elias Klemm	Core course, lecture course	[11]
Germany: TU Ilmenau, Jens Köhler and Mike Günther	Hands-on experiments	[20, 28]
Germany: TFH Wildau, Andreas Foitzik and Franz Wildenauer	Hands-on experiments	[17]
Japan: IIS, University of Tokyo, Teruo Fujii	Hands-on microfluidics course, undergraduate.	
Netherlands: TU Eindhoven, Evgeni Rebrov and Jaap Schouten	Elective course	
United States: Mississippi State University, Adrienne Minerick	Analytical Microdevice Technology Special Topics Course: lecture, a Survivor Game, student-led discussions of technical articles, and a semester-long concept development project	[56]

Continued

TABLE 12.1 Continued

Location, Teacher(s)	Course Description	Remarks
United States: University of Tennessee at Chattanooga (UTC), Frank Jones	Three-stage approach: (1) Assistance in microbioreactor simulation and experimentation efforts, (2) Teaching modules incorporated into an existing fluid mechanics lab course, and (3) Teaching modules for the simulation of channel-based bioreactors incorporated into the advanced level chemical and environmental processes lab course	[25, 26]
United States: University of Washington, Seattle, Bruce A. Finlayson	Optional course, undergraduate; described as "transport effects in microreactors"; Student assignments in CFD	[12–14]
United Kingdom: University of Hull, Paul Watts	Optional course, lecture course on microreactors for chemical synthesis for final-year students	

issues. This is a valuable step towards a better communication between chemists and chemical engineers. An even wider scope of issues is pursued in the IMPULSE project funded by the European Union (EU) (see Chapter 14). In this program, onsite training as well as distance learning courses including a Web-based tool (WBT) is available as a web platform [55] on the results of the IMPULSE project concerning sophisticated process design with multiscale structured elements. A crucial role in developing proper educational material has been played by IMPULSE member Britest [6]. Based in the United Kingdom, Britest is a nonprofit organization that optimizes processes and manufacturing strategies using a set of proprietary methodologies.

IMPULSE also supplies material to the joint master's degree program of URV-ETSEQ (Tarragona, Spain), ICTP (Prague, Czech Republic), and INPL (Nancy, France), with Warsaw, Manchester, and Aachen as associated partners. In Germany, INPL in Nancy and RWTH Aachen University are looking into using the public teaching platform Moodle [see, e.g., 21, 35]. A Web-based platform offering course materials, additional materials, a forum, a chat rooms, and more, Moodle will be used for teaching courses in microreaction technologies or process design. In the public portion of micro-Moodle, the fundamentals of this technology can be made readily available to just about anyone interested in studying the subject. Such an approach can easily be used by an increasing number of students without a corresponding increase in personnel or related resources (scalability). Yet another approach to scalable education has been followed by MIT. There, in the MIT OpenCourseWare project, it has been decided to openly publish teaching material [33, 38].

Finally, as we live in times of tight schedules and reduced financial resources, trade magazines have proven a very reliable source for a first, yet often very comprehensive, overview of the topic. To give just one example, *Chemical Engineering* (with a circulation of over 60,000) reported on the budding field of microreactors as early as 1994 [36, 40] and continues to cover the field on a regular basis (see, e.g., [41, 42]).

12.5 FUTURE OUTLOOK

More and more organizations as well as governmental departments are recognizing the immense potential of microreactors. They are thus either joining their efforts to propagate this technology or offering substantial support for its implementation.

Hence, in a coordinated action costing 1.4 million euro, three German organizations have set out to support educational projects focused on microreaction technology. Grants by the German Federal Department of Education and Research of up to 1.5 million € are aimed at developing affordable miniaturized lab-scale apparatuses and components for hands-on courses at universities. The Fonds der Chemischen Industrie will spend 200.000 € to supply chemistry departments with miniaturized components, and the German Environmental Foundation (DBU) has supported the development of a teaching handbook with another 200.000 €.

In addition, a number of projects have also been initiated by European companies and supported by the EU: One of them is the European Technology Platform for Sustainable Chemistry (Ref SusChem), which aims to foster and focus attention on European research in chemistry [48]. Founded in 2004, the initial partnership has constantly expanded, with many other stakeholders having joined since, helping to implement SusChem's goals, especially on a local level. SusChem also has a strong focus on the educational aspects surrounding new technologies.

In summary, the advent of microreactors may well change the way chemists learn their craft, as it has prompted the demand for the wide-ranging education in this budding field. Education in microreaction technology will eventually cover an extensive range of disciplines, ages, and walks of life.

BIBLIOGRAPHY

1. Armenante, P. M., and Manfredi, J. J. (2006). Pharmaceutical engineering programs and courses for the working pharmaceutical professionals. Paper presented at the AIChE Annual Meeting, San Francisco, CA, November 17, 2006.

2. Baatar, B., Wanko, E., and Liauw, M. A. (2009). Hands-on experiments on RTD in microreactors. Paper in preparation.

3. Baerns, M., Behr, A., Brehm, A., Gmehling, J., Hofmann, H., Onken, U., et al. (2006). *Technische Chemie.* Weinheim: Wiley VCH.

4. Bazzanella, A., and Steinbach, C. (2005). *DECHEMA Report: μVT Guide.* Leitfaden Industrielle Nutzung der Mikroverfahrenstechnik, Frankfurt/M.

5. Bradley, J. D. (2001). UNESCO/IUPAC-CTC global program in microchemistry. *Pure Appl. Chem.*, 73(7): 1215–1219.

6. Britest Web site. http://www.britest.co.uk.

7. Cohen, J. (1994). Stephen Thompson: Call him the czar of small-scale chemistry. *Science*, 266(5186): 889.

8. ASME Continuing Education Institute, Las Vegas, USA. http://catalog.asme.org/Education/ShortCourse/MICRO_PROCESS_TECHNICAL.cfm.

9. Dietrich, T. R., and Liauw, M. A. (2004). Mikroverfahrenstechnik in der Lehre und Ausbildung an Hochschulen. *Chem. Ing. Tech.*, 76(5): 517–519.

10. Dröscher, M. (2003). New innovation concept at Degussa. Project houses and new business development. Paper presented at the IUPAC CHEMRAWN XVI Consultation Forum Innovation: The Way from Pure to Applied Chemistry, August 9, 2003, Ottawa, Canada. http://www.iupac.org/symposia/conferences/chemrawn/crXVI/crXVI-N13-Droescher.pdf.

11. Emig, G., and Klemm, E. (2005). *Technische Chemie.* Revised 5th ed. Berlin: Springer. Chapter 16, pp. 444–467.

12. Finlayson, B. A. (2006). *Introduction to Chemical Engineering Computing*: Hoboken, NJ: Wiley.

13. Finlayson, B. A., Drapala, P. W., Gebhardt, M., Harrison, M. D., Johnson, B., Lukman, M., et al. (2006). Micro-component flow characterization. In M. Koch, K. VandenBussche, and R. Chrisman, (eds.), *Micro-Instrumentation for THE and Process Intensification*. Hoboken, NJ: Wiley.

14. Bruce A. Finlayson, University of Washington, Course listings at Web site: http://faculty.washington.edu/finlayso/

15. Fogler, H. S. (2006). *Elements of Reaction Engineering*. 4th ed. Upper Saddle River, NJ: Pearson Education.

16. Galilei, G. (1638). *Discorsi e dimostrazioni matematichi intorno a due nuove scienze attenenti alla mecanica & i movimenti locali.* Leiden: Elsevier.

17. Gast, F.-U., Dittrich, P. S., Schwille, P., Weigel, M., Mertig, M., Opitz, J., et al. (2006). The microscopy cell (MicCell). A versatile modular flowthrough system for cell biology, biomaterial research, and nanotechnology. *Microfluid Nanofluid*, 2(1): 21–36.

18. Gorges, R., Klemm, W., Kreisel, G., Ondruschka, B., Scholz, P., and Taubert, T. (2004). Microreaction technology in the technical chemistry curriculum at Friedrich Schiller University of Jena—Status and prospects. *Chem. Ing. Tech.*, 76(5): 519–522.

19. Gorges, R., Taubert, T., Klemm, W., Scholz, P., Kreisel, G., and Ondruschka, B. (2005). Integration of microreaction technology into the curriculum. *Chem. Eng. Tech.*, 28(3): 376–379.

20. Günther, M., and Köhler, J. M. (2004). Electrochemical microflow setup for standard experiments in university courses. *Chem. Ing. Tech.*, 76(5): 522–526. See also http://www.tu-ilmenau.de/fakmn/Miniaturisierung-che.1378.0.html.

21. Hall, J. M., and Moore, John W. (2006). Nanoscience for teachers: An online course at UW-Madison created with Moodle. Paper presented at 37th Great Lakes Regional Meeting of the American Chemical Society, Milwaukee, WI, May 31–June 2, 2006, Abstract GLRM-221.

22. Hessel, V., and Schulz, M. (2004). VDI (seminar on) Mikroverfahrenstechnik-Eine intelligente Lösung für organische Synthesen und Mischprozesse. *Chem. Ing. Tech.*, 76(5): 516.

23. Huether, A., Geisselmann, A., and Hahn, H. (2005). Process intensification: A strategic option for the chemical industry. *Chem. Ing. Tech.*, 77(11): 1829–1837. See also http://www.degussa.com/degussa/en/innovations/creavis/project_houses/.

24. Incroprera, F. P., DeWitt, D. P., Bergman, T. L., and Lavine, A. S. (2007). *Fundamentals of Heat and Mass Transfer.* 6th ed. Hoboken, NJ: Wiley.

25. Jones, F. J., and Elmore, B. B. (2000). Incorporating chemical process miniaturization into the ChE curriculum. *Chem. Eng. Ed.*, 34(4): 316–319, 325.

26. Jones, F., and Bailey, R. T. (2003). Simulation and experimentation in microbioreactor design: Incorporation into engineering education. Paper presented at Proceedings of the 2003 ASEE Southeastern Section Annual Meeting, Macon, GA, April 6–8, 2003.

27. Kaneda, S., Matsunaga, M., Takagi, N., and Fujii, T. (2004). Microfluidic devices for educational purposes. *Nikkei Shinbun*, 06/21: 17.

28. Kirner, T., Jaschinsky, P., and Kohler, J. M. (2004). Spatially resolved detection of miniaturized reaction-diffusion experiments in chip reactors for educational purposes. *Chem. Eng. J.*, 101(1–3): 163–169.

29. Kirschning, A. (2002). Chemistry in flux—New microreactors for the laboratory, Prax. Naturwiss. *PdN-ChiS*, 51(4): 28–30.

30. Klemm, W., Ondruschka, B., Köhler, M., and Günther, M. (2006). Laboratory applications of microstructured devices in student education. In: O. Brand, G. K. Fedder, C. Hierold, J. G. Korvink, and O. Tabata, (eds.), *Advanced Micro & Nanosystems, Micro Process Engineering, Vol. 5: Fundamentals, Devices, Fabrication and Applications*, (N. Kockmann, ed.) pp. 463–495, Weinheim: Wiley-VCH.

31. Leitner, W. (2004). Focus on education in green chemistry, *Green Chem.*, 6: 351.

32. Liauw, M. A. (2005). Introducing microreaction engineering in education. Paper presented at AIChE Annual Meeting, Cincinnati, OH, October 30–November 5, 2005.

33. Margulies, A. H. (2004). A new model for open sharing: Massachusetts Institute of Technology's OpenCourseWare initiative makes a difference. *PLoS Biol.*, 2(8): E200.

34. Matlack, A. S. (1999). Teaching green chemistry. *Green Chem.*, 1(1): G19–G20.

35. Moodle Web site. http://www.moodle.org.

36. Moore, S., Fouhy, K., and Samdani, G. S. (1994). Miniaturization reaches the CPI. *Chemical Engineering*, October, p. 41.

37. Müller, D. H., Liauw, M. A., and Greiner, L. (2005). Microreaction technology in education: Miniaturized enzyme membrane reactor. *Chem. Eng. Tech.*, 28(12): 1569–1571.

38. OCW Web site: http://ocw.mit.edu/.

39. OECD (ed.) (2006). *Starting Strong II: Early Childhood Education and Care*, OECD Publishing, Paris.

40. Ondrey, G. (1995). Microreactor engineering: Birth of a new discipline? *Chemical Engineering*, March, p. 52.

41. Ondrey, G. (ed.) (2009). Microreactors widen their reach for increased capacities. *Chemical Engineering*, April, p. 18.

42. Ondrey, G. (ed.) (2009). Multiscale reactors come of age for making fine chemicals. *Chemical Engineering*, March, p. 14.

43. Ondruschka, B., Scholz, P., Gorges, R., Klemm, W., Schubert, K., Halbritter, A., et al. (2002). Micro-scale heat exchangers in a practical course in chemical engineering. *Chem. Ing. Tech.*, 74(11): 1577–1582.

44. Rem_exp_Leip Project, Virtual lab course, Prof. Helmut Papp, University of Leipzig, Germany. http://leipzig.vernetztes-studium.de/index_engl.html.

45. Rem_exp_Tenn Project, Engineering Laboratories on the Web, Prof. Jim Henry, University of Tennessee, Chattanooga, USA. http://chem.engr.utc.edu/.

46. Stange, T., Kiesewalter, S., Russow, K., Hessel, V., Provence, M., Balsalobre, C., Boulon, P., and Mounier, E. (2002). *PAMIR—Potential and Applications of Microreaction Technology.* IMM, Mainz.

47. Stief, T., and Langer, O.-U. (2004). DECHEMA-Weiterbildungskurs "Mikroverfahrens-technik." *Chem. Ing. Tech.*, 76(5): 516–517.

48. SusChem Web site. http://www.suschem.org/.

49. van Roon, A., Govers, H. A. J., Parsons, J. R., and van Weenen, H. (2001). Sustainable chemistry: an analysis of the concept and its integration in education. *Int. J. Sustainability Higher Ed.*, 2(2): 161–180.

50. Verband der Chemischen Industrie (ed.) (2006). *Industriewirtschaft in Zahlen.*, Frankfurt/ M.: VCI.

51. Wilkes, J. O. (2005). *Fluid Mechanics for Chemical Engineers with Microfluidics and CFD* (2nd Edition). Upper Saddle River, NJ: Prentice-Hall.

52. Pieters, B., Andrieux, G., and Eloy, J. C. (2006). Technologies and market trends in micro-reaction. *Chim. Oggi,* 24(2): 41–42.

53. Web-based hands-on training: www.mikropraktikum.de/index-e.html

54. Müller, C. M., Schlüter, M., and Räbiger, N. (2009). µPr@ktikum: web-based practical hands-on training in microprocess engineering through North German Universities in partnership with the industry. Paper presented at ACHEMA 2009, May 13, 2009, Frankfurt/M.

55. IMPULSE educational Web Platform: www.impulse.inpl-nancy.fr.

56. Minerick, A. (2009). Creative Learning in a Microdevice Research-Inspired Elective Course for Undergraduate and Graduate Students. *Chem. Engng Educ.,* Submitted.

CHAPTER 13

MICROREACTION SYSTEMS FOR LARGE-SCALE PRODUCTION

ANNA LEE Y. TONKOVICH and ERIC A. DAYMO

13.1 OVERVIEW OF LARGE-SCALE OPPORTUNITY AND CHALLENGES

The decision to use microreaction systems in the production of large volumes of chemicals is based on economics, rather than on plant footprint or reductions in reactor volume. The three economic drivers for modifying existing plants or for building new plants are (1) reducing the capital cost of existing unit operations, (2) reducing the capital cost of the total production facility based on flowsheet changes that microreaction technology might allow, and (3) operating cost advantages. Operating cost savings typically come from the use of an alternative and less expensive feedstock, better use of an existing feedstock by improving selectivity, and better thermal integration to reduce utility consumption per production volume. In industrial manufacturing applications, the economic driving force and process requirements must be established first, and then technical feasibility evaluated. The evaluation approach is to conduct a top-down analysis to define the technical and economic performance level required of a microreaction system, and then determine whether this performance can be achieved.

Several themes emerge while considering dozens of chemistries for potential application in large-scale microreaction systems. First, no two systems are completely identical; one should not expect a design created for one product to be applicable to a second product. Although this may appear to be counterintuitive for cost control, in fact, it enables greater cost advantage for a cost-sensitive process. Much like consumer goods, the designs for different applications vary but the manufacturing methods for microreaction units must be standardized to make a profit. Second, although the

Microchemical Engineering in Practice. Edited by Thomas R. Dietrich
Copyright © 2009 John Wiley & Sons, Inc.

design for each application is uniquely optimized, developing families of reactor designs appears highly profitable. This approach allows reactor performance to be tuned to maximize economics, while maintaining an economy of mass production. Third, the economics for large-scale processing are driven by several factors and each must be considered. The difference of a few cents per pound of product may make or break the viability of an application. For example, feedstock cost often is usually a strong driver in achieving target economics, so that increasing reaction selectivity by a few points or by changing to a slightly less expensive feedstock may lead to project success.

Several companies are working to develop microreaction systems for commercial applications. These technologies may be classified by their scale: small, mid, large, and mega. The small-scale systems [$<1,000$ metric tons (mt)/yr] tend to focus on liquid reactions at a laboratory scale or at a small pilot scale, or on small-capacity gas-phase reactions for fuel processing applications. The mid-scale applications (1,000–10,000 mt/yr) are geared toward specialty chemicals and typically require only a single reactor to achieve production capacity. Large-scale microreaction systems (10,000–1 million mt/yr) include gas, liquid, and or multiphase reactions. Each application may require multiple reactors in parallel to achieve the required plant production capacity. The mega-scale application area (>1 million mt/yr) is not yet developed; only two examples are given in the open literature: a proposed 1.4 million mt/yr plant to produce Fischer–Tropsch liquid and a 1 million mt/yr ethylene plant. In a mega-scale process, the microreaction technology faces stiffer competition from conventional technology because of economies of scale. Conventional reactors generally scale according to a six-tenths power law, whereas microreaction systems have a less dramatic economy of scale.

The importance of scale relationships will be more evident in the discussion of scale-up integration to compete with conventional technology.

Many companies are actively developing microreaction technology at different scales and for different applications. Although many active development efforts have been described in the literature, it is likely that many have not. For example, developments in academic or government research are not listed among commercial applications of microreaction technology unless they are involved in a partnership with industry. Furthermore, where mass production rates were not explicitly given, reasonable assumptions of fluid density or process efficiency were made to create consistent basis categories.

Examples of demonstrated microreaction products, or those in development, are listed below as a function of capacity:

Small-Scale

- Sigma Aldrich: 3 mt/yr using a CYTOS system [1]
- Heatric: 8 mt/yr for hydrogen production [2]
- Merck: 15 mt/yr for specialty intermediate production [3]
- CPC: 20 mt/yr CYTOS pilot system [4]

- Lonza: 44 mt/yr, which is average to large-scale for fine chemical/pharmaceutical campaign [5]
- Little Things Factory: 63 mt/yr [6]
- Mikroglas: 105 mt/yr [7]
- Clariant: 130 mt/yr demonstrated for the synthesis of red, yellow azo pigments [8]
- Xi'an Huain Chemical Company/IMM: 130 mt/yr, nitroglycerin plant operational in China [9]
- Chart: 160 mt/yr H_2 reported production for refueling station with SINTEF and the University of Warwick [10]
- Syntics: 260 mt/yr [11]

Mid-Scale

- Siemanns Axiva: 1,500 mt/yr [12]
- Ehrfeld Microtechnik: 8,800 mt/yr [13]

Large-Scale

- Heatric: 12,000 to 60,000 mt/yr methanol production [14]
- DSM/Karlsruhe: 15,000 mt/yr of liquid chemical products [9]
- Velocys: 16,000 mt/yr production of transportation fuels [15]
- IMM: 26,000 mt/yr starlaminator mixer [8]
- Velocys: 30,000 mt/yr hydrogen production [16]
- Compact GTL (formerly GTL Microsystems): 45,000 mt/yr for remote gas-to-liquids (GTL) applications [17]
- Degussa/UDH: 150,000 mt/yr vinyl acetate synthesis [18]
- UOP/IMM: 162,000 mt/yr direct hydrogen peroxide plant in development [19]

Mega-Scale

- Velocys: 1 million mt/yr of ethylene [20]
- Velocys: 1.4 million mt/yr of FT liquids [21, 22]

13.2   SCALE-UP CONSIDERATIONS FOR LARGE-SCALE SYSTEMS

Large-capacity microreaction systems require special consideration of scale-up parameters for integration into a commercial plant. These parameters include manifolding within the microchannel reactor, manifolding to multiple reactors, manufacturing to achieve an economy of mass production, robust catalyst integration, and plant integration. Each parameter is discussed in further detail below.

13.2.1 Manifolding Within the Microreaction Unit

Dividing flows to several hundred, thousand, or more parallel microchannels for large-capacity systems creates a special challenge. The flow must be sufficiently uniform to all channels without adversely impacting the system pressure drop or cost. Many approaches have been described in the literature for manifolding microreaction systems in general; however, additional constraints may be required for large-scale systems. The three main approaches to manifolding large-scale systems are (1) cross-flow, where each stream has an open reactor face; (2) cocurrent or countercurrent schemes, where multiple fluids must be manifolded in close proximity; and (3) orifice restrictions. The key consideration for large-scale production units is defining sufficient flow distribution in terms of volume, process performance, and pressure drop cost. Typically, as the manifold gets larger, the pressure drop gets smaller, and the corresponding flow maldistribution decreases. Unfortunately, as the volume of the manifold increases, economics are challenged.

The simple manifolding scheme for large units is cross-flow. Each array of channels is open in a face and a large plenum or external manifold may be directly attached (via welding, brazing, gasket, and the like). The manifold acts as a large duct and adds only a modest amount of pressure drop and economic penalty. The velocity in the manifold is typically reduced to less than 10% of the velocity in the connecting microchannels to ensure sufficient flow distribution. This approach is useful primarily for two-stream unit operations or selected three-stream unit operations, where fluids A and B interact in one portion of the device and fluids A and C in the other. For some processes, however, the simple cross-flow manifold is not an option.

Countercurrent and cocurrent unit operations are more challenging to manifold. Each of two or more fluids must be connected at the same end of a reactor. For these designs, one or more of the fluids must first move laterally across a layer to distribute evenly before turning to enter a parallel array of process microchannels. Increasing the fluid pressure or manifold volume may not be a luxury that economics will allow to ease the design and manufacturing challenge.

Two primary approaches have been described in the literature for manifolding counterflow (or coflow) large-scale microchannel reactors—either with or without internal subdivision of flow. The approach without internal flow subdivision has been used commercially by Heatric for more than 15 years in heat exchange applications, and is described in their commercial literature. Fluids are divided into discrete flow passageways at the unit face that mate uniquely with individual process microchannels. Flow travels in a z-shape flow path within the microchannel: first across the shim, then turning orthogonally to flow along the shim, and then finally turning again to exit the unit on the opposing side. The cross-flow sections experience less efficient heat transfer and are a substantial portion of the device—as shown in technical literature and images of large-scale devices. The large advantage of this approach rises from minimal flow maldistribution. The approach may be cost disadvantaged from the increase in hardware volume and pressure drop, the requirement for only partially etched methods of manufacturing, and challenging internal catalyst integration.

A second approach to manifolding large-scale systems for counter- or cocurrent flow has been described by Velocys [23]. In this approach, flow undergoes a series of subdivisions within the microchannel reactor to both limit the volume of the manifold to a small fraction of the overall volume and to limit the manifold pressure drop. Pressure drop is precisely calculated at all points along the flow circuit, and ranged based on dimensions in as-manufactured reactors, to predict the amount of flow in each flow path. Flow always follows the path of least resistance. The modest addition of flow resistance is included in some circuits to improve flow uniformity to all flow circuits [23]. An example of this manifold approach is shown in Figs. 13.1 and 13.2, where sides A and B are subsequently joined to create passive flow control to 72 parallel channels as illustrated.

A final approach to manifolding large-scale systems is the integration of orifices in each microchannel to provide a sufficiently high inlet pressure drop to control flow distribution. This approach has the disadvantage of increasing overall system pressure drop, but it does not add significantly to overall volume.

13.2.2 Manifolding to Multiple Units and Plant Integration

For most large- and mega-scale applications, a single reactor unit is not sufficient to achieve full plant capacity. Multiple full-scale reactors must be connected in parallel to achieve the desired capacity—analogous to a multitubular conventional reactor. As the capacity of the reactor block increases, the number of external connections decreases. For a full-scale microreaction system, there are three possible dimensions to make a block, and thus capacity, larger. Increasing the reactor length increases

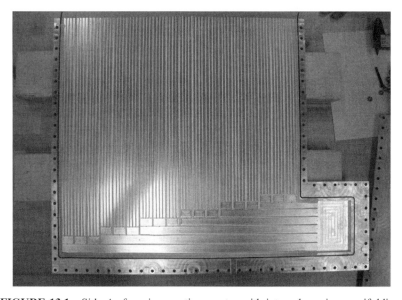

FIGURE 13.1 Side A of a microreaction reactor with internal passive manifolding.

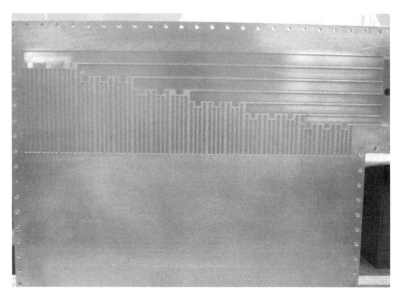

FIGURE 13.2 Side *B* of a microreaction reactor with internal passive manifolding. Sides *A* and *B* are joined to form the manifold.

capacity, but at the expense of pressure drop. Increasing the reactor height or stack height, as made by laminates, is a consideration, but may face manufacturing constraints. Increasing the reactor width also increases the capacity, but does so by increasing the reactor flow length for cross-flow applications—which may negatively affect pressure drop or performance. For counter- (or co-) flow operations, as the reactor becomes wider, the overall portion of a block devoted to internal manifolding also increases, thereby reducing the ratio of active volume to total volume and potentially affecting plant economics.

Several approaches are possible for manifolding multiple full-scale reactors. Each individual reactor must accept or reject multiple fluids. A flange is preferred to connect flow to and from the plant to the microreaction system. There may be multiple microchannel reactors connected to a single flange. Flow pipes and fluid connections may further be housed within a thermal enclosure, assembly, or pressure containment shell (Fig. 13.3). Alternatively, the connections may not have additional external containment.

Methods for attaching manifolds to the microreaction blocks include welding, brazing, and gaskets. Each has distinct advantages and disadvantages. Welding is the most permanent approach, but requires careful treatment along bond or braze joints of the stack. Brazing requires careful alignment of the reactor surface and header. Gaskets represent the most flexible approach, but may be prone to leaking in some service environments. Each approach must be carefully evaluated for a particular application.

The single flange connection to an assembly eases the integration challenges of large-scale microreaction systems to a chemical plant. The reactor assemblies are each skid-mounted and easy to integrate.

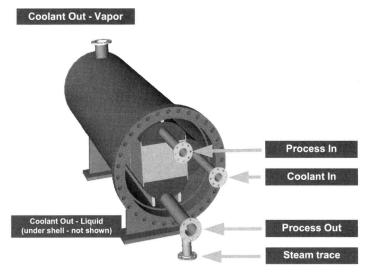

FIGURE 13.3 Schematic of a reactor housed in a thermal shell.

13.2.3 Mass Manufacturing Requirements for Large-Scale Microchannel Units

All current known work for manufacturing large-capacity microreaction systems is based on a laminate or sheet assembly method. Thin metal sheets are stacked to form a unit. No large-scale system made from a material other than a metal is known of at this time. Metals have a natural advantage for large-scale systems as they often reflect existing materials of construction and have proven durability, corrosion resistance, and ease of integration with pipes and connections.

The process for manufacturing large-scale microreaction systems involves several sequential steps, as follows:

- Selecting shims (sheets and laminates) of the appropriate materials of construction and thickness
- Feature forming
- Catalyst substrate integration
- Alignment and stacking
- Joining (brazing, bonding, other)
- Machining and finishing

A summary of these steps is shown in Fig. 13.4. In general, the manufacturing methods are a combination of industrially proven methods, albeit often for other applications, and support an economy of mass production.

Shim Selection Materials of construction must be selected that are compatible with the process chemistry and mechanical rigors of the operating conditions. Most

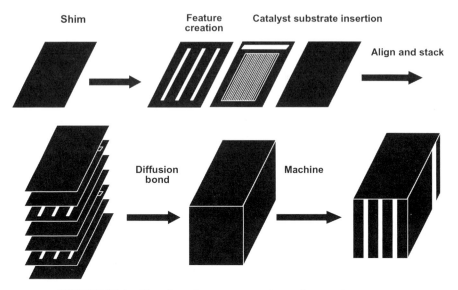

FIGURE 13.4 Manufacturing steps to produce microreaction reactors.

large- and mega-scale applications are based on metal shims made of either stainless steel or high-temperature alloys. The thickness of the material relates to the final microreaction dimensions, in that the critical microreaction dimension is the material thickness or a fraction thereof.

Feature Formation Features formed in the shim create the final microchannels. Multiple methods for channel creation include photochemical machining (PCM), electrochemical machining (ECM), electrical discharge machining (EDM), laser cutting, water jet cutting, and stamping. In a study on the mass manufacturing of microchannel devices [24], it was shown that above a threshold of parts formed per year, stamping was the least expensive method of feature formation. Below the threshold, PCM and ECM were cost-competitive. As cited, the cost of stamping tool formation required a high part volume per year to be cost-competitive. The use of PCM and ECM allows for more design flexibility when the production volumes are relatively low.

Catalyst Substrate Integration For some applications, it may be advantageous to integrate a catalyst substrate into the stack prior to bonding. Examples of different substrate forms are discussed in a subsequent section. The advantage of integrating catalyst substrates prior to reactor joining is the elimination of the catalyst substrate insertion and the associated thermal contact resistance. The added thermal resistance may be acceptable for some reactions, but unacceptable for others. Careful analysis is required.

Alignment and Stacking Each shim must be stacked with the correct orientation and order prior to bonding. Quality-control schemes are necessary to validate the location and order of all shims during the stacking process for large- and mega-scale microreaction systems. Alignment is maintained through the use of a series of alignment rods to minimize relative shim-to-shim distortion. An example is shown in Fig. 13.5.

Joining The stack may be joined either by diffusion bonding or brazing to form fluid-tight internal passageways. Although under some special circumstances gaskets may be considered to join together sheets, it would be challenging to maintain internal seals between parallel passageways. For both diffusion bonding and brazing, metallic interlayers may be required to assist with the process. In general, the temperature is higher for bonding processes than brazing to accommodate the formation of grain growth across the shim boundaries. Brazing is a lower-temperature process and relies on the use of interlayers to connect two adjacent shims.

A method for diffusion bonding metallic microchannel plates was first described by Martin et al. in 1999 [25] for copper, and stainless steel. The required bonding temperature for each, respectively, was 630, and 920°C, with bonding pressures of 41 MPa, and 28 MPa, respectively. All metallic shims were described as being well cleaned to create good contact between the like metallic faces and pressed in a vacuum ram press to facilitate diffusion bonding. In a subsequent publication [26], it was further noted that diffusion bonding stainless required 4 h at the described conditions. Another method of bonding has been described by Alfa Laval [27] with the use of transient liquid-phase interlayers to assist the joining process and create superior material properties. This interlayer melts at the bonding conditions to reduce the metal joining temperature.

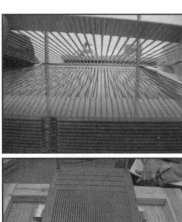

FIGURE 13.5 Examples of stack alignment and orientation.

Machining and Finishing Access to the internal passageways may be created by removing perimeter metal through machining. For some applications, the channels may extend all the way to the device perimeter, and thus a machining step may not be required. After finishing, the device is carefully cleaned to remove all residual fluids and materials and prepare the device for subsequent steps to add active catalyst layers.

An example of a large-capacity manufacturing scale-up device that represents all the required manufacturing steps is shown at one-third commercial scale in Fig. 13.6. The device contains more than 1,500 shims, with thicknesses ranging around a mean of 0.4 mm. Each shim was roughly 60 cm by 20 cm, and the stack height was 60 cm. Each shim was made from a nickel alloy as required for the high-temperature methane reforming reaction.

13.2.4 Catalyst Integration

For catalyst integration in large-scale microreaction systems, there are three approaches: before reactor assembly, after reactor assembly, or a hybrid of the two. Each approach has distinct challenges.

Coating a catalyst directly to the microchannel laminate prior to assembly allows for line of sight application and more options to control quality and catalyst

FIGURE 13.6 Manufactured microreaction device (one-third commercial scale).

uniformity. However, this approach requires the catalyst to withstand the rigors of metal joining, primarily temperature, and may make refurbishment a challenge. High-temperature resistance is not required for gasket-sealed reactors.

A second option for robust commercial application is applying the active catalyst to the reactor after joining the reactor laminates. A nonactive catalyst substrate may be optionally integrated into the shim stack prior to bonding to ease subsequent active catalyst integration efforts. By decoupling the joining conditions from the catalyst stability, more catalyst options are made available. In addition, the methodology for applying a catalyst after initial joining can be reused during catalyst refurbishment cycles. The challenge with this approach is maintaining catalyst uniformity.

A third option is to insert removable catalyst structures after joining the reactor. The catalyst is applied *ex situ* and may undergo substantial quality control prior to insertion. Multiple options for insertable catalyst forms have been described in the literature, including felts, foils, and foams. Figure 13.7 shows some practical examples.

Although there is simplicity in the insertion approach of an *ex situ* prepared catalyst within a fully joined reactor, there are also challenges—namely, the formation of a contact resistance between the catalyst and microchannel wall. For endothermic reactions, such as reforming, a contact resistance limits the rate at which the reaction can proceed. For exothermic reactions, including oxidation reactions, increased thermal resistance to heat removal from the catalyst may deleteriously impact selectivity. Regardless, this must be carefully considered before selecting a final catalyst form.

For systems where a catalyst is applied as a wash-coat solution, the catalyst must be both filled and drained uniformly among many parallel channels. This challenge

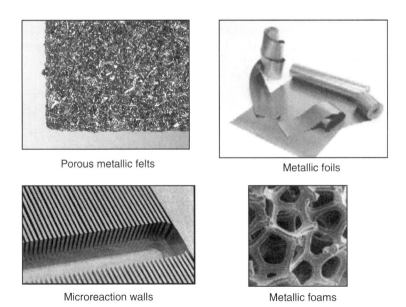

Porous metallic felts

Metallic foils

Microreaction walls

Metallic foams

FIGURE 13.7 Catalyst forms for microreaction systems.

becomes greater as the size of the reactor and number of channels increase. The catalyst must also be uniformly, or nearly so, retained along the reactor length. Fortunately, it has been shown that the requirement for axial catalyst uniformity may not be strict for some reactions. Some recent modeling results [28] suggest that the overall performance for one microchannel reactor scales with the total amount of catalyst in a channel and can tolerate some axial catalyst maldistribution. At any rate, this is a chemistry-specific question that must be addressed for each reaction in question.

13.3 COMPETING WITH CONVENTIONAL TECHNOLOGY AND ECONOMY OF SCALE

The literature suggests that processes which are inherently limited by heat and/or mass transfer, constrained by the size/weight of process equipment, or challenged by safety issues are good candidates for the implementation of microreaction systems. The conventional chemical industry, however, has had decades (or longer) to optimize chemical processes. Particularly when only cost benefits are at stake, there are many factors that can favor conventional systems. In some cases, the microreaction solution may be limited to particular market segments and in others the competing technology may be less expensive. The factors that should be considered when selecting microreaction process technology to replace conventional chemical equipment are discussed below.

13.3.1 Microreaction System Advantages

Step-out Chemistries Microreaction components have the greatest potential to replace conventional process equipment when the inherent heat and mass transfer advantages allow the application of new catalysts and/or reaction pathways to improve the yield (e.g., eliminating side-products and/or increasing conversion). Directionally, the greatest impact will be for nonequilibrium systems (e.g., oxidation reactions). For equilibrium products (e.g., syngas), conventional and microreaction systems will produce the same quantity of product at the same operating conditions (temperature, pressure, and feed composition) and thus compete solely on cost.

Thermal Integration and Lower Operating Costs For many commodity chemicals, the economics are driven by operating rather than capital costs. Microreaction components typically demonstrate greater thermal integration and thus often create an operating cost advantage over conventional systems. In some cases, however, the waste heat can be recovered (e.g., steam generation); for conventional plants where waste heat value can be reclaimed, the operating cost advantage for microreaction systems narrows.

Modular Construction and Lower Installation Factors Microreaction components are good candidates for modular (skid) construction as opposed to on-site "stick-built" construction. The smaller footprint and integrated heat transfer systems (e.g., no box furnaces) typically result in components that can fit on standard tractor trailers. By utilizing modular construction, labor and quality can be more carefully controlled. In addition, weather and other site-specific political and labor factors may have less impact on the cost and schedule. Installation factors for microreaction systems can be as low as one-fifth the value used for conventional process equipment, reflecting the significant savings associated with intensified, mass-produced systems [29]. Although most large chemical plants have stick-built (on-site constructed) components, recently some smaller-scale chemical units have been packaged and sold as skid-mounted systems (e.g., Howe-Baker steam methane reformers [30] and the Davy process technology compact reformer modularized GTL systems [31], among others [32]).

13.3.2 Conventional System Advantages

Intrinsic Advantage not Available   Perhaps the greatest advantage for conventional systems occurs when the intrinsic advantage needed to justify a microreaction process is not available. Catalytic systems may not be selective and/or active enough to justify a plant modification.

Presence of Solids Incompatible with the Scale of Microreactions   A study by Lonza [5] of the pharmaceutical and fine chemical reactions considered for process intensification found that over 60% were deemed incompatible with microreaction systems because of the presence of solids (either as a mobile catalyst, product, or reactant). Processes where the inherent sizes of solids (especially products or reactants) are greater than the intrinsic scale of microchannels would clearly not be candidates for microreaction systems.

Lower Capital Cost per Unit Equipment Volume   Although process intensification reduces the footprint of chemical equipment, it usually results in a higher cost per unit vessel volume (e.g., $\$/m^3$). Often, this trend toward higher capital intensity is based on the high metal volume (typically greater than 50%) of the microreaction equipment. Assumptions on the capital intensity of a microreaction unit can be used to derive a maximum unit volume. If the estimated microreaction equipment volume exceeds this critical volume derived from estimates of capital intensity, the conventional system will likely be the lower-cost alternative.

Fewer Changes to Balance of Plant Equipment   Once it is established that a microreaction design has an advantage over the conventional unit (e.g., operating cost, capital cost, safety, etc.), the impact on balance of plant (BOP) and outside battery limits (OSBL) equipment should be considered for a final assessment of economic viability. Example considerations include:

- The start-up procedure may be different, requiring new equipment as compared to the conventional process.
- Site-specific equipment spacing guidelines may erode the footprint advantage of the microreaction process.

- Higher pressure drop through the microreaction equipment may require a transition from a blower to a compressor, which has higher capital and operating costs.
- Number/size/cost of guard beds and particulate filters may increase costs.

Process Equipment Cost vs. Overall System Cost   For some chemical units, such as water gas shift (WGS) in a hydrogen plant, if the candidate process equipment for replacement by microreaction technology is a relatively small portion of the overall system cost, the development effort to miniaturize the unit may not be justified from a pure capital cost basis.

Other factors that should be considered when evaluating microreaction vs. conventional process equipment include:

- Corrosion (microreaction components have thinner walls)
- ASME certification (standard pressure vessel methods may or may not apply)
- Catalyst refurbishment

In the end, a careful analysis of the economic drivers—both operating and capital—will show that some, but not all, chemical applications may achieve a significant return on investment by transforming a process using microreaction systems.

13.4   EXAMPLES OF LARGE-SCALE PRODUCTION

Several companies are working on large- or mega-scale demonstrations of microreaction technology. To date, no plants are known to have operated with a scale of production exceeding 100,000 mt per year. Several companies have completed technology development to manufacture and distribute technology to service the chemical industry for this scale of production, including Degussa, with Uhde GmbH and TU Chemnitz as partners; UOP; and Velocys.

13.4.1   DEMiS® Project

A team of researchers from Degussa and Uhde GmbH, together with several partners from academia under Demonstration Project for the Evaluation of Microreaction technology in industrial Systems (commonly known as the DEMiS® project), supported by the German Federal Ministry of Education and Research, described the demonstration of a pilot-scale unit and plan implementation of microreaction technology at a large scale. The first application was a pilot plant unit for the gas-phase epoxidation of propene using hydrogen peroxide at a several mt/yr scale [33–35]. Development of the pilot-scale unit focused on addressing several key technology risks and demonstrating the selected solutions. Specifically, the design and method of manufacturing addressed flexible methods to integrate catalysts, the mechanical integrity of the hardware, thermal management, safety considerations, including avoiding flammable mixtures in pipes and connections and equal flow distribution.

Development of the laboratory-scale unit focused on the design of the industrial-scale reactor, which in turn defined the design of the pilot plant reactor. The channel design of the laboratory reactor, as shown in Fig. 13.8, mimics the commercial reactor with respect to fluid dynamics, reaction engineering, and heat management. In another key design aspect, the channels were slits and thus only in the micrometer range in one dimension. The defined capacity was created by stacking plates, each with defined and identical reaction slits. The authors note that one scale-up parameter was modified in the pilot unit: the reactor length. It was cited that this change did not influence the critical mass transfer in the microchannel lateral dimension at equal residence time and thus the reactor performance in the pilot unit will approach that in the commercial reactor.

The dimensions of the pilot-scale unit were a height of 4 m, diameter of 1.4 m, and weight of 9 tons. The reactor modules stacked within the pilot unit were 1.0 by 1.6 m in dimension. The dimension of the microchannel was less than 1 mm, such that the reaction was safely operated in the explosive regime.

The catalyst (titanium silicalite) for the propene epoxidation reaction was integrated within the microchannels before stacking the plates to allow for line of sight application of the catalyst and excellent quality control. The reproducible catalyst thickness was found to be on the order of several hundred microns. The

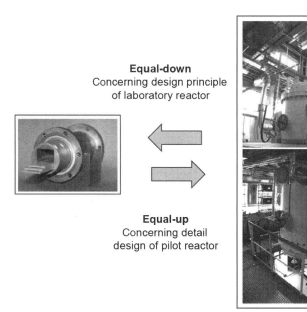

FIGURE 13.8 Design approach for microreactors, where the design of the test unit is derived from the full-scale reactor design and subsequent scale-up is achieved by numbering up reaction channels [34].

coated catalyst plates were stacked in a reactor housing to achieve both the target capacity and equal flow distribution. An example of the catalyst-coated plates is shown in Fig. 13.9, where uniformity has been demonstrated with an external application approach to both single-channel test reactors and large multichannel plates.

After the successful demonstration for the production of propene oxide, the industrial partners Uhde and Degussa have continued work on implementing microreaction technology for large-scale gas-phase applications, for example, vinyl acetate, with capacities in the range beyond 100,000 mt/yr. The development partnership consists of three main groups, each bringing unique core competencies to address development challenges. Degussa is developing the catalyst and coating techniques for metal microchannels. Reliability and feasibility of the coatings at the laboratory scale are being developed by the Chemnitz University of Technology. Uhde GmbH is developing the microreactor process and system engineering.

One obstacle that micro-reaction technology has to overcome on its way to chemical production is the industrial manufacture of mechanically and chemically stable catalyst coatings. Conventional vinyl acetate catalysts coated in the laboratory (extreme right, or right) and coated DEMiS® pilot reactor (below)

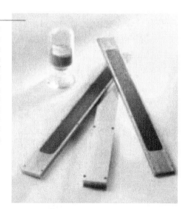

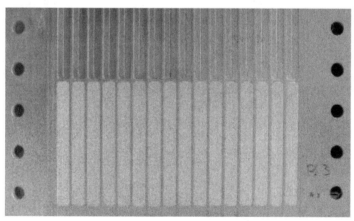

FIGURE 13.9   Robust catalyst integration protocols have been developed for the production of vinyl acetate in a microreactor. The coatings of several hundred microns in thickness are externally applied to reaction plates for quality control prior to reactor assembly [18].

The oxidative coupling of ethylene and acetic acid to vinyl acetate is highly exothermic and well suited for improved selectivity of a few percentage points over the current 92% with tight temperature control—thus addressing a key advantage of microreaction systems, *step-out chemistries*. The conventional multitubular reactor has a temperature exotherm well over 100 K. The large-scale microreaction system will target a 10% cost reduction in the plant capital cost requirement and a few percent lower operating costs. A pilot plant operated at a production capacity of 1 kg/h in 2007, and the marketing of the commercial-scale technology is expected to commence. The pilot plant to commercial unit capacity represents a scale-up factor of roughly 17,000X—a feat that is possible only because the channels are numbered up and, unlike with conventional technologies, all relevant parameters influencing heat and mass transport are kept constant.

13.4.2   DSM/Karlsruhe Reactor

Background   The development and fabrication of highly efficient micro heat exchangers and microreactors for chemical applications began at the Karlsruhe Research Center around 1990 [36]. In this design approach, metal foils with micro-channels are stacked on top of each other, joined by a diffusion welding process, and welded into a housing. Vacuum tightness is achieved between the reaction channels and the channels for temperature control as well as tightness to the outside and pressure resistance up to several hundred bar. Reactor scale-up to obtain technically relevant throughputs was based on increasing the number of channels inside the device until the required throughputs, residence times, and heat transfer capacities were reached. By this "internal numbering up," thousands or ten thousands of micro-channels can be integrated in a single reactor unit [37].

Karlsruhe Research Center and the company of DSM Fine Chemicals entered in a cooperative agreement to achieve the commercial synthesis of an organic intermediate product using a microreactor instead of a stirred tank in order to enhance the yield and to obtain a better product at lower cost [38]—again exploiting the advantage of a step-out chemistry in a microreaction system. The reactants for the synthesis are concentrated sulfuric acid and a mixture of organic liquids. The reaction is very fast and highly exothermic. The reactants have large differences in viscosity. The throughput of the production microreactor was 1,000 to 2,000 kg/h of reaction mixture.

Manufacturing the Production Microreactor   Before starting the design of the production microreactor, the potential advantages of microreaction technology for the described synthesis reaction and the optimal operating conditions were investigated using a laboratory-scale microreactor with reaction mixture throughputs of 1 kg/h. The lab-scale microreactor was equipped with an exchangeable micromixer, a passage with microchannels for the reaction, and a passage in cross-flow for strict temperature control using water as a fluid. A nickel-based alloy was used as material.

The experiments with the lab-scale microreactor showed a yield increase of 20% compared to the values of the stirred tank reactor; the residence time in the lab reactor was around several seconds. The scale-up of the lab reactor with 1 kg/h to the production reactor with a throughput of 1,000 to 2,000 kg/h was done by "internal numbering up" according to the Karlsruhe concept. The number of reaction channels was chosen according to the desired throughput, ensuring the same residence time as in the lab reactor.

The production microreactor is 70 cm long and has a weight of 300 kg. It has two inlets for the reactants and one outlet. In the inlet zone, the two reactants are mixed very rapidly and intensively. A device designed with the aid of computational fluid dynamics (CFD) was integrated downstream of the mixing part to distribute the reaction mixture equally over the inlets of the downstream reaction channels.

The microreactor contains several thousands of reaction channels and several tens of thousands of cross-flow cooling channels for temperature control. The volume of the reaction passage is about 3 L. The cooling capacity is several hundred kW. Before application, the production reactor was certified by the German licensing organization TÜV.

Integration in the DSM Plant and Results   The microreactor was integrated in DSM's production plant and operated with throughputs between 1,000 and 1,700 kg/h over several months in two production campaigns in 2005 and the beginning of 2006. Approximately 600 tons of product were obtained. As in the lab reactor, the yield was improved by 20% compared to the stirred tank used previously. The amounts of raw materials and of the waste streams were lowered, and plant operation safety was improved. The reactor is shown in Figs. 13.10 and 13.11.

Future Development   Based on the valuable experience gained so far in the production campaigns of DSM, the Karlsruhe Research Center will focus future development work on improving the production microreactor and developing new production microreactors for other applications in chemical production.

13.4.3   Velocys Hydrogen and FT Plants

Velocys is developing a steam methane reformer (SMR) to produce industrial hydrogen. The reactor is designed for a natural gas process feed between 20 and 30 atm inlet pressure and 3 : 1 steam-to-carbon ratio. The combustion fuel is from the pressure swing adsorption (PSA) tail gas (hydrogen-depleted reformate). A conventional medium-temperature WGS reactor will operate between the SMR reactor and PSA unit. Each commercial-scale reactor generates approximately 1-MM SCFD hydrogen (PSA product basis). To scale up, 1-MM SCFD reactors can be coupled in "assemblies" that can accommodate up to seven reactor blocks. Further increases in plant capacity can be accommodated through the use of multiple assemblies. The entire reforming unit—including nonmicroreaction technology, such as compressors and filters—is skid-mounted and shop-assembled.

A demonstration reactor shown in Fig. 13.12 produced 9,000 standard cubic feed per day (SCFD) hydrogen at a scale slightly less than 1% of a commercial reactor. The design and operation of each channel mimic the commercial-scale reactor. Cold reactant, air, and fuel were fed to one end of the reactor, where they subsequently

FIGURE 13.10 Microreactor of the Karlsruhe Research Center, successfully used in commercial chemical production at DSM Fine Chemicals, Austria (length of reactor: 70 cm, throughput: 1,000–2,000 kg/h).

recuperated heat with hot product and combustion exhaust streams before reacting near 850°C to form synthesis gas in less than 10-ms contact time. The unit has very low emissions and produced less than 10 ppm of NOx (nitrogen oxides). As can be seen in Fig. 13.13, the footprint of the Velocys reformer is substantially smaller (90% volume reduction) than an equivalently scaled conventional counterpart. Despite the substantial footprint advantage, the Velocys SMR is projected to reduce the capital cost for hydrogen production by 20%. The technology builds on the advantages of microreaction systems with both *improved thermal integration* and *modular construction* to compete with conventional technology. The low NOx emissions are a step-out chemistry that reduces additional plant emission abatement equipment and thus capital cost.

Velocys is also developing a Fischer–Tropsch (FT) reactor for the conversion of syngas to middle distillates from stranded natural gas with Total S.A. In a related project, Velocys is developing a FT reactor for a U.S. Army synthetic fuel production facility. Like the SMR application, the FT reactor has substantial CAPEX advantages over conventional systems through the advantage of step-out chemistry. Designed for operation at around 230°C with a temperature rise of a few degrees Celsius (as compared to 25°C for a conventional reactor), the Velocys FT reactor is designed to maximize carbon efficiency and selectivity. During a bench-scale demonstration, a Velocys microreaction FT reactor was operated for more than 500 h at less than

FIGURE 13.11 Microreactor of the Karlsruhe research center successfully used in commercial chemical production at DSM, Austria. This view shows the front of the stirred tank used for the reaction so far (length of reactor: 70 cm, throughput: 1,000–2,000 kg/h).

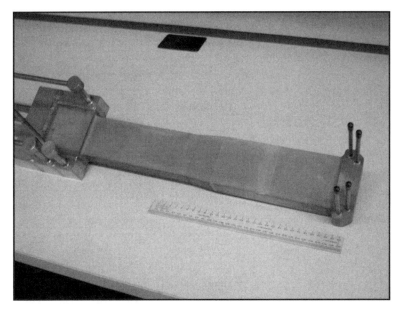

FIGURE 13.12 Pilot-scale demonstration with methane steam reforming reactor with integral heat generation from catalytic combustion to produce 9,000 SCFD of hydrogen.

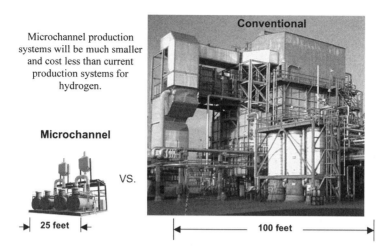

Microchannel production systems will be much smaller and cost less than current production systems for hydrogen.

Microchannel

VS.

Conventional

→| **25 feet** |← |← — — — — — **100 feet** — — — — — →|

FIGURE 13.13   The 20-MM SCFD Velocys SMR is substantially smaller than the equivalent conventional reactor.

300-ms contact time (CT). For the duration of this run, the average CO conversion was 70% and the average CH_4 selectivity was about 8%. The CAPEX of a greenfield 30,000-BPD (barrel per day) GTL plant was estimated given the observed reactor performance, a detailed P&ID, and quotes for major equipment. All equipment is skid-mounted and shop-assembled. Like the SMR, individual reactors are installed in reactor assemblies that are 8 ft in diameter and 28 ft in length. Each reactor assembly contains 12 Velocys FT reactors with manifolds for products, reactants, and cooling. Both inside battery limits (ISBL) and outside battery limits (OSBL) costs

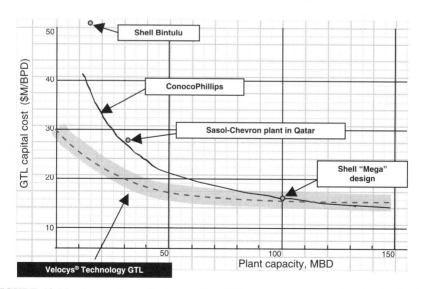

FIGURE 13.14   Comparison of microreaction GTL process vs. conventional technologies.

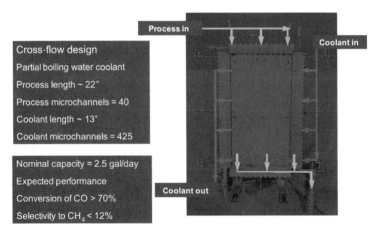

FIGURE 13.15 Bench-scale FT reactor scheduled for start-up in 2007.

were calculated. As compared to publicly available cost, there is a substantial cost advantage associated with the Velocys microreaction technology up to a plant capacity of about 50,000 BPD (Fig. 13.14). Between 50,000 and 100,000 BPD, microreaction technology is expected to have a smaller CAPEX advantage over conventional systems. Above 100,000 BPD, economy of scale makes conventional systems less expensive than the microreaction FT system. The OPEX of the Velocys FT system also creates a modest cost advantage.

Velocys is currently preparing to test a pilot reactor that is full-scale in length and a repeating unit of the full-scale reactor. The pilot-scale reactor, as shown in Fig. 13.15, is designed to produce 2.5 gal/day of FT products. A successful pilot reactor operation was completed in 2007. A second pilot reactor operated at 2 gas/day of FT products in 2008. The second generation pilot reactor validated commercial manufacturing methodology.

13.5 FUTURE OF LARGE-SCALE PRODUCTION OPPORTUNITIES

The development of microreaction technology commenced in the 1990s, intensified in the new millennia, and is poised for widespread commercial integration within the next decade. The level of interest across the chemical industry is high, and the pace of implementation will, in part, follow the lead of the early adopters. As the clear economic advantage is demonstrated at a large production volume for one company, the perceived and real technology risk will be lower for other companies—thus raising the risk of not adopting microreaction technology and losing market share.

Beyond the economic drivers for integrating the technology into the industry, other forces may accelerate or delay the introduction of new technology and are challenging to predict. Chemicals are in large part tied to the use of and cost of oil. Continued high prices for crude oil may enhance the implementation of new technology. Geopolitical forces that disrupt supplies of oil may enhance the development and implementation of technology based on alternate feedstocks, including coal, biomatter, or stranded reserves in politically accessible locations.

Although the timeline of adoption is uncertain, it is clear that new technologies that offer a cost advantage, reduced usage of feedstock per pound of product, or enable the use of alternative and more attractive feedstocks will always have a place in the chemical industry. Microreaction technology has the potential to be a step-change technology that moves the industry forward.

BIBLIOGRAPHY

1. Rouhi, A. M. (June 28, 2004). Microreactors eyed for industrial use. *Chemical and Engineering News*, 82(27): 18–19. http://www.pubs.acs.org/isubscribe/journals/cen/82/i27/html/8227sci2.html.

2. Seris, E., Abroamowitz, G., Johnston, A., and Haynes, B. (2005). Demonstration plant for distributed production of hydrogen from steam reforming of methane. *Chemical Engineering Research and Design*, 83(A6): 619–625.

3. O'Driscoll, C. (January 2004). Small is bountiful. *Chemistry World*, 1(1) http://www.rsc.org/chemistryworld/Issues/2004/January/bountiful.asp.

4. CPC (2006). *CYTOS Pilot System*. http://www.cpc-net.com/cytosps.shtml.

5. Roberge, D., Ducry, L., Bieler, N., Cretton, P. H., and Zimmermann, B. (2005). Microreactor technology: A revolution for the fine chemical and pharmaceutical industries? *Chemical Engineering Technology*, 28(3): 318–323.

6. Micro Residence Tool (2003). *Products: High Throughput Microfluidics 2003*. http://www.ltf-gmbh.de/en/products/microfluidic/reaction/mic_residence_tool.html.

7. mikroglas chemtechnik GmbH (2006). *Microreaction Technology, Microreactors, and Microreaction System Data*. http://www.mikroglas.com/index_e.html.

8. Hessel, V., and Lowe, H. (February 2005). Microreactor technology: Applications in pharma/chemical processing. Online *The Pharmaceutical Technology Journal*. http://www.iptonline.com/pdf_viewarticle.asp?cat=5&article=269.

9. ACHEMA (2006). Press release on IMM production capacity. 28th International Exhibition-Congress on Chemical Engineering, Environmental Protection and Biotechnology, Frankfurt am Main, Germany, May 15–19, 2006. http://www.achema.de/data/achema_/tb_08_e_Process%20Intensification.pdf.

10. Wood, M. (2005). Compact reactor technology for gas processing. Paper presented at the Microchannel Heat Exchangers Workshop, Houston, TX, December 7, 2005.

11. SYNTICS GmbH (2006). *Turbulence Mixing Reactor Description and Properties*. http://www.syntics.de/data-live-syntics-two/docs/pdf/20050822%20Produktdatenblatt%20TMR-englisch.pdf.

12. Ackerman, U. (2006). *New Chemistry with Micro Process Engineering*. http://www.mstonline.de/mikrosystemtechnik/mst-germany/industrial_applications_I/09_Ackermann_Bayer.pdf.

13. Ehrfeld Microtechnik BTS (2005). *Specification/Materials Data*. http://www.ehrfeld-shop.biz/shop/catalog/specification.php.

14. Banister, J., and Rumbold, S. (2005). A compact gas-to-methanol process and its application to improved oil recovery. Paper presented at the Gas Processors Association Europe Annual Conference, Warsaw, Poland, September 2005.

15. Jarosch, K., Mazanec, T., McDaniel, J., Tonkovich, A. L., and Fitzgerald, S. (2006). Compact mobile synthetic fuel unit. Paper presented at the American Institute of Chemical Engineers Spring National Meeting, Orlando, FL, April 23–27, 2006.

16. Tonkovich, A. Y., Perry, S., Wang, Y., Qiu, D., LaPlante, T., and Rogers, W. A. (2004). Microchannel process technology for compact methane steam reforming. *Chemical Engineering Science*, 59: 4819–4824.

17. Accentus (2003). *FMC Technologies, Accentus Form Gas-To-Liquids Joint Venture*. Press release. http://www.accentus.co.uk/press_releases/gtl_microsystems_press_release.pdf.

18. Albrecht, J., Becht, S., Geisselmann, A., Hahn, H., Caspary, K. J., Schirrmeister, S., et al. (2006). Degussa and Uhde develop a production scale microstructured reactor for use in commercial chemical plants. *Elements* (Degussa science newsletter), 15: 18–31.

19. Hessel, V., Löwe, H., Müller, A., and Kolb, G. (2005). *Chemical Micro Process Engineering: Processing and Plants*. Weinheim, Germany: Wiley-VCH Verlag and GmbH & Co.

20. Mazanec, T. (2006). Catalytic selective oxidation in microchannel reactors. Paper presented at the ACS Fall Symposium, San Francisco, CA, September 12, 2006.

21. Wang, Y., Jun, J., Cao, C. and Mazanec, T. J. (2005). Microprocess technology for Fischer–Tropsch gas-to-liquids. *American Chemical Society Division of Petroleum Chemistry*, 50(1): 69–70.

22. Jarosch, K., Tonkovich, A. L., Perry, S., Kuhlmann, D., and Wang, Y. (2005). Microchannel reactors for intensifying gas-to-liquid technology. In *Microreactor Technology and Process Intensification*, Wang, Y. and Holliday, J. (eds.), ACS Symposium Series, Washington, DC, Vol. 914: pp. 258–273.

23. Tonkovich, A. L., Arora, R., and Fitzgerald, S. (2005). Commercial scale microchannel technology methodology and capabilities. Paper presented at Group TK-IMRET 8, 8th International Conference on Microreaction Technology, AIChE Spring National Meeting, Atlanta, GA, April 10–14, 2005. Session 133: Modularization and Multiscale Design.

24. Werner, T. M., Schmitt, S. C., Daymo, E. A., and Wegeng, R. S. (1999). Microreaction gasoline vaporizer unit manufacturing cost study. PNNL-12226, Pacific Northwest National Laboratory, Richland, WA.

25. Martin, P. M., Matson, D. W., and Bennet, W. D., Microfabrication methods for microchannel reactors and separation systems. *Chemical Engineering Communication*, 173(1), June 1999, 245–254.

26. Martin, P. M., Matson, D. W., and Bennett, W. D. (1999). Microfabrication methods for microchannel reactors and separations systems. *Chemical Engineering Communications*, 173: 245–254.

27. News from Alfa Laval: Super Steel Bonding-Hall 4.0, stand H2-T8 (2006). *Achema Preview*. http://www.worldcoal.com/Hydrocarbon/HE_Achema_AlfaLaval.htm.

28. Tonkovich, A. L. Y., Yang, B., Perry, S. T., Fitzgerald, S. P., and Wang, Y. (2007). From seconds to milliseconds to microseconds through tailored microchannel reactor design of a steam methane reformer. *Catalysis Today*, 120, 21–29.

29. Hoare, R., and Gerhard, S. (2001). Building blocks for capital projects. *The McKinsey Quarterly*, 2: 56–61.

30. Hoitsma, K., and Snelgrove, P. (July–August 2002). Effective reformer design & erection. *World Refining*, 24–28.

31. Hensman, J. R., and Ashley, M. (Winter 2004). Modularized technology for GTL applications. *Process Technology Quarterly*, 131–135.

32. Bryden, R. T. (2000). To skid or not to skid. AACE International Transactions, 44th, EST.01.1–EST.01.8.

33. Klemm, E. J., Dietzsch, E., Schüth, F., Becker, F., Markowz, G., Döring, H., et al. (2006). DEMiS®: Results from the development and operation of a pilot-scale micro reactor on the basis of laboratory measurements. Paper presented at Group TK-IMRET 8, 8th International Conference on Microreaction Technology, AIChE Spring National Meeting, Atlanta, GA, April 10–14, 2005, Session 131: Microstructured Reactor Plant Concepts.

34. Klemm, E., Schwarz, T., Döring, H., Markowz, G., Becker, F., Geißelmann, A., et al., (2006). Catalyst coating in lab- and technical-scale for microreactors of the DEMiS®-type. Paper presented at 9th International Conference on Microreaction Technology (IMRET 9), Potsdam, Germany, September 6–8, 2006, Topic 4: Materials Aspects, Nanostructures and Nanoparticles.

35. Markowz, G. (2004). DEMiSTM project enters critical phase: New impetus for microreaction engineering. *Elements* (Degussa science newsletter), 4: 4–8.

36. Bier, W., Keller, W., Linder, G., Seidel, D., and Schubert, K., (1990). Manufacturing and testing of compact micro heat exchangers with high volumetric heat transfer coefficients. *DSC-Microstructures, Sensors and Actuators*, 19: 189–197.

37. Schubert, K., Brandner, J., Fichtner, M., Schygulla, U., and Wenka, A. (2001). Microstructure devices for application in thermal and chemical process engineering. *Microscale Thermophysical Engineering*, 5: 17–39.

38. Vorbach, M., Bohn, L., Kotthaus, M., Kraut, M., Pöchlauer, P., Wenka, A., et al. (2006). First large-scale application of microreaction technology within commercial production of DSM. Paper presented at ACHEMA 2006, Frankfurt, Germany, May 15–19, 2006.

CHAPTER 14

PROCESS INTENSIFICATION

MICHAEL MATLOSZ, LAURENT FALK, and
JEAN-MARC COMMENGE

14.1 INTRODUCTION

This chapter reviews the methods of intensification by microstructuring. A short introduction outlines the objectives of intensification and takes a quick look at current related activities throughout the world and in Europe.

Process intensification is a rather difficult domain and there is, strictly speaking, no general method showing how it should proceed. On the one hand, several successful examples of intensification may be found in the literature but without a view toward a transposable procedure. On the other hand, there are failed examples on which silence persists, without our knowing exactly why they failed. In fact, all processes cannot benefit from the advantages of microreactors and microstructured systems because of the slowness of certain phenomena (rate-limited reactions, e.g.).

In this chapter, we attempt to fill the methodological gap of intensification by proposing a simple method of analysis of the process based on its characteristic times. This method enables us to identify the times of important phenomena and their couplings. Very often, a limiting phenomenon exists, with the characteristic time relatively higher compared to the characteristic times of other phenomena. To intensify a process thus means, from the technical point of view, to act on the limiting phenomenon to decrease the characteristic time and thus accelerate the rate of the global process. Several case studies are presented here for the purpose of illustration.

14.2 DEFINITIONS AND OBJECTIVES

Process intensification involves the development of innovative methods and devices that, in comparison to existing approaches, offer the chance of a dramatic

Microchemical Engineering in Practice. Edited by Thomas R. Dietrich
Copyright © 2009 John Wiley & Sons, Inc.

improvement in the quality of production, substantial reduction in the ratio of equipment size to production capacity, and significant drop in the consumption of energy and production of waste [1]. Intensified process equipment and production systems are therefore key enabling factors for step–change improvement in process/plant efficiency, with respect to space, time, energy, raw materials, safety, and the environment; microstructured devices and components have an important role to play in this context. If successful, intensified process technologies should not only be cheaper but also globally more sustainable.

By doing more with less, intensified processes and devices should enable higher throughputs with equivalent or even smaller–volume devices, greater production with lower energy consumption, smaller quantities (or even absence) of solvents, substantially lower risk, reduced environmental impact, and higher selectivity for similar or even higher conversions of reactants. Process intensification should also enable a reduction in the number of process steps as well as the use of novel and more eco–efficient synthesis routes and thereby contribute directly to the following desirable outcomes for sustainable chemistry:

- Highly efficient, inherently safe, and environmentally benign technologies
- Smaller-sized facilities with maximum reuse of materials
- Multipurpose equipment with increased flexibility and decreased plant cost

Despite substantial academic interest and a number of successful industrial applications in recent years, highly intensified micro and/or meso–structured components, process equipment, and devices have yet to achieve their true potential on a large scale in the process industries. Additional research and development efforts are needed and must be focused on overcoming the barriers to widespread implementation in industry. Among the efforts required is the development of a methodology for the integration of microstructured equipment and devices into large–scale production facilities. For that purpose, a structured multiscale approach is essential, capable of adapting length and timescales to production requirements. The development of such a structured multiscale design lies at the heart of current research activity, including a major academic–industrial research initiative in Europe, the IMPULSE project [2–4].

14.3 INTERNATIONAL STATE-OF-THE-ART

Over the last 15 years, substantial research on small–scale structured devices for chemical applications has been undertaken, and a host of academic studies, as well as eight successful editions of the proceedings from the International Conference on Microreaction Technology (IMRET), have established a solid scientific basis for the fabrication and analysis of individual (generally unconnected) units. A number of reference books are now available [5–7], a substantial contribution to the already well-established general area of process intensification [8].

Although much of the research efforts in this area have been undertaken in Europe, significant contributions have also emanated from other regions—in particular, the United States and Japan. Although the American effort on microstructured devices was initially oriented largely toward transportation and military applications, the academic work of K. Jensen at MIT [9] has clearly addressed chemical production applications. An ambitious program on microchemical systems has also been initiated in Japan. The project, begun in 2002 and led by J. I. Yoshida, has led to very significant progress in close collaboration between academic institutions and the Japanese chemical industry [10]. Among the examples cited is the use of intensified devices for the oxidation of aromatics with peroxides under severe conditions [11].

A number of processes utilizing small–scale structured devices have been employed for process intensification on the production scale, most notably by Siemens, Merck, Clariant, and Degussa [12–15] in Europe, as well as the recent industrial developments at Velocys, FMC, and Dow Chemical [16] in the United States. In addition, the Chinese firm of Trustchem in Shanghai, in collaboration with the Institut für Mikrotechnik (IMM) in Mainz, Germany, has recently used microstructured process components for intensified operation for the synthesis of an azo pigment [17]. The Forschungszentrum Karlsruhe (FZK) in Karlsruhe, Germany, has reported its collaboration with DSM Fine Chemicals GmbH in Linz, Austria, in manufacturing a product for the plastics industry [18]. Additional industrial applications have also been reported by Johnson & Johnson R&D in the United States for highly exothermic reactions, unstable intermediates, and hazardous reagents [19] and by Ube Industries in Japan on a microscale tube reactor system for the production of pharmaceutical intermediates [20].

These successful efforts have generally been initiated by "technology push," and widespread industrial "market pull" has not yet appeared to be comprehensive or systematic on the part of chemical producers. The state-of-the-art in recent years has tended to focus on individual small-scale structured components and devices on the laboratory scale, rather than complete production systems. The major challenge facing the chemical industry today is therefore not the further development of individual locally structured units but instead the effective integration of those units in complete production systems.

In this connection, it should be noted that the seductively simple ideas of direct "numbering-up" or "scale-out" of microreactor systems initially envisioned in the 1990s are now being brought into question (see, e.g., the critical remarks in this regard in [21] and [22]). It now appears clear that the principle of numbering-up through the direct interconnection of individual small-scale units into large-scale production systems does not, in fact, solve the scale-up problem as initially intended. On the contrary, numbering-up displaces the (well-known) chemical engineering problem of process scale-up with a (for the moment, essentially unsolved) problem of multiscale process interconnection. In a similar manner, the true impact of individual process intensification units on whole-process performance (including reactant work-up and related process logistics) has not been completely explored. Basic design principles for process layout and process performance evaluation are clearly needed, and these issues will require substantial research efforts in view of the development of a truly generic multiscale design methodology.

As a complement to thorough research and development efforts, it should be noted that a sophisticated methodology for detailed equipment design, interconnection, and layout is only of use for industrial application once an initial, preliminary decision has been made to explore new technological options. The availability of approximate, short-cut methods and principles, derived from more complete, rigorous research results, is an additional challenge for emerging innovation that cannot be ignored.

Examination of the state-of-the-art reveals that for a thorough evaluation of true technological opportunities for the use of small-scale structured components in chemical production, a comprehensive and systematic protocol is required as an aid to decision making and for ultimate design and exploitation. Whether for the retrofit of structured components in an existing plant or for the new design of future plant facilities, a new methodological approach is an urgent need and will make a clear contribution to future industrial competitiveness in chemical production technologies. Comparable to a pinch analysis for heat integration or hazard and operability (HAZOP) for safety issues, the structured multiscale design methodology should be developed in such a way as to permit reliable qualitative and quantitative technoeconomic evaluation of structured multiscale process systems for both existing and potential production processes.

Although the direct numbering-up of microreactors is generally not an appropriate technological solution, a targeted, integrated approach to the use of microstructured devices and process components sets the stage for a true paradigm shift in the principles of chemical process engineering. Rather than adapting the operating conditions and chemistry to available equipment, the process structure, architecture, and equipment can now be adapted to the physicochemical transformation. Production units can be created by the integration of diverse, small–scale structured units into large–scale macroproduction devices. A key feature of the resulting structured chemical devices is local process control (through integrated sensors and actuators), leading to enhanced global process performance. This new multiscale design, characterized by the construction of large–scale systems with small–scale inner structuring, is a major step forward in the quest for a truly competitive and sustainable chemical industry.

The analysis of process performance in terms of length and timescales as presented in [23] and illustrated below is therefore an important step forward in the development of the structured multiscale design methodology required for the widespread use of microstructured process components for process intensification in industrial practice.

14.4 GUIDELINES

In order to take advantage of the potential of microstructured systems for process intensification, appropriate tools and methodologies are required for a proper design and relevant evaluation of their performance. Unfortunately, the recent development of these tools has occurred more quickly than the development of related design methodologies. As a consequence, this section presents a tool based on the analysis of the characteristic times of the physical and chemical processes involved in the system studied.

At first, this approach can be used to identify the limiting phenomenon, by comparing the individual processes influencing the global performance of the system. In a second step, the further analysis of these times enables us to develop strategies for a selective intensification of the process.

14.4.1 Characteristic Time Analysis

A general property of chemical engineering processes is the fact that their global performance is the result of a competition between physical and chemical phenomena (chemical reactions, mixing, heat or mass transfer, etc.). Nevertheless, even under steady-state conditions, these phenomena proceed at different rates, such as the chemical equilibrium of a reversible reaction that can be interpreted as the result of two opposing reactions. In opposition to this time dependence, the first specific feature of microstructured reactors is related to their typical dimensions, commonly admitted to be lower than a millimeter.

The characteristic time analysis introduced here seeks to combine both these features, by relating the individual characteristic times of the fundamental phenomena to the dimensions of interest in the process. These times can then be compared to the space-time or local convection time, respectively defined as

$$\text{Space-time:} \quad \tau = \frac{V}{Q}$$

$$\text{Local convection time:} \quad \tau_{\text{conv}} = \frac{R}{u}$$

where V denotes the system volume, Q the volume flow rate, R the characteristic dimension, and u the average velocity.

The expressions for the fundamental characteristic times can be obtained by various methods. They can be extracted from a list of physical and chemical properties involved in the process using a blind dimensional analysis similar to the Buckingham method utilized for dimensionless numbers. Unfortunately, this method may yield varying expressions for a unique phenomenon, which can be difficult to use thereafter. A more efficient procedure consists of extracting the characteristic times from the governing equations.

Table 14.1 presents a nonexhaustive list of the most commonly used characteristic times, covering homogeneous and heterogeneous reactions, as well as transfer phenomena, gravity, and surface tension effects. These times and their expressions are presented here in ascending order of their power dependence on the characteristic geometrical dimension. What should be noticed first is the large variation of the scale dependence with respect to the phenomenon. Homogeneous reactions exhibit no dependence on the characteristic dimension, whereas transfer phenomena are strongly influenced by the presence of walls and obstacles.

TABLE 14.1   Expressions of Various Elementary Characteristic Times and Their Dependence on the Characteristic Dimension R

Characteristic Time	Expression	Dependence on R
Nth-order homogeneous reaction	$t_{hom} = \dfrac{C_0}{r_0} = \dfrac{1}{k \cdot C_0^{n-1}}$	0
Gravity	$t_{grav} = \sqrt{\dfrac{2R}{g}}$	1/2
Apparent first-order heterogeneous reaction	$t_{het,1} = \dfrac{R}{2k_s}$	1
General heterogeneous reaction	$t_{het} = \dfrac{C}{r}$	Varying
Surface tension	$t_{surf} = \sqrt{\dfrac{\rho R^3}{2\sigma \cdot \cos(\theta)}}$	3/2
Viscosity	$t_{visco} = \dfrac{\rho R^2}{\mu}$	2
Diffusive mass transfer	$t_{diff} = \dfrac{R^2}{D_m}$	2
Convective mass transfer at constant Sherwood number	$t_{mass} = \dfrac{R^2}{Sh \cdot D_m}$	2
Heat conduction	$t_{cond} = \dfrac{\rho \cdot Cp \cdot R^2}{\lambda}$	2
Convective heat transfer at constant Nusselt number	$t_{heat} = \dfrac{\rho \cdot Cp}{\lambda} \dfrac{R^2}{Nu}$	2

Before discussing these dimension dependencies and the potentials for selective intensification, attention must be paid to the fact that a small characteristic time represents a fast phenomenon: A second-responding system runs faster than a minute-responding system. The information presented in Table 14.1 therefore clearly shows how microstructured reactors can selectively influence the rates of these combined phenomena. For example, reducing the characteristic dimension accelerates transfer phenomena, whereas the rates of homogeneous reactions remain unchanged.

Although particular expressions are presented in Table 14.1, a few general features may be noted. As far as transfer phenomena are concerned, a general expression of their characteristic times can be proposed by considering the transfer analogies. Indeed, heat transfer, mass transfer, and momentum transfer are similar since they describe the transfer of a particular extensive property that can diffuse in a specified medium. A general expression can be written as

$$t_{transfer} = \dfrac{R^2}{D_{phen} \cdot N_{dim}}$$

where R denotes the characteristic dimension, D_{phen} the diffusivity of the transferred extensity through the considered medium, and N_{dim} a dimensionless number required when convection effects are taken into account.

The dimensionless number N_{dim} does not appear when the transfer is purely diffusive or conductive. Under these conditions, a square-power dependence relates the characteristic time to the geometrical dimension. If the transfer process includes convection effects, the corrective dimensionless number is introduced, enabling us to take into account the influence of the velocity profiles. Table 14.1 presents two characteristic times when this number is taken into account: The convective heat and mass transfer times include, respectively, the Nusselt and Sherwood numbers. As soon as velocity and extensity profiles are fully developed, this dimensionless number can be considered constant and the square dependence remains valid. Under developing conditions, the dependence is more complex.

Chemical reactions can also be considered using a very general expression of their characteristic times. The construction principle is similar and consists of relating the concentration of a reactant of interest C_0 at initial conditions to the rate of its consumption r_0 at these same initial conditions (index 0 refers to the initial conditions):

$$t_{reaction} = \frac{C_0}{r_0}$$

This very general expression that can be used for homogeneous reactions as well as heterogeneous reactions may be simplified for particular cases, as shown in Table 14.1 for an nth-order homogeneous reaction and an apparent first-order heterogeneous reaction. Unfortunately, many reactions of interest exhibit complex kinetics that prevent one from using simplified expressions and require numerical estimations of the characteristic reaction times. Estimations can also be obtained from empirical kinetics measurements under similar operating conditions.

Particular attention must be paid to heterogeneous reactions, whose apparent kinetics may vary as a function of the operating conditions or reaction conversion. As an example, the usual Langmuir–Hinshelwood kinetics appears as a first-order reaction under low partial pressures, and as a zeroth-order reaction under high partial pressures of the reactant. Biological and enzymatic reactions present similar difficulties.

A second difficulty for heterogeneous reactions, which makes previous empirical kinetics measurements difficult to use, is the knowledge of their intrinsic kinetics. The reaction rate employed in the general expression above must be carefully chosen. Confusion between the intrinsic rate and an apparent reaction rate can have severe consequences on the analysis of intensification potentials. The comparison of the Thiele modulus and Weiss modulus enables us to understand possible mistakes. Their expressions are

$$\text{Thiele modulus:} \quad \varphi_s^2 = \frac{r_s L^2}{D_e C_s}$$

$$\text{Weiss modulus:} \quad \varphi_s' = \frac{\bar{r} L^2}{D_e C_s}$$

where r_s denotes the intrinsic reaction rate, $\bar{r}$ the apparent reaction rate, L the characteristic dimension, D_e the effective diffusivity in the catalyst, and C_s the surface concentration.

In addition to the possible misuse of the intrinsic or apparent reaction rate, these expressions demonstrate the relation that exists between the characteristic times presented here and the commonly used dimensionless numbers. Indeed, the Thiele modulus and Weiss modulus are the ratios of the effective diffusion time in the catalyst layer to the intrinsic or apparent reaction time. As a consequence, their values enable us to find the limiting phenomenon.

The usual dimensionless numbers can also be translated as the ratio of characteristic times:

$$\mathrm{Re} = \frac{\rho u d}{\mu}, \quad \mathrm{Da}_I = \frac{k_s R}{D_m}, \quad \mathrm{Bo} = \frac{\rho g d^2}{\sigma}, \quad \text{and} \quad \mathrm{We} = \frac{\rho d u^2}{\sigma}$$

For example, the Reynolds number relates the viscosity time to the local convection time. The first Damköhler number compares the diffusive mass-transfer time to the reaction time of a first-order heterogeneous reaction. The Bond number includes the ratio of the surface tension time to the gravity time. The Weber number relates the surface tension time to the local convection time.

To conclude this section, one should note that the characteristic time analysis proposed here is absolutely similar to usual methodologies based on dimensionless numbers. The main advantage to be highlighted is that this approach enables us to classify all the involved phenomena along a unique scale: the timescale. This characteristic time methodology is an approach that preserves good physical intuition when one analyzes the coupled phenomena of a complex process. As a result, a hierarchy of the physical and chemical processes can be established. This hierarchy is the basis of the selective intensification that can be performed thereafter using microstructured reactors.

14.4.2 Coupled Processes

Each characteristic times described above corresponds to a different fundamental phenomenon. The performance of a particular process that may be expressed as the conversion of a reaction or heat-transfer efficiency results from the coupling between these phenomena. As a consequence, the apparent characteristic time of this process is a combination of these individual times. The way these times are combined differs depending on their interactions.

To illustrate how the individual times combine, a plug-flow reactor where two parallel homogeneous first-order reactions occur is considered. The reactions and their kinetics are

$$A \;\longrightarrow\; B \qquad r_1 = k_1 C_A$$
$$A \;\longrightarrow\; C \qquad r_2 = k_2 C_A$$

The evolution of the concentration C_A along the reactor with respect to the space-time τ is described by

$$C_A = C_A^0 \exp\left(-(k_1 + k_2)\tau\right)$$

As a consequence, the performance of this system, that is, the conversion of reactant A, is

$$X_A = 1 - \exp\left(-\frac{\tau}{t_{\text{glob}}}\right)$$

where t_{glob} denotes the global time observed experimentally. This time is related to the reaction times as

$$\frac{1}{t_{\text{glob}}} = \frac{1}{t_1} + \frac{1}{t_2} \quad \text{with} \quad t_1 = \frac{1}{k_1} \quad \text{and} \quad t_2 = \frac{1}{k_2}$$

For this case, the global time is a harmonic combination of the individual times of the parallel phenomena. As a consequence, the global time will be on the same order of magnitude as the smaller individual time, that is, the fastest phenomenon.

It is interesting to note that this combination does not depend on the type of reactor used. Indeed, if these chemical reactions are considered in a continuously stirred tank reactor, the outlet concentration and conversion are, respectively,

$$C_A = \frac{C_A^0}{1 + (k_1 + k_2)\tau} \quad \text{and} \quad X_A = \frac{(\tau/t_{\text{glob}})}{1 + (\tau/t_{\text{glob}})} \quad \text{with} \quad \frac{1}{t_{\text{glob}}} = \frac{1}{t_1} + \frac{1}{t_2}$$

These expressions show that the individual times combine in the same way in both reactors. As a result, this combination is only influenced by the interactions between the fundamental phenomena.

This first feature can be developed by performing a similar analysis on consecutive phenomena. Unfortunately, purely consecutive phenomena are rare in chemical engineering. A case study may then consist of considering the performance of a first-order heterogeneous reaction $A \rightarrow B$ coupled with mass transfer in a fluid phase. In a cylindrical plug-flow reactor, under steady-state conditions, the mass-transfer rate from the fluid phase to the catalytic wall J_A and the apparent reaction rate r_s are

$$J_A = k_g(C_A^g - C_A^s)$$
$$r_s = k_s C_A^s$$

where J_A is the transferred flux, k_g the mass-transfer coefficient, C_A^g the reactant concentration in the bulk fluid, C_A^s the reactant surface concentration, and k_s the apparent first-order heterogeneous rate constant. The reactant concentration along the reactor decreases with respect to the space-time as

$$C_A^g = C_A^{g,0} \exp\left(-\frac{a k_g k_s}{k_g + k_s}\tau\right) \quad \text{with a conversion} \quad X_A = 1 - \exp\left(-\frac{\tau}{t_{\text{glob}}}\right)$$

Developing this expression enables us to relate the global characteristic time to the reaction and mass-transfer times:

$$t_{glob} = \frac{1}{ak_g} + \frac{1}{ak_s} = \frac{R}{2k_s} + \frac{R^2}{Sh \cdot D_m} = t_{het,1} + t_{mass}$$

The combination is very different from the first case: The global time for consecutive phenomena is the sum of the individual times. This relation implies that the observed performance will result from the largest individual time, that is, the slowest phenomenon.

The analysis of both these elementary cases demonstrates the complexity that can be expected during the analysis of complete chemical engineering studies of interest. Nevertheless, these examples also show that the individual times combine similarly to electrical resistances: The equivalent resistance to resistances in series is the sum of the resistance, whereas the reciprocal resistances are used for parallel resistances.

As far as process intensification is concerned, the global time of the heterogeneous reaction coupled with mass transfer is a basis to discuss the hierarchy of involved phenomena and to highlight the intensification potentials using microstructured reactors. Indeed, the individual reaction and mass-transfer times exhibit a different dependence on the characteristic dimension R. The reaction time is, here, linear with R, whereas the mass-transfer time presents a squared power dependence. As a consequence, reducing this dimension will intensify mass transfer faster than the reaction rate, enabling a mass-transfer limitation to vanish.

The complexity described above can be demonstrated by studying a physical process that involves three coupled phenomena. Indeed, the ascension of a liquid meniscus in a vertical capillary involves gravity, surface tension effects, and viscous flow. The efficiency of this process is defined as the ratio of the current height of liquid in the capillary to the height reached by the liquid at equilibrium. A simplified force balance enables us to extract the expression of a global characteristic time as a function of the elementary times:

$$t_{glob} = \frac{t_{grav}^4}{t_{surf}^2 \cdot t_{visco}}$$

Nevertheless, this complexity is counterbalanced by a certain homogeneity of the relations between the efficiency of the coupled processes and the global characteristic time of the process. Indeed, in most cases, the efficiency of the process can be related to the number of operation units (NOU), which is defined as the ratio of the space-time of the fluid in the device to the global operation time of the process carried out in the device:

$$NOU = \frac{\tau}{t_{glob}}$$

The relations between the efficiency and the global characteristic time can generally be explicitly derived from mass or force balances. The examples above, dealing with parallel and consecutive phenomena, have established the two main relations between the efficiency η and the NOU:

$$\text{For a plug-flow reactor:} \quad \eta = 1 - \exp(-\text{NOU})$$

$$\text{For a continuously stirred tank reactor:} \quad \eta = \frac{\text{NOU}}{1 + \text{NOU}}$$

More complex couplings, such as the capillary ascension or enzymatic reaction, do not yield explicit expressions, but implicit algebraic or even differential equations relating the efficiency and the NOU. In spite of this difficulty, these relations exhibit common features that can be seen in Fig. 14.1. This illustration compares the η-NOU relations for first-order reactions in a plug-flow reactor and in a continuously stirred tank reactor, for an enzymatic reaction following Michaelis-Menten kinetics in a plug-flow reactor and for the capillary ascension. These curves do not perfectly overlap but present similarities that can be advantageously exploited to study the intensification potentials of a system. Indeed, even if the η-NOU relation is not explicitly known, measurement of the efficiency and comparison with space-time enable us to obtain an order for magnitude for the global characteristic time. Further comparison of this apparent characteristic time to the elementary times of the phenomena involved allow us to detect which phenomenon is actually limiting the efficiency, and what sort of intensification strategy should be considered.

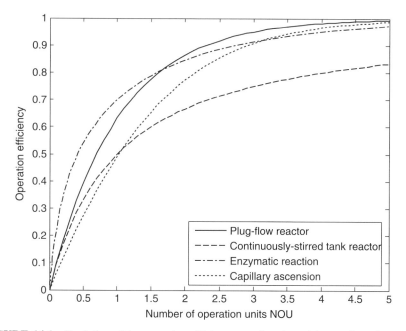

FIGURE 14.1 Evolution of the operation efficiency as a function of the number of operation units, for various operations.

This primary analysis then enables us to reduce the global characteristic time of the system. As a consequence, the space-time can also be reduced, whereas the NOU, which is equivalent to the efficiency, can be maintained as a constant. Finally, a decrease in the space-time induces a decrease in the reactor volume at a fixed flow rate, that is, miniaturization of the system, which is one of the objectives of process intensification.

14.4.3   Uniformity vs. Dispersity

Microreactors and microstructured reactors are essentially based on the principle of parallel channels that are all supposed to operate identically. Based on this assumption of an ideal similarity between the channels, it is possible to easily design and estimate the performances of an industrial microstructured system extrapolated by numbering-up. The identical operating conditions are based on the equal repartition of the flow between the channels, which is carried out thanks to a distribution chamber at the inlet and a collecting chamber at the outlet of the channels set.

The design of an optimal geometry for the multistructured system is difficult, and furthermore, its manufacturing may not be perfect. In particular, geometrical defects may appear during catalyst coating or be induced by chemical deposit during operation, which can partially or totally block specific channels. As a result, channels may not operate identically because of flow maldistribution. The reasons for and quantification of this loss of performance are given below.

As presented in the previous paragraph, the efficiency of an operation (heat and mass transfer or reaction conversion) is an increasing function of the residence time of the fluid in the channel, assuming as a plug-flow system [Fig. 14.2(a)]. In the case of a reactive system implying multiple reactions, the yield or selectivity of one of the intermediary products is a function of the residence time that presents an optimal value [Fig. 4.2(b)]. In both cases, these functions are concave.

Let us consider two parallel channels A and B of the same length in which a certain operation is carried out. If the flow passing in both channels were strictly identical, the residence time would be the same and the operating efficiency [Fig. 14.4(a)]

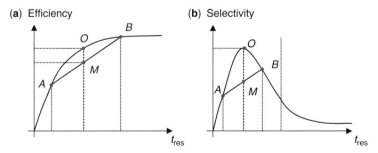

FIGURE 14.2   Evolution of the efficiency (a) and selectivity (b) as a function of the residence time.

and selectivity [Fig. 14.2(b)] would be optimal, represented by the operating point O in the figure. Because of the maldistribution, the residence time in channel B is larger than in channel A, with each one characterized by a specific efficiency and selectivity. Because of the concavity of the operating curve, the mean operating point M has a lower efficiency [Fig. 14.2(a)] and lower selectivity [Fig. 14.2(b)]. More generally, any deviation in flow uniformity in the channels induces a decrease in performance.

For any feature $Z(t_{red})$, a function of the residence time t_{red} and characteristic of an operation, it is possible to calculate the mean value $\bar{z}$ over all the channels by the average value weighted by the local flow rate q_i in the channels:

$$\bar{Z} = \frac{\sum_{i=1}^{n} q_i Z(t_{red,i})}{Q}$$

where Q denotes the total flow rate being subdivided in n channels.

In fact, the flow in small channels is very often laminar and the velocity profile parabolic. This makes the assumption of a plug-flow not totally valid and it is necessary to consider the real flow field. Therefore, it is preferable to generalize the previous concept by using the residence time distribution concept $E(t_{red})$:

$$\bar{Z} = \int_{0}^{\infty} Z(t_{red}) E(t_{red}) \, dt_{red}$$

Example   The following aims to quantify the impact of the flow nonuniformity on the performance of a multistructured reactor in which the consecutive reaction $A \rightarrow R \rightarrow S$ is carried out. Kinetics are first order: $r_1 = k_1 C_A$ and $r_2 = k_2 C_R$. In the case of a plug-flow, the relative yield on R is given by

$$\frac{C_R}{C_{A0}} = \frac{k_1}{k_2 - k_1} \left(e^{-k_1 t_{red}} - e^{-k_2 t_{red}} \right)$$

The maximal yield value is obtained for an optimal residence time:

$$t_{opt} = \frac{1}{k_2 - k_1} \ln\left(\frac{k_2}{k_1}\right)$$

Because of the maldistribution between the channels and the actual parabolic flow in every channel, the residence-time distribution in the entire reactor (constituted by a set of parallel channels) is not a perfect plug-flow. It is assumed here to be represented by a normal distribution, where the mean is equal to the optimal residence time (maximal yield).

Figure 14.3 represents the impact of the maldistribution on the yield of intermediate product R for different kinetics constants. The maldistribution is characterized by the normalized standard deviation (i.e., standard deviation of the normal distribution divided by the mean residence time). For a normalized standard deviation

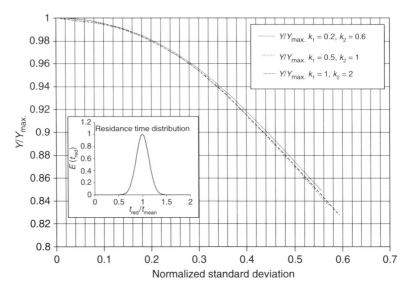

FIGURE 14.3 Impact of maldistribution on the yield as a function of the normalized standard deviation of the residence time distribution.

of zero, the flow is ideally equidistributed between every channel and the yield is maximal($Y/Y_{max.} = 1$). For a normalized standard deviation higher than 0.2, the performance can be affected sensitively and efficiency loss may reach several percent.

14.5 CASE STUDIES

14.5.1 Continuous Gas–Liquid Reactor

We show in this example how it is possible from the mass balance equations on a reactor to derive a simple equation relating the reactor performance to the characteristic times of the process.

In this particularly simple reaction case (first-order kinetics $A + B \rightarrow$ Products in an ideal continuous mixed-flow reactor), an analytical equation can be obtained. The analysis of this equation indicates the most relevant characteristic times, that is, the operating parameters to be modified in order to intensify the process. In the case of a more complex system where no analytical solution can be found and only numerical solutions may be obtained, the present approach still remains valid and can be used to select the best operating parameters for intensification.

Let us consider a gas–liquid continuous perfectly mixed reactor schematically presented in Fig. 14.4. Reactant A present in the gas phase is absorbed in the liquid phase and reacts with reactant B.

In the case of a slow reaction, the gas reactant A that is transferred into the liquid phase has time to diffuse far from the gas–liquid interface in the entire volume of liquid before reacting with the reactant B. The concentration gradient of species A in the liquid is very flat, and one may consider that the reaction takes place in the bulk volume of liquid. Three mass-balance equations, one per phase and per reactant, can be written as

Mass balance on reactant A in the gas phase:

$$Q_g C_{A,g}^{in} - k_{gl}a\left(C_{A,g}^{out} - \alpha C_{A,l}^{out}\right)V = Q_g C_{A,g}^{out}$$

Mass balance on reactant A in the liquid phase:

$$k_{gl}a\left(C_{A,g}^{out} - \alpha C_{A,l}^{out}\right)V - kC_{A,l}^{out}\varepsilon V = Q_l C_{A,l}^{out}$$

Mass balance on reactant B in the liquid phase:

$$Q_l C_{B,l}^{in} - kC_{A,l}^{out}\varepsilon V = Q_l C_{B,l}^{out}$$

where additional relations are considered:

Equilibrium condition:

$$C_{A,g}^{eq} = \alpha C_{A,l}^{eq}$$

Reaction rate in the liquid phase: $r = k\,C_A$ and specific parameters used:

k_{gl} = the mass-transfer coefficient
a = the specific interfacial area per volume of the reactor
ε = the volume fraction of the liquid phase

According to the previous chapter, one can introduce the following characteristic times:

Characteristic time of reaction:

$$t_{reaction} = \frac{1}{k\varepsilon}$$

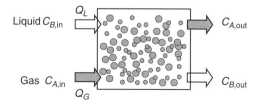

FIGURE 14.4 Schematic view of a gas–liquid continuous perfectly mixed reactor.

Characteristic time of mass transfer:

$$t_{\text{mass}} = \frac{1}{k_{gl}\, a}$$

Space-time of the liquid phase:

$$\tau_l = \frac{V}{Q_l}$$

Space-time of the gas phase:

$$\tau_g = \frac{V}{Q_g}$$

and the ratio of specific times:

The Damköhler number defined as the ratio of the space-time of the liquid to the characteristic time of reaction:

$$\text{Da} = \frac{\tau_l}{t_{\text{reaction}}}$$

The number of transfer unit (NTU) defined as the ratio of the space-time of the liquid to the characteristic time of mass transfer:

$$\text{NTU} = \frac{\tau_l}{t_{\text{mass}}}$$

The performance of the reactor is given by the conversion of reactant B:

$$X_B = \frac{1 - C_{B,l}^{\text{out}}}{C_{B,l}^{\text{in}}}$$

The combination of the three mass balances and rearrangement yield a single expression of the conversion as a function of the characteristic times of the process:

$$X_B = \frac{C_{A,g}^{\text{in}}}{C_{B,l}^{\text{in}}} \left[\frac{\text{Da}}{(1 + \text{Da})(t_{\text{mass}} + \tau_g)/\tau_l + \alpha} \right]$$

From this equation, one can see that the conversion increase (intensification) is obtained for large values of the Damköhler number and low values of the ratio $(t_{\text{mass}} + \tau_g)/\tau_l$ (in the limit of the model used here, i.e., slow reaction compared to mass transfer). These conditions corresponds to:

High concentration of reactant A at the reactor inlet

High value of the residence time of the liquid phase that can be obtained by increasing the reactor volume and/or decreasing the liquid flow rate

Low value of the characteristic time of mass transfer that can essentially be obtained by increasing the G/L interfacial area

Low value of the residence time of the gas phase obtained by decreasing the reactor volume and/or increasing the gas flow rate.

Some operating parameters have the opposite effects on the conversion. Furthermore, all these operating parameters do not have the same relative effect, with some having a more pronounced impact. A method for analysis consists of evaluating the sensitivity of each parameter by calculating the elasticity criteria.

Elasticity is a measure of the incremental percentage change in the variable X_B with respect to an incremental percentage change in another variable Y:

$$E_{X/Y} = \frac{\Delta X_B / X_B}{\Delta Y / Y}$$

Figure 14.5 and Table 14.2 present the values of the elasticity of the conversion with respect to key operating parameters and can be considered "intensificability" charts. From these elasticity values, it is possible to determine a capability criterion for intensification. The two parameters having a great influence (elasticity equal to 1) are the residence time of the liquid phase and the concentration of reactant A at the inlet. The residence time of the gas phase, with an elasticity value of -0.1 (a decrease of 10% of the residence time leads to an increase of 1% of the conversion) is a much less sensitive parameter. As for the mass-transfer characteristic time, with a quasi-null elasticity value (-2.10^{-4}), intensification is almost impossible. In the present case of a

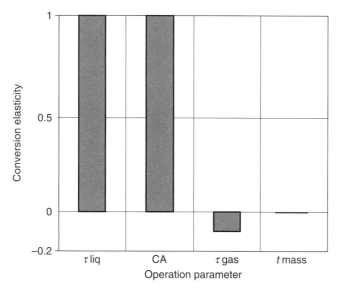

FIGURE 14.5 Values of the elasticity of the conversion with respect to key operating parameters.

TABLE 14.2   Values of the Elasticity of the Conversion with Respect to Key Operating Parameters

Parameter	τ_{liquid}	$C_{A,inlet}$	τ_{gas}	t_{mass}
Elasticity	1	1	-0.1	-2.10^{-4}
Intensification	Easy	Easy	Moderate	Nonrelevant

gas–liquid kinetics-limited reaction, it would be necessary for a 10% conversion increase to use a microstructured reactor, which would present an increase of the specific interfacial area by a factor of 500 (assuming a constant mass-transfer coefficient).

14.5.2 Transposition of a Fast Exothermic Reaction from a Semi-Batch Stirred Tank to a Continuous Microreactor

In fine chemistry, there are numerous exothermal fast reactions carried out in batch reactors. Because of the weak cooling capacities of the stirred tank reactors, these reactions are often slowed down by diluting the concentration of the reagents and by slowly feeding the reactor, so that the thermal power released by the reaction can be safely evacuated by the device.

Krummradt et al. [13] studied experimentally in various scales and in various structures of processes the exothermal synthesis ($\Delta_R H = -300$ KJ/mol) of an organic compound of fine chemistry. This organometallic Grignard synthesis includes side reactions (parallel and consecutive) that are thermally activated. The following main reaction is fast but the reaction rate is not exactly known:

These same authors showed experimentally that the reaction yield could be increased while diminishing considerably the operation time by a factor of 2,000, by replacing a conventional stirred tank by a set of microreactors (Table 14.3).

This a priori very spectacular result can be explained very easily thanks to the characteristic time analysis.

TABLE 14.3   Experimental Results from Krummradt et al. [13]

Reactor Type	Reactor	Operation Time	Yield (%)
Semibatch stirred tank	Laboratory (0.5 L)	Feed time 0.5 h	88
	Industrial (6,000 L)	Feed time >5 h	72
Continuous plug-flow	IMM microreactor	Space-time <10 s	95
	Minireactor	Space-time <10 s	92

This analysis is developed from the mass and heat balances on two configurations of reactors that are the semibatch reactor and the microreactor, considered here as a plug-flow reactor [25]. In both configurations, the reactive liquid is cooled by an external medium assumed to be at a constant temperature of the wall.

Semibatch Reactor   The expressions of heat and mass balances are, respectively,

$$UA(T_{wall} - T) = (V\rho c_P + \Gamma_R)\frac{dT}{dt} + Q\rho c_P(T - T_{in}) + \Delta_R H \cdot r_P V$$

$$r \cdot V = -V\frac{dC_A}{dt} - Q(C_A - C_{A0})$$

For simplification, it is assumed that the reactor operates in isothermal conditions, which implies $\frac{dT}{dt} = 0$. This restrictive condition has the advantage of avoiding the determination of the reaction rate r, which is not precisely known here.

The combination of equations yields the differential equation of the concentration in the semibatch reactor:

$$(at_f + t)\frac{dC_A}{dt} + C_A = C_{A0} - \frac{UA(T_{wall} - T)}{Q\Delta_R H} - \frac{\rho c_p(T_{in} - T)}{\Delta_R H}$$

where $a = \dfrac{V_0}{V_f}$ is the ratio of the initial volume in the semibatch reactor and the fed volume.

After integration, the concentration of reactant A in the reactor can be written as

$$C_A = C_{A0}\left(1 + \frac{t_f}{J \cdot t_{heat}} \cdot (T_{wall} - T) + \frac{(T_{in} - T)}{(a+1) \cdot J}\right) \cdot \left(\frac{t}{a \cdot t_f + t}\right)$$

where

$$J = \left(\frac{(-\Delta_R H) C'_{A0}}{\rho c_P}\right) \text{ is defined as the adiabatic temperature rise with}$$

$$C'_{A0} = \frac{C_{A0}}{1 + a}$$

$$t_{heat} = \frac{m \cdot c_P}{U \cdot A} = \frac{\rho \cdot V \cdot c_P}{U \cdot A} \text{ is the characteristic time of heat exchange.}$$

$$t_f = \frac{V_f}{Q} = \left(\frac{1}{1+a}\right)\frac{V}{Q} \text{ is the feeding time.}$$

The operation's characteristic time is defined as the feeding time required to ensure the isothermal behavior of the reactor:

$$t_f = \left(\frac{T_{in} - T}{J \cdot (a+1)} + X\right) \cdot \frac{t_{heat}J}{(T - T_{wall})}$$

where X is the conversion $X = \dfrac{C'_{A0} - C_{Af}}{C'_{A0}}$ calculated at the concentration C_A (time $= t_f$).

From the previous relation, it can be noted that the operation's characteristic time is directly proportional to the characteristic time of heat exchange, which is the key parameter for intensification.

Plug-Flow Reactor   The mass and heat balances resolution leads to the following expression of the space-time required to ensure the isothermal conditions of the reactor. The space-time is defined here as the characteristic operation time:

$$\tau = \frac{J \cdot X \cdot t_{\text{heat}}}{T - T_{\text{wall}}}$$

where $t_{\text{heat}} = \dfrac{\rho c_p D^2}{4\lambda \text{Nu}}$ is the characteristic time of heat transfer with λ the thermal conductivity of the fluid, and D the diameter of the channel, and Nu the Nusselt number.

It should be noted that this characteristic time, proportional to the square power of the channel dimension, strongly decreases when the diameter in the mini- or micro-reactor diminishes. This explains the strong impact of miniaturization on intensification, compared to classical stirred tanks.

The characteristic operation time in both configurations exhibits the same relationship that is proportional to the characteristic time of heat transfer and to a function ψ_{process} that is characteristic of the structure of the process as follows:

$$\tau_{\text{operation}} = \psi_{\text{process}}(J, X) \cdot t_{\text{heat}}$$

Numerical application shows that function ψ_{process} has almost the same order of magnitude for the two types of reactors investigated here, so the key parameter to modify the operation time is the time for heat transfer and not the process structure.

Generally, this relation illustrates how it is possible to reduce the operation's characteristic time by reducing the characteristic time of heat exchange. Therefore, the reduction of the process size, which decreases the characteristic time of heat exchange, explains the very weak values for the residence time in the microstructured reactor compared to the large vessel.

We estimate the characteristic operation time in the different structures thanks to the previously determined relations. For a given conversion rate, we assumed the following conditions to calculate the characteristic parameters of the process:

$$T = T_S, \quad T - T_{\text{in}} = T - T_{\text{wall}} = 20\,\text{K}, \quad C_A = 1\,\text{Mol/L}$$

The surface-to-volume ratio of the stirred tank $\left(\dfrac{A}{V}\right) = (4.3) \cdot V^{-1/3}$ and the global heat exchange coefficient is $U = 200\,\text{Wm}^{-2}\text{K}^{-1}$ for the stirred tank (the heat-transfer limitation in the jacket).

To simplify the case of the plug-flow reactor, the theoretical value of 3.66 is assumed for the estimation of the Nusselt number for fully developed temperature profiles. The results presented in Table 14.4 indicate that the best efficiency in

TABLE 14.4   Predicted Values for the Operation's Characteristic Time

Reactor	Diameter (m)	U $(W \cdot m^{-2} \cdot K^{-1})$	$\dfrac{A}{V}$ (m^{-1})	Characteristic Time of Heat Transfer t_{heat}	Operation Characteristic of Time
Fed batch	2	200	2.4	2.5 h	7 h
Plug-flow reactor (minireactor)	0.001	2,200	4,000	5×10^{-1} s	1.5 s
Plug-flow reactor (microreactor)	0.0001	22,000	40,000	5×10^{-3} s	1.5×10^{-2} s

terms of characteristic operation time is obtained in the plug-flow reactor, followed by the continuous and semibatch stirred tank.

We highlight the effect of the miniaturization that allows a significant reduction in the characteristic operation time in the plug-flow configuration. In the semibatch process, we calculate a feed time of 7 h. By comparison, the adopted experimental feed time, 5 h, does not enable us to trust the isothermal criteria; it must be checked for good agreement between the experimental data and the results predicted by the analysis.

The case of the plug-flow reactor, with a channel diameter equal to 100 μm, leads to a very short space-time ($5 \cdot 10^{-3}$ s), which may not be realistic. Indeed, the chemical regime will probably be reached and the reaction rate must be the limiting phenomenon that will fix the space-time as a consequence.

14.6   OTHER INTENSIFICATION PRINCIPLES

The previous sections have presented the possibilities offered by microstructured reactors to selectively accelerate phenomena in order to acheive the objectives of process intensification. Unfortunately, microstructured reactors may present drawbacks that can become detrimental from an operational or technical point of view. For example, channel clogging with solid particles, or the technical impossibility of largely parallel numbering-up, may lead one to search for other intensification technologies.

Stankiewicz and Moulijn [1] give an overview of the numerous equipments and technologies to intensify a process, depending on the phenomena involved. Among these equipments, one can cite as examples emblematic spinning-disk reactors, monolithic reactors, static mixers, or high-gravity systems, whose applications can be identified easily by comparison with conventional technologies.

In addition to these equipments, intensification methodologies probably offer the most significant improvements since they directly intend to act on couplings between phenomena. Multifunctional reactors such as reactive distillation, reactive extraction, membrane reactors, or reactive extrusion have already demonstrated their potentials. Hybrid separations and the use of alternate energy sources are under continuous development and may enable users to develop totally new applications.

Nevertheless, in spite of this high number of possibilities, the choice of one technology instead of another remains very difficult. For a given application, several equipments or technologies appear to be potential candidates for intensification, but qualitative and quantitative selection criteria are still lacking for a proper choice from an operational and technical point of view. Work is still required on this particular step toward a unified decision-making approach, one that will enable us to properly choose the most appropriate intensification technology. Simplifying this decision-making step will accelerate the development and operation of these technologies at the industrial scale and make it possible to fulfill the objective of sustainable production.

BIBLIOGRAPHY

1. A. Stankiewicz and J. A. Moulijn (January 2000). *Chemical Engineering Progress*, 96: 22–34.

2. T. Bayer, J. Jenck, and M. Matlosz (May 2004). IMPULSE—Ein neuartiger Ansatz für die Prozessentwicklung. *Chemie Ingenieur Technik*, 76(5): 528–533.

3. T. Bayer, J. Jenck, and M. Matlosz (April 2005). IMPULSE—A new approach to process design. *Chemical Engineering and Technology*, 28(4): 431–438.

4. Impulse, http://www.fzk.de/fzk/idcplg?IdcService=FZK&node=0739&document=ID_007705& lang=en, DECHEMA e. V.

5. M. Matlosz, W. Ehrfeld, and J.-P. Baselt (eds.) (2001). *Microreaction Technology. IMRETS: Proceedings of the 5th International Conference.* Berlin: Springer-Verlag.

6. V. Hessel, S. Hardt, and H. Löwe (2003). *Chemical Micro Process Engineering: Fundamentals, Modelling and Reactions.* Weinheim: Wiley-VCH.

7. V. Hessel, H. Löwe, A. Müller, and G. Kolb (2005). *Chemical Micro Process Engineering: Processing and Plants.* Weinheim: Wiley-VCH.

8. A. I. Stankiewicz and J. A. Moulijn (2004). *Re-engineering the Chemical Processing Plant: Process Intensification.* New York: Marcel Dekker.

9. K. F. Jensen (1999). Microchemical systems: Status, challenges and opportunities. *AIChE Journal*, 45: 2051–2054.

10. J.-I. Yoshida (2003). Highly selective reactions using microstructured reactors. Keynote lecture, *7th International Conference on Microreaction Technology*, Lausanne, Switzerland, September 2003. http://www.mcpt.jp.

11. K. Yube and K. Mae (2005). Efficient oxidation of aromatics with peroxides under severe conditions using a microreaction system. *Chemical Engineering Technology*, 28: 331–336.

12. T. Bayer, D. Pysall, and O. Wachsen (2000). Micromixing effects in continuous radical polymerization. In *Proceedings of the 3rd International Conference on Microreaction Technology* (IMRET 3), pp. 165–170. Frankfurt/Main, Germany, April 19–21, 2000. Berlin: Springer-Verlag.

13. H. Krummradt, U. Koop, and J. Stoldt (2000). Experiences with the use of microreactors in organic synthesis. In *Proceedings of the 3rd International Conference on Microreaction*

Technology (IMRET 3), pp. 181–186. Frankfurt/Main, Germany, April 19–21, 2000. Berlin: Springer-Verlag.

14. U. Nickel, et al. (February 7, 2002). Conditioning of pigments. U.S. Patent Application US2002/0014179. Wille, C., Autze, V., Kim, H., Nickel, U., Overbeck, S., Schwalbe, T., et al. (2002). Progress in transferring microreactors from lab into production—an example in the field of pigments technology. In *Proceedings of the 6th International Conference on Microreaction Technology* (IMRET 6), AIChE Pub. No. 164 (2002), pp. 7–17. New Orleans, LA, March 11–14, 2002.

15. R. Schuette, et al. (March 7, 2002). Process and device for carrying out reactions in reactor with slot-shaped reaction spaces. U.S. Patent Application US2002/0028164.

16. Microchannels: Reactors with parallel microscale channels offer commercial benefits. *Chemical and Engineering News*, October 11, 2004, 82(41): 39.

17. H. Pennemann, S. Forster, J. Kinkel, V. Hessel, H. Löwe, and L. Wu (2005). Improvement of dye properties of the azo pigment yellow 12 using a micromixer-based process. *Org. Process Research Development*, 9(2): 188–192.

18. Industrial applications of micro heat exchangers, http://www.fzk.de/fzk/idcplg? IdcService=FZK&node=0909&lang=en, DECHEMA e. V.

19. X. Zhang, S. Stefanick, and F. J. V. Villani (2004). Application of microreactor technology in process development. *Org. Process Research Development*, 8(3): 455–460.

20. T. Kawaguchi, H. Miyata, K. Ataka, K. Mae, and J.-I. Yoshida (2005). Room-temperature Swern oxidations by using a microscale flow system. *Angewandte Chemie International Edition*, 44(16): 2413–2416.

21. K. F. Jensen (2001). Microreaction engineering—Is small better? *Chemical Engineering Science*, 56: 293–303.

22. M. Matlosz and J.-M. Commenge (2002). From process miniaturization to multiscale design: The innovative, high-performance chemical reactors of tomorrow. *Chimia*, 56: 654–656.

23. J.-M. Commenge, L. Falk, J.-P. Corriou, and M. Matlosz (2005). Analysis of microstructured reactor characteristics for process miniaturization and intensification. *Chemical Engineering Technology*, 28(4): 446–458.

24. J. M. Engasser and C. Horvath (1974). A simple additivity relation for analysis of heterogeneous catalytic reactors with first order reaction. *Chemical Engineering Science*, 29: 2259–2262.

25. S. Lomel, L. Falk, J. M. Commenge, J. L. Houzelot, and K. Ramdani (2006). The microreactor: A systematic and efficient tool for the transition from batch to continuous process? *Chemical Engineering Research and Design*, 84(A5): 363–369.

CHAPTER 15

STANDARDIZATION IN MICROPROCESS ENGINEERING

ALEXIS BAZZANELLA

15.1 INTRODUCTION

A large variety of components and equipment for microprocess engineering is nowadays offered by different manufacturers and suppliers all over the world. Although this clearly demonstrates the advances and evolution of this technology, it presents particular challenges and/or opportunities for both manufacturers and users. From a manufacturer's perspective, a portfolio of different components (e.g., reactors, mixers, etc.) offers the possibility of designing and constructing standardized elements to be used in different components. The repeated use of such elements in different components increases the manufacturing lot of identical parts and is an effective means to reduce manufacturing costs. This is, in particular, relevant for toolboxes containing different components with identical footprint, connectors, and housing. In such cases, microstructured elements can be designed as inlays of an otherwise standardized housing, which can be produced in higher numbers. Figure 15.1 gives an example of this approach; several microprocess engineering toolboxes are commercially available.

From a user's perspective, the availability of a wide variety of components holds the possibility of matching a selected piece of equipment to the physicochemical requirements of the intended chemical process. Parameters such as material selection, required mixing time, heat dissipation capability, residence time, pressure, temperature, or viscosity can be defined or determined and provide a basis for selecting the most appropriate equipment for a given process step. Subsequently, for example, lab-scale plants for process development can be erected using appropriate equipment pieces for the different process steps. This scenario is usually supported by the commercially offered toolboxes and systems, which allow the more or less rapid

Microchemical Engineering in Practice. Edited by Thomas R. Dietrich
Copyright © 2009 John Wiley & Sons, Inc.

FIGURE 15.1 Standardized heat exchanger module with microstructured inlay; component of the modular microreaction system (courtesy Ehrfeld Mikrotechnik BTS).

set-up of different modules to a complete system, including, for example, sensors, pumps and other actuators, and control instrumentation.

However, it is generally much more difficult or even prohibitive to couple components from different suppliers due to the lack of compatibility of components in terms of connectivity. This results from not only the large variety of used fluidic connectors (e.g., Swagelok, UNF threads, and HPLC fittings) but also different component geometries, sealing approaches, materials, and mechanical requirements. Although this problem can, in principal, be circumvented by using different fluid adapters and tubing, the resulting connectors have a rather large hold-up compared to the connected microstructured components, and the process of building the set-up is time-consuming and tedious.

Toolbox and system manufacturers have partly reacted to customers' demands to couple components from other suppliers to their systems by providing "gateways" based on adapter modules or by specifically designed modules to adopt OEM equipment. However, these measures are far from a desirable generic and flexible approach to component connectivity.

Subsequently, first steps toward the vision of flexibly constructed microplants using different microstructured components irrespective of the origin are described. These are based on the MicroChemTec initiative in Germany.

15.2 THE MICROCHEMTEC STANDARDIZATION INITIATIVE

The strategic research project Modular Micro Chemical Engineering (MicroChemTec [1]) was a standardization initiative funded by the German Ministry of Education and Research. This project, running from October 2001 to January 2005, aimed to establish a uniform standard and modular approach to microstructured process technology, thus providing a solution to the connectivity problem of different microstructured components. The project, conducted by research institutes with expertise in microprocess engineering such as IMM and Forschungszentrum Karlsruhe, was steered by a large panel of industrial experts that included both equipment manufacturers and representatives of the chemical and pharmaceutical application sectors. Three years of research and development resulted in a manufacturer-spanning toolbox of compatible microprocess components provided by different manufacturers. The core concept of the project was a modular fluidic connection system, the *backbone*, which allows both commercial and demonstration-type microstructured devices to be coupled in all three dimensions in a flexible and easy manner. Micro heat exchangers, reactors, and mixers of different manufacturers are surface-mounted onto this backbone. Standardized interfaces ensure that devices can be exchanged easily, for example, to evaluate different types of mixers. In addition, the project targeted the integration of sensors and control equipment.

15.2.1 The Backbone Concept: Fluidic Platform for Surface-Mount Microstructured Components

The backbone concept is based on the idea of a standardized fluidic bus system, in which the flow is passed through a central spine. Figure 15.2 shows a microplant with different component modules connected via the backbone. The backbone consists of individual blocks with a defined footprint and standardized fluid connectors at defined positions. Microstructured components, such as micromixers, micro heat exchangers, miniaturized pumps, etc., with fluid connectors matching the backbone can be surface-mounted to one or more backbone elements, depending on their size.

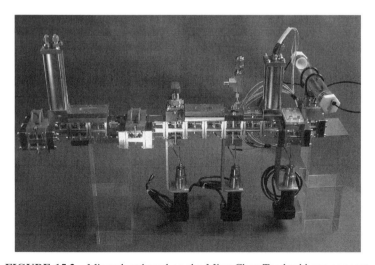

FIGURE 15.2   Microplant based on the MicroChemTec backbone concept.

The backbone thus provides fluid paths between different modules and the integration of heating/cooling fluids. Modules are always connected via the backbone and not directly to each other. The fluidic connectors of the modules and the backbone are complementary and follow a defined system of flow inlets and outlets. This is essential to achieve full compatibility and to ensure that different components with identical functionality can be exchanged without compromising the flow path scheme of the microplant.

15.2.2 Backbone Elements

The backbone consists of several cubic backbone elements, each following a footprint of 45 mm. As depicted in Fig. 15.3, the backbone blocks are modular components, each incorporating a number of pipes and housing parts. For the fluidic connections, three types of pipes are available: a straight pipe, an elbow, and a T-piece. These standard pipes are sufficient to build up all required fluid connections between the backbone and surface-mount module. Three different pipe diameters allow for different flow rates and throughputs, corresponding to the capabilities and requirements of the modules attached. This way, volume flows up to 100 L/h are feasible in the microplant.

The individual backbone blocks are connected by screwing the front plates of two blocks using appropriate seals. This way, the backbone is built up. Figure 15.4 schematically shows the front plate and top views of a backbone block. The front plates can accommodate up to four pipes, thus allowing four fluids to be conveyed in parallel in one backbone. The top plate has eight optional positions to accommodate pipes to be coupled to the microstructured module. These eight optional fluid ports allow the connection to modules of different functionality (e.g., mixing, and heat exchange), which require, respectively, different number and position of fluid inlets and outlets.

In addition to the standard backbone, full block elements of different materials with channels can be used for different applications. Ceramic or PEEK (polyether ether ketone) blocks can be employed to achieve the thermal decoupling of adjacent process steps in the microplant at a different temperature level. Stainless steel blocks can be used as heat exchanger modules, in which cooling or heating fluids are

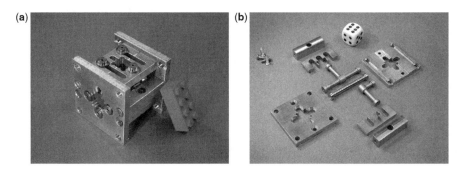

FIGURE 15.3 (a) Assembled backbone element. (b) Backbone parts: housing and piping elements.

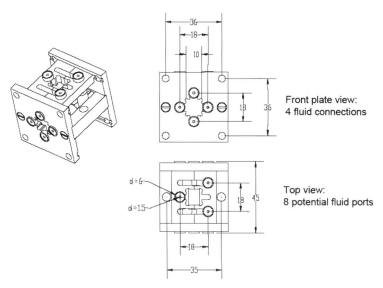

Front plate view:
4 fluid connections

Top view:
8 potential fluid ports

FIGURE 15.4 Schematic depiction of a backbone block, indicating the different optional fluid ports.

provided in one or more channels. Figure 15.5 shows a PTFE (polytetrafluoro-ethylene) backbone block for thermal decoupling and its use in a microplant including process steps at different temperatures.

15.2.3 Compatibility of Modules

The particular standardization effort of the MicroChemTec project included the rationale of the connection interface between the backbone blocks and the manufac-turers' microstructured modules. This required the predefinition of the geometry and function of the fluid ports (inlets and outlets) as described subsequently. The com-patibility of equipment modules to the backbone is achieved either by designing the module to directly match the position and geometry of the fluid ports to the back-bone interface, or by using an adapter plate to adjust such a position and geometry. Examples of both cases are depicted in Fig. 15.6. The redesign of modules to match the standardization specifications is more complex and costly, but is the preferential option in terms of reliability and hold-up, as adapters cause more sealed fluid connec-tions and additional volume.

In order to achieve the compatibility of a large number of modules from different manufacturers without imposing too many restrictions on the module design, the module size is not limited to the backbone block footprint of 45 mm × 45 mm. Depending on the size, modules can be coupled to one, two, or more backbone blocks. In Fig. 15.6, for instance, module (a) is coupled to two backbone blocks, while module (b) is coupled to one backbone block.

In any case, the position and function of fluid ports with respect to the backbone have to be maintained, to keep full compatibility. This is schematically depicted in

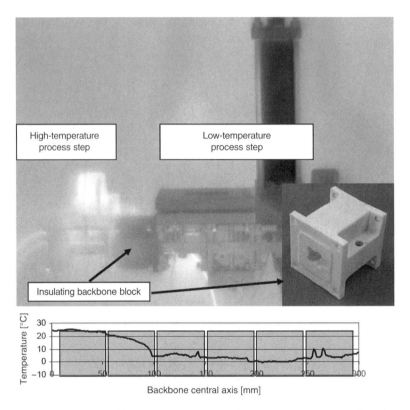

FIGURE 15.5 Use and effect of a thermal insulating backbone block in microplant with different temperature levels [2].

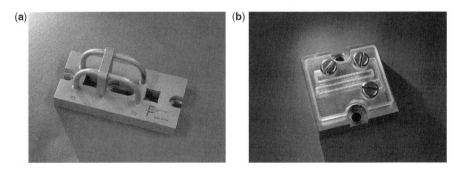

FIGURE 15.6 Backbone compatible modules: (a), heat exchanger with standardized fluid ports (Forschungszentrum Karlsruhe); (b), glas micromixer with steel adapter plate (Little Things Factory).

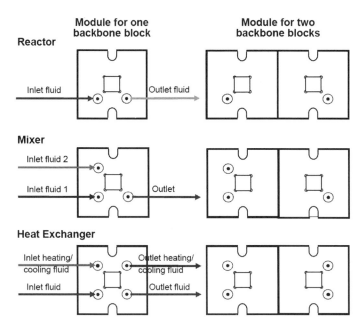

FIGURE 15.7   Allocation of fluid ports for different equipment modules: The systematic of fluid inlets and outlets is mandatory to ensure full compatibility with the backbone.

Fig. 15.7 for three types of modules connected to one or two backbone blocks, respectively. Fluid inlets and outlets occur at defined positions, the allocation depending on the type and function of the module. A very simple reactor possesses one inlet and one outlet at predefined positions. These two standard ports are used in every module for conveying reaction medium from the backbone into the module and back into the backbone. A mixer has an additional inlet for the second fluid; a heat exchanger has both an additional inlet and outlet for the heating/cooling fluid. This fluid port allocation is mandatory to ensure that a module at a given position on the backbone can be exchanged by another module of identical function without compromising the fluid distribution in the backbone.

15.2.4 Building a Microplant Based on the Backbone

Based on the schematic flow diagram of the process, the construction of a microplant starts with the selection of appropriate standardized modules meeting the respective process requirements (manufacturers specifications or experimental characterization). Subsequently, a schematic piping and installation plan is created for the microplant that is then used to build up the backbone. Finally, the modules are surface-mounted to the backbone and peripheric equipment is installed. Figure 15.8 shows a microplant for the sulfonation of toluene with gaseous SO_3, one of four demonstration plants that were erected and extensively tested during the MicroChemTec project for the proof-of-concept of the toolbox. This plant consists of 10 modules from 7 different

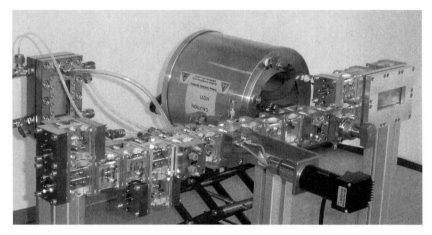

FIGURE 15.8 Microplant for the sulfonation of toluene with gaseous SO_3; the microplant includes the following modules: microgear pumps (HNP Mikrosysteme), heat exchanger (Forschungszentrum Karlsruhe), microfalling film reactor (IMM Mainz), micromixer and microheat exchanger (Little Things Factory), residence time module (CPC GmbH), microreactor (mikroglas chemtech), and sampling valves (Swagelok).

manufacturers. Furthermore, several sensors (temperature, pressure, and mass flow controllers) and a control unit have been integrated (not depicted here).

15.3 CONCLUSION AND OUTLOOK

Standardization is a common step in the development process that transform a new technology into a mature one. In this sense, the MicroChemTec initiative is a first step toward the vision of truly compatible modules irrespective of the manufacturer. With the more widespread implementation of microprocess engineering as a tool of process development, the further harmonization of systems and toolboxes for lab applications across manufacturers can be expected, driven by customers' demand for high flexibility. Related standardization initiatives such as the new sampling/sensor initiative (NeSSI) [3] aimed at the modularization and miniaturization of process analyzer sample system components have to be taken into account or may even be harmonized with standardization activities in microprocess engineering.

In the future, the standardization of fluid connectors and industry-accepted standards on how to rationally connect and distribute fluid streams—in particular, between components of different characteristic dimensions—will become an increasingly important issue for the implementation of microstructured components in industrial pilot plants and on the production level. Accordingly, current research and development activities with respect to industry-scale microstructured equipment for high tonnages, and subsequent concepts of equaling-up and scaling-out of microstructures will finally lead to corresponding standardization efforts on the plant level.

BIBLIOGRAPHY

1. Web site: http://www.microchemtec.de, DECHEMA e.V., Frankfurt, Germany.
2. A. Müller, et al. (2005). *Chemical Engineering Journal*, 107(1–3): 205–214.
3. Web site: http://www.cpac.washington.edu/NeSSI/NeSSI.htm. Center for Process Analytical Chemistry, University of Washington.

PART V

APPLICATIONS

CHAPTER 16

POLYMERIZATION IN MICROFLUIDIC REACTORS

EUGENIA KUMACHEVA, HONG ZHANG, and ZHIHONG NIE

16.1 INTRODUCTION

Recent progress in developing new microfabrication techniques and microreaction technologies has raised exciting opportunities in reaction engineering [1]. A broad range of new strategies has been developed to achieve fast throughput synthesis of organic and inorganic compounds, to conduct efficient studies of reaction mechanisms, and to carry out online characterization of reaction products. Synthesis in microreactors is characterized by high rates of heat and mass transfer (and thus high reaction yields), and it can be performed under conditions more aggressive than in conventional reactors [2, 3]. Generally, it is possible to achieve good control of reaction temperature, especially for exothermic reactions. New reaction pathways too difficult for conventional reactors can be conveniently explored in microreactors [4]. Furthermore, these reactions are safe: If a small amount of a harmful chemical is released, the failed microreactor can be easily isolated.

The literature is burgeoning with reports on the microfluidic syntheses of organic, bioorganic, and inorganic materials [5–11]. Microfluidic synthesis of polymers is beginning to evolve: From 2004 through 2006 several research groups reported on polymerization reactions carried out in continuous microfluidic reactors [12–19]. This chapter focuses on the recent accomplishments in the microfluidic synthesis of polymers. The reports on microfluidic polymerization can be divided into two groups: (1) polymerization in a continuous solution flowing through the microchannels [12], and (2) polymerization in droplets emulsified in a continuous phase moving through the microchannels [14].

Microchemical Engineering in Practice. Edited by Thomas R. Dietrich
Copyright © 2009 John Wiley & Sons, Inc.

16.2 POLYMERIZATION IN A CONTINUOUS SOLUTION FLOWING THROUGH MICROCHANNELS

Polymerization in a continuous flow has been demonstrated for living cationic polymerization [20], free-radical polymerization [21], and the atom transfer radical polymerization (ATRP) of synthetic polymers [13]. Polypeptides and polyaminoacids have been successfully polymerized in a microfluidic reactor [22]. Recently, free-radical polymerization conducted in microfluidic reactors has been modeled using numerical simulations [23]. Typically, a laminar microfluidic reactor consists of a micromixer (where premixing of reactants takes place) and a reaction microchannel (where the monomers undergo continuous polymerization). Polymerization is initiated by heating [21] or photoirradiation [22]. The molecular mass of polymers is typically controlled by varying the polymerization time through the time of residence of the reactants on the chip. The latter variable is manipulated by changing the design of the microfluidic device and the flow rates of liquid phases.

Figure 16.1 shows a typical microfluidic reactor for the temperature-controlled free-radical polymerization of vinyl monomers (such as butyl acrylate, methyl methacrylate, or styrene), which is initiated using 2,2-azobis(isobutylronitrile) (AIBN) [21]. Following mixing of AIBN and the monomers at the T-junction (M1) and in the microreactor (R1), the temperature increases from room temperature to $100°C$ (R2) and AIBN initiates polymerization. The extent of polymerization is controlled by the flow rates of the liquid monomers. The polymerization is terminated by forcing the mixture into the third compartment (R3), where the temperature decreases to $0°C$.

Maeda et al. performed polymerization of amino acid N-carboxyanhydride (NCA) in a microfluidic reaction system and compared the properties of microfluidically synthesized polymers with those obtained in a conventional polymerization reaction [22]. The authors used ring-opening polymerization under a dry argon atmosphere

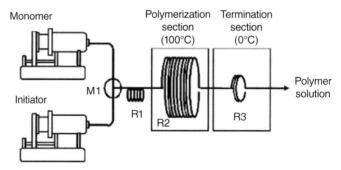

FIGURE 16.1 Schematic of microfluidic reactor for polymerization in continuous flow. The solutions of a monomer and an initiator are introduced into the T-shape micromixer (M1). The mixture passes through a microreactor (R1) at room temperature where complete mixing takes place. The mixture then enters the second microreactor (R2) where upon heating to $100°C$, the initiator decomposes and polymerization takes place. The polymerization is stopped by cooling the mixture to $0°C$ in the third microreactor (R3). The final solution is collected at the exit of the channel. (Adapted, with permission, from [21].)

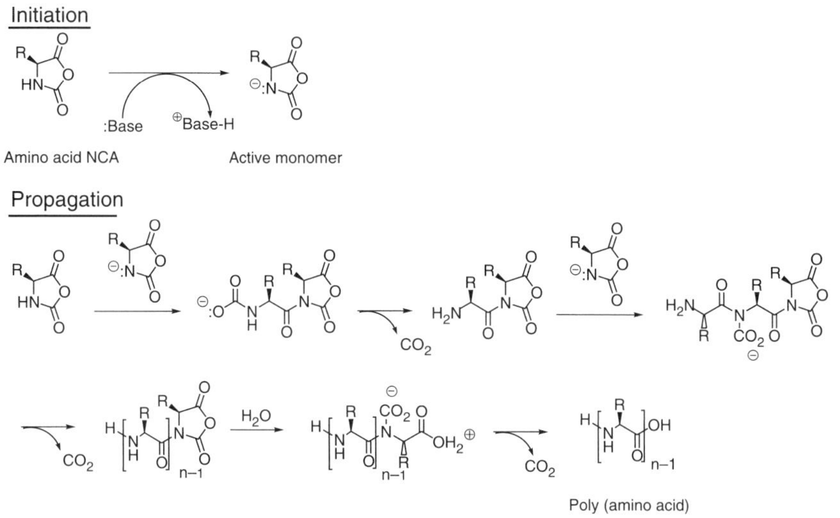

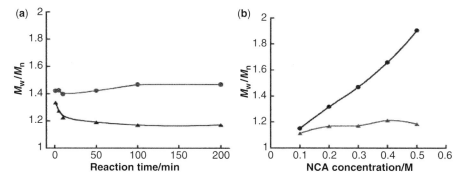

FIGURE 16.2   Mechanism of polymerization of *N*-carboxy anhydrides (NCA). The reaction is initiated by a tertiary amine. Propagation proceeds via either the carbamate or the amine mechanism. (Adapted, with permission, from [22].)

(Fig. 16.2). A *N,N*-dimethylformamide solution of NCA and *N,N*-dimethylformamide solution of an initiator triehtylamide (TEA) were introduced in two separate channels of the microfluidic device and mixed in the micromixer. By changing the flow rates of the solutions of NCA and TEA, the residence time of the liquids in the microchannels was changed from 1 to 200 min. The molecular weight of the resulting polymer gradually increased with increasing residence time. In comparison with polymerization of the same monomers in a bulk system, the polymers synthesized in the microfluidic reactor had a narrower molecular mass distribution achieved for the same 60% conversion (Fig. 16.3).

FIGURE 16.3   Variation in polydispersity of polypeptide molecules synthesized in a microreactor (▲) and in a bulk system (●) plotted as a function of (a) the reaction time and (b) the concentration of NCA. (Adapted, with permission, from [22].)

The ability to synthesize polymers with a narrow distribution of molecular masses is a characteristic feature of microfluidic polymerization, which originates from the following factors.

1. *Efficient Mixing of Reactants*   Due to the very small length scales of micro-reaction systems, diffusion-controlled mixing occurs very rapidly, in comparison with diffusion in conventional reactors [24]. Efficient mixing can further be enhanced by changing the design of a microfluidic reactor by, for example, focusing the liquid streams or by creating vortices in wavy channels.

2. *Control of Reaction Temperature*   Large surface-to-volume ratio of fluid entities in microreaction synthesis leads to an efficient heat transfer between the reactants and minimizes temperature fluctuations in a reaction mixture.

The main problem of microfluidic polymerization in a continuous flow is a gradual increase in the viscosity of the polymer solution with the increasing conversion of a monomer to a polymer. This feature may lead to the change in the characteristics of flow and can ultimately cause clogging of microchannels. Although the limitation can be partly overcome by, for example, increasing the pressure in the channel [25], the molecular mass of polymers synthesized *via* microfluidic polymerization in continuous solutions typically does not exceed 40,000 g/mol [21].

16.3 POLYMERIZATION IN DROPLETS

In the second approach, polymerization is conducted in droplets dispersed in the liquid flowing through the microchannels, so that every droplet acts as a polymerization "nanoliter reactor." The process includes two stages: a microfluidic emulsification and polymerization of monomer(s). The droplets can be hardened to form polymer particles or, alternatively, each droplet can act as a small container for solution polymerization.

Two important aspects of polymerization in droplets have to be stressed. First, polymer synthesis in droplets allows control over viscosity and heat transfer in the system. Second, microfluidic emulsification yields droplets with an extremely narrow polydispersity and an excellent control over droplet size, shape, and morphology. If these features are preserved in the solid state, polymer particles with predetermined dimensions, shapes, and morphologies can be obtained [14, 16–18].

Below we describe the characteristic features of the microfluidic synthesis of polymer particles: the emulsification and polymerization of liquid monomers, the production of particles with controlled shapes and morphologies, and the synthesis of particles with various compositions. We focused on the particles obtained by polymerization and omitted the microfluidic production of polymer colloids by, for example, evaporating solvents or ionic crosslinking. We also limited our discussion to continuous in situ polymerization, that is, when both the emulsification and polymerization processes are conducted on a microfluidic chip.

16.3.1 Formation of Droplets with Controlled Sizes

Polymer colloids in the size range from tens to hundreds of micrometers have important applications in separation and purification technologies, in the immobilization of bioactive species, in toners, and as calibration standards. In many of these applications, a narrow distribution of sizes of polymer particles is of great importance. Currently, the preparation of monodisperse polymer colloids in the designated size range is either material-specific or time-consuming (as it includes a multistep procedure).

Microfluidic polymerization produces particles with diameters from several microns to hundreds of microns, and polydispersity below 5% (and under particular conditions below 1%) [14]. Most common geometries and microfluidic devices used for emulsification include T-junctions [26], flow-focusing devices [27], and terrace-like microchannels (Fig. 16.4) [28]. The formation of droplets in each of these devices has its own characteristic features. However, the size of droplets is generally controlled by varying the ratio of flow rates of the continuous and droplet phases. Typically, the dimensions of droplets decrease with the increasing flow rate of the continuous phase and decreasing flow rate of the liquid to be dispersed. The variation in droplet size is a function of a dimensionless Reynolds number $Re = \rho R U / \mu$ and a Capillary number $Ca = \mu U / \gamma_{12}$, where U is the average velocity of the liquid; R is the characteristic length scale of the system; ρ and μ are the density and viscosity of the liquid, respectively; and γ_{12} is the interfacial tension between the liquids.

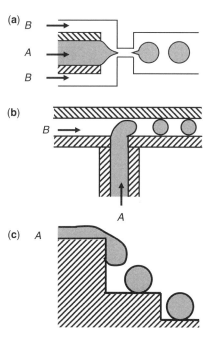

FIGURE 16.4 Schematic illustration of various designs of microfluidic droplet generators: (a) T-junction, (b) flow focusing, and (c) terrace like geometry. A and B are the dispersant and continuous phases, respectively.

At the moment, no mechanism exists that allows the prediction of droplet size as a function of the macroscopic properties of monomers. However, experimental results indicate that for liquids with similar surface energies, larger droplets with a narrower size distribution are obtained from liquids with higher viscosities.

The size of droplets is also determined by the geometry of a microfluidic device. For example, for emulsification at the T-junction, the ratio of the width of the orthogonal inlet channel to that of the main channel plays an important role in determining the modes of formation of the droplets and thus their size [29]. Finally, emulsification strongly depends on the wetting properties of liquids: A liquid to be dispersed should have a low affinity to the material of the microfluidic device [16, 30, 31], while the continuous phase should wet the device.

16.3.2 Polymerization of Monomer Droplets

Droplets generated by microfluidic emulsification are typically solidified by free-radical polymerization [14, 16] or polycondensation reactions [32, 33]. The latter process occurs at the interface between the continuous and droplet phases and yields polymer capsules.

Polymer particles obtained by a photoinitiated free-radical polymerization were produced from a variety of vinyl monomers, for example, tripropyleneglycole diacrylate, dimethacrylate oxypropyldimethylsiloxane, styrene, ethyleneglycole diacrylate, and pentaerythritol triacrylate [14]. To conduct on-chip photopolymerization, the microfluidic droplet generator was connected to an extension wavy microchannel where the droplets were exposed to UV-irradiation. The resulting particles were collected at the exit of the channel. Conversion of a monomer to a polymer depended on the concentration of a photoinitiator, the time of residence of droplets in the microfluidic reactor (i.e., the time of UV-irradiation), and the intensity of UV-irradiation. It was demonstrated that the concentration of a photoinitiator in a monomer is an important factor in particle synthesis [30]. Figure 16.5 shows the effect of the concentration of the photoinitiator 1-hydroxycyclohexylphenyl ketone on the polymerization of tripropyleneglycole diacrylate (TPGDA). Low conversion was obtained at low concentrations of the photoinitiator, whereas at a high concentration of the photoinitiator, a large amount of heat released in the exothermic polymerization reaction "exploded" polymer particles. The monomer was fully converted to the polymer when the concentration of photoinitiator was optimized. Following polymerization, particle dimensions were about 90 to 95% of the size of the corresponding droplets. This feature helped to prevent clogging of the microfluidic reactor with polymer particles whose size was comparable to the microchannel dimensions.

Interfacial polycondensation was conducted at the interface between the droplets and continuous phase, yielding capsules of nylon 6,6 [32]. Polymerization occurred between the bifunctional comonomers (1,6-diaminohexane and adipoyl chloride) that were dissolved in a hexadecane/dichloroethane mixture and in water, respectively. In order to avoid the potential problems associated with rapid polymerization in the orifice, the aqueous droplets of 1,6-diaminohexane were first obtained in hexadecane. Then a stream of adipoyl chloride in the mixture of dichloroethane and hexadecane

FIGURE 16.5 SEM images of polyTPGDA polymer particles produced by continuous polymerization in the microfluidic reactor at concentrations of photoinitiator, C_{in}, of (a) 2, (b) 4, and (c) 6 wt%. (d) SEM image of particles obtained via polymerization of pentaerythritol triacrylate obtained. Scale bar is 100 μm. (Adapted, with permission, from [30].)

was introduced into the continuous phase, resulting in the formation of the nylon-6,6 membrane engulfing a liquid core.

In an alternative procedure, a mixture of sebacoyl and trimesoyl chloride was emulsified in an aqueous solution of poly(ethyleneimine) The reaction between amine and carboxyl groups yielded polyamide capsules [33].

16.3.3 Synthesis of Polymer Particles with Controlled Shapes

Microfluidic Synthesis of Nonspherical Particles using Confinement
Polymer particles with unconventional shapes can serve as the building blocks in the self-assembly of colloids in new functional structures. Polymeric particles produced in microfluidic reactors can acquire nonspherical shapes by (1) confining droplets in microchannels and (2) by trapping their nonequilibrium shapes by *in situ* photopolymerization [30, 34]. The diameter d_0 of an undeformed droplet

(and the corresponding particles) is determined as $d_0 = (6V/\pi)^{1/3}$, where V is the droplet volume. The shape of droplets is governed by the ratio between the value of d and the dimensions of the microchannel such as its width w and height h. Figure 16.6 shows the schematics of the production of droplets with various shapes and the optical microscopy images of the corresponding polymer particles. For $w > d$, and $h > d$, the droplets minimize their surface energy by acquiring a spherical shape [Fig. 16.6 (a, a')]. For $w > d$ and $h < d$, the droplets experience a one-dimensional confinement and transform into disks [Fig. 16.6(b, b')]. For $w < d$ and $h < d$, the droplets undergo two-dimensional confinement and form rods [Fig. 16.6(c, c')]. The aspect ratio of the rods is controlled by changing the volume of droplets and the dimensions of the microchannels. The dimensions, polydispersity, and shapes of droplets were preserved in the corresponding polymer particles.

Microfluidic Synthesis of Microfibers and Microtubes Microfluidics-based polymerization provides a route to the preparation of polymer microfibers and microtubes [35]. Figure 16.7 shows the experimental design for the preparation of these objects. In Fig. 16.7(a), a monomer 4-hydroxybutyl acrylate (4-HBAL), pre-mixed with a photoinitiator, is sheared by the sheath flow of a mixture of poly(vinyl alcohol) (PVA) and a deionized water. The laminar flow supports the cannular shape of the monomer thread. Solidification of the liquid thread by photopolymerizing it

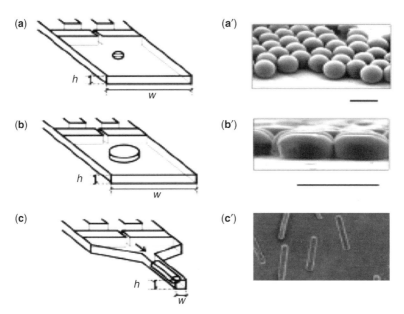

FIGURE 16.6 Schematic (a–c) and optical microscopy images (a'–c', d) of polyTPGDA particles in various shapes: microspheres (a, a'), disks (b, b'), rods (c, c'), and (d) ellipsoidal particles obtained via photopolymerization of the droplets produced at $Q_w = 8$ mL/h, $Q_m = 0.1$ mL/h, and $C_{in} = 4$ wt%, h and w are the height and width of the channel, respectively. Scale bar is 50 μm. (Adapted, with permission, from [30].)

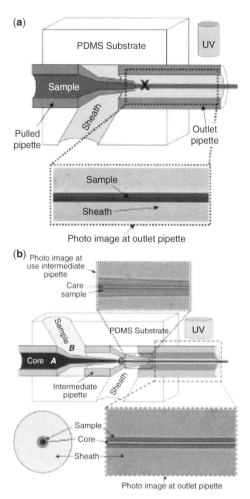

FIGURE 16.7 Schematic of the apparatus for the production of microfibers (a) and micro-tubes (b). In (a), a monomer 4-HBAL premixed with a photoinitiator forms a tread in an aqueous solution of PVA. The resulting thread is photopolymerized in the outlet pipette. In (b), a coaxial thread of 4-HBAL and an aqueous solution of PVA are surrounded by the sheath flow of an aqueous solution of PVA. Photopolymerization and removal of the PVA-water core produces a microtube. (Adapted, with permission, from [35].)

"on the fly" produces a microfiber. Microtubes were obtained by using a coaxial flow of two liquids: a 50/50 vol% aqueous PVA solution (an inner phase) and 4-HBAL premixed with a photoinitiator (used as a "cladding") [Fig. 16.7(b)]. The continuous phase was an aqueous solution of PVA. After the "cladding" was photopolymerized, the inner liquid was removed. The diameter of the microtube cores and thickness of microtube walls were controlled by changing the flow rates of liquids [35].

The advantage of the microfluidic production of microfibers and microtubes provides efficient control of their structures by changing the flow rate of liquids. Since the microobjects were obtained under ambient conditions and exposed to UV-irradiation for a short time, this is useful for the production of microstructures loaded with bioactive species, for example, enzymes [36].

Polymerization using Projection Lithography In contrast to the two methods described above, a continuous-flow photolithography-based microfluidic synthesis does not utilize the break-up of liquid threads (Fig. 16.8). In this method, a polymerizable liquid, for example, poly(ethylene glycole diacrylate), mixed with a photoinitiator flows through the microchannel. The liquid is photopolymerized by UV-irradiating it through a mask. The shapes of the resulting solid microstructures are determined by the features of the mask. This method simultaneously produces many particles surrounded by the flowing liquid. Synthesis is conducted in a micro-fluidic reactor fabricated in the material permeable for oxygen, for example, PDMS. The presence of a thin oxygen inhibition layer between the walls of the microfluidic device and the polymerized objects provides lubrication between the moving particles and microchannel surface.

Figure 16.9 shows particles synthesized using this method: polygons, high-aspect-ratio objects, and nonsymmetric objects. All the microstructures exhibited a good correspondence to the features of the original mask [37].

16.3.4 Synthesis of Polymer Particles with Controlled Morphologies

The polymerization of particles in continuous microfluidic reactors provides a versatile strategy for the reproducible and scalable production of polymer particles with

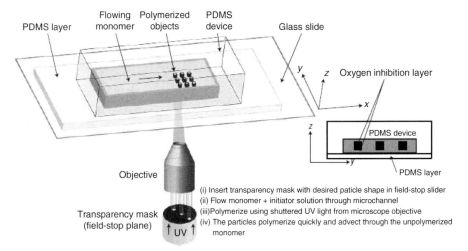

FIGURE 16.8 Schematic of the generation of microstructures via a mask-defined photopolymerization. The monomer flowing through a PDMS device is irradiated through a mask defining the shape of the spot on the monomer layer. The resulting polymer particles are surrounded with an unreacted monomer. (Adapted, with permission, from [37].)

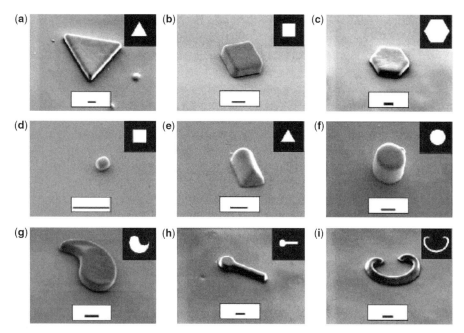

FIGURE 16.9 SEM images of polymer microstructures obtained using a continuous-flow photolithography-based microfluidic method. Scale bar is 10 μm. (Adapted, with permission, from [37].)

various morphologies. The microfluidic synthesis of polymer colloids uses hydrodynamic means to control particle morphologies, in addition to the thermodynamic variables [14, 16, 38–42]. Below we describe recent advances in the synthesis of particles with a core-shell (capsular) morphology and polymer colloids with a Janus or ternary structure.

Synthesis of Polymer Particles with a Core-Shell (Capsular) Structure Microencapsulation has many important applications in the cosmetics, food, nutrient, and pharmaceutical industries. The distribution in sizes of microcapsules is an important property that determines the loading and release of functional ingredients from the interior of particles. Conventional polymerization methods produce polymer microcapsules with polydispersities in the range from 10 to 50%. Microfluidic techniques can circumvent this drawback by producing microcapsules with polydispersity below 5%. Furthermore, microfluidic synthesis yields polymer capsules with precise control over the size of capsular core, thickness of shells, and overall droplets per capsule [16, 38, 39]. The synthesis of polymer capsules with a thin rigid wall and liquid core using condensation reactions has been described above [32, 33, 43]. Alternatively, polymer capsules and core-shell particles with rigid cores can be obtained by generating double emulsions and polymerizing one or both liquids in the core-shell droplets.

Figure 16.10 shows a schematic of the generation of double emulsions by breaking up a coaxial jet formed by two immiscible liquids. The jet is formed by two monomers or a monomer and nonpolymerizable liquid. A pressure gradient acting along the long axis of the microfluidic flow-focusing device forces three liquids (the encapsulated phase, "cladding" phase, and continuous phase) into the narrow orifice. The coaxial jet extends into the downstream channel and subsequently breaks up, releasing core-shell droplets [Figs. 16.10(a) and (c)]. The droplets flow through the extension wavy channel where they are solidified by a photoinitiated free-radical polymerization [Fig. 16.10(b)].

The microfluidic strategy allows precise control over the size of core-shell droplets, diameter of cores, and thickness of droplet shells by varying the ratio of the flow rates of three liquid phases [16, 38]. The radius of the core-shell drops, d_0, can be predicted as $d_0 = (1.5\lambda_{\text{breakup}}d^2)^{1/3}$, where λ_{breakup} is the interfacial capillary wavelength, that is, the length of the last wave within the coaxial jet before it breaks up into droplets, and d is the average diameter of the coaxial jet in the equilibrium region. The average diameter of the coaxial jet is calculated as $d = [(4/\pi)(Q_{\text{drop}}/V_{x,\text{cont}})]^{1/2}$, where $V_{x,\text{cont}}$ is the velocity of the continuous phase in the center of the channel, $V_{x,\text{cont}} = 1.5 \, Q_{\text{cont}}/A_{\text{channel}}$; Q_{drop} and Q_{cont} are the flow rates of the

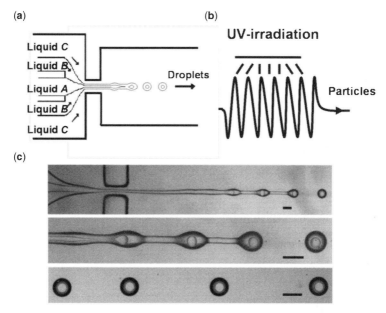

FIGURE 16.10 (a) Schematic of production of droplets by laminar coflow of silicone oil (A), monomer (B), and aqueous (C) phases. The orifice had a rectangular shape with a width and height of 60 and 200 μm, respectively. (b) Schematic of the wavy channel used for photopolymerization of the monomer in the core-shell droplets. (c) Typical optical microscopy images of the break-up of a liquid jet formed by coaxial thread of silicone oil (10 cst) and EGDMA, and the resulting core-shell droplets flowing in the downstream channel. (Adapted, with permission, from [16].)

droplet and continuous phases, respectively; and $A_{channel}$ is the area of cross section of the microfluidic channel.

This approach provided a way to control the number of cores per capsule [16, 38] (Fig. 16.11) by varying the relative position and lengths of interfacial capillary waves on the two jets. Core-shell droplets with multiple cores formed for $\lambda_m > \lambda_o$, that is, when the frequency of break-up of the oil thread exceeds the frequency of break-up of the monomer thread (λ_m and λ_o are the interfacial capillary wavelengths of the monomer and oil threads, respectively). The cores were uniform in size when the value of λ_m was commensurate with the integer number of λ_o; otherwise, the cores were polydispersed.

Polymer particles were obtained by polymerizing a monomer phase in the droplets and, if needed, removing an oil phase. Figure 16.12 shows the SEM images of the resulting polymer colloids.

In the second approach, the core-shell droplets (and particles) were produced by using two mechanisms of microfluidic emulsification—namely, the dripping and jetting mechanisms [38, 39]. Figure 16.13(a) shows the schematic of the production and the optical microscopy image of double emulsions [38]. In this approach, the innermost fluid is pumped through a tapered cylindrical capillary tube, and the middle fluid (a prepolymer Norland Optical Adhesive) is injected through the outer coaxial region.

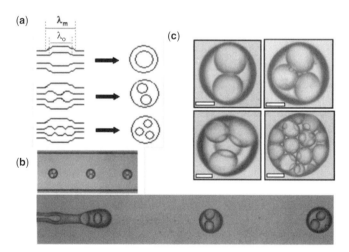

FIGURE 16.11   (a) Schematics of the formation of droplets with multiple cores. Inner phase: silicone oil; outer phase: monomer liquid. (b) Optical microscopy image of core-shell droplets with two cores flowing through the microfluidic device. (c) Break-up of the coaxial jet into two-core droplets. (d) Isolated core-shell droplets containing a different number of SO cores engulfed with TPGDA shell. Conditions for the formation of core-shell droplets with two cores: Q_w, 8 mL/h; Q_m, 0.11 mL/h; Q_o, 0.052 mL/h. Core-shell droplets with three cores: Q_w, 12 mL/h; Q_m, 0.16 mL/h; Q_o, 0.05 mL/h. Core-shell droplet with four cores: Q_w, 9 mL/h, Q_m, 0.155 mL/h, Q_o, 0.054 mL/h. Core-shell droplet with multiple cores: Q_w, 10 mL/h, Q_m, 0.165 mL/h, Q_o, 0.052 mL/h. Scale bar is 40 μm. (Adapted, with permission, from [16].)

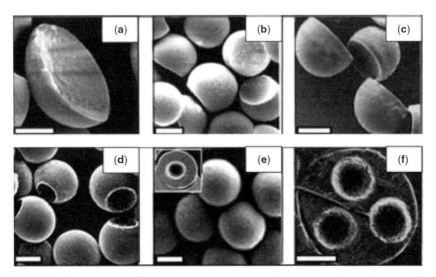

FIGURE 16.12 Scanning electron microscopy images of polymer microbeads obtained by polymerizing tripropyleneglycole diacrylate (TPGDA) in TPGDA- silicone oil droplets, after removing silicone oil. (a) Polymer plates; (b) truncated microspheres; (c) polymer hemispheres; (d) polymer "bowls"; (e) polymer capsules with a single core; (f) cross section of a polymer capsule with multiple cores. The particle is embedded in epoxy glue. Scale bar is 40 μm. (Adapted, with permission, from [16].)

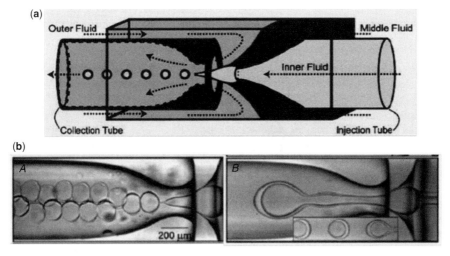

FIGURE 16.13 Design of microfluidic devices for generating double emulsions. (a) Schematic of the coaxial microcapillary fluidic device. The collection tube (a round tube on the left) can be a simple cylindrical tube with a constriction as shown here, or it can be tapered into a fine point (not shown). (b) Optical microscopy images of the generation of core-shell droplets by combining a dripping (A) and ajetting (B) mechanisms of droplet formation. (Adapted, with permission, from [38].)

The outermost fluid (a continuous phase) is introduced through the outer coaxial region from the opposite direction. All fluids are then forced through the exit orifice [Fig. 16.13(a)]. Polymer microcapsules were produced by photopolymerizing the shells in the core-shell droplets. A similar microfluidic device consisting of two coaxial nozzles was reported by Canan Calvo et al. for the preparation of core-shell droplets that upon the solidification of shells yielded polymer capsules [39].

Synthesis of Janus and Ternary Particles Janus particles (JPs) consist of two hemispheres with different compositions and properties. Janus particles have potential applications in studies of particle self-assembly, in the stabilization of water-in-oil or oil-in-water emulsions, and as dual-functionalized optical, electronic, and sensor devices [44]. Hydrodynamic techniques provide a simple and scalable route to the preparation of JPs [19, 40–42, 45]. A microfluidic approach to Janus and ternary (three-phase) particles utilizes the emulsification of monomer droplets with a corresponding multiphase structure, which is followed with a rapid thermo- or photoinitiated polymerization [19, 42, 45]. (Alternatively, Janus droplets can be formed by two polymer solutions: Rapid evaporation of the solvent traps the two-phase structure of the particles in the solid state [40, 41].

The multiphase droplets were obtained by using an electrohydrodynamic [40, 41] or hydrodynamic [19, 42, 45] method (Fig. 16.14). In the electrohydrodynamic method, the liquid drop at the tip of the nozzle transforms into a nanometer-thick thread that breaks up into nanodroplets. This method is applicable to the production of droplets from polar (e.g., aqueous) polymer solutions, and thus a limited range of liquids can be used for the preparation of Janus particles. Furthermore, although this process may yield droplets (and particles) in the submicrometer size range, it yields colloids with a broad size distribution [40, 41].

In the hydrodynamic flow focusing, the laminar thread of monomer streams flowing side by side breaks up into droplets under the shear force imposed by the continuous phase [Fig. 16.14(b)] [19, 42, 45]. In this strategy, both polar and nonpolar monomers and liquid polymers can be used for the production of Janus and ternary particles. The multiphase droplets have a very narrow size distribution, but the size of particles falls in the range of approx. 20 to 200 μm, that is, it is significantly larger than that obtained by using electrohydrodynamic emulsification.

In both cases, diffusive mixing between the two adjacent streams is dominant; therefore, no convective transport occurs across the interface between them. However, in order to have a sharp interface between the phases in the particles, it is imperative to use highly immiscible liquids and/or conduct a fast solidification of the liquid phases.

By using a microfluidic emulsification, the volume of each compartment in the Janus and ternary droplets (and in the corresponding particles) can be precisely manipulated by varying the flow rates of each phase. Figure 16.15 shows the approach to the manipulation of the morphology of Janus particles [42]. Assuming that each phase forms a truncated sphere, a simple relationship was derived that related the ratio of flow rates Q_{M1}/Q_{M2} to the ratio of volumes V_{M1}/V_{M2} (or to the

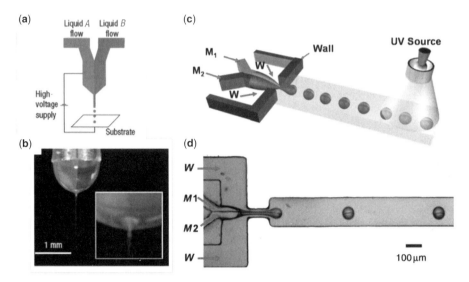

FIGURE 16.14 (a) Schematic of the experimental set-up used for electrohydrodynamic emulsification. When exposed to an applied electric potential, the bipolar jetting liquid experiences an electrical field formed between the tip of the liquid and the counterelectrode (collecting substrate). (b) A digital image of a typical biphasic Taylor cone with jet. A 2 wt% aqueous solution of poly(ethylene oxide) ($M_w = 600,000$ g mol^{-1}) was used for both jetting fluids. Each phase was dye-labeled to produce red and green colors. The inset shows an image of the swirllike jet ejection point [40]. (c) Schematic of the hydrodynamic generation of Janus droplets from immiscible monomers $M1$ and $M2$, emulsified in a 2 wt% aqueous solution of sodium dodecul sulfate (W). The droplets are irradiated with UV-light in the downstream channel. (d) Optical microscopy image of formation of Janus droplets. (Adapted, with permission, from [40].)

ratio of heights of each hemisphere h_{M1}/h_{M2}) of the monomers $M1$ and $M2$ in the Janus droplet [Fig. 16.15(a)]:

$$\frac{Q_{M1}}{Q_{M2}} = \frac{V_{M1}}{V_{M2}} = \frac{h_{M1}^2(3R - h_{M1})}{h_{M2}^2(3R - h_{M2})}$$

where R is the radius of the spherical Janus droplet. Good correlation was found between the predicted and calculated variation in V_{M1}/V_{M2} (and h_{M1}/h_{M2}) with an increasing ratio of flow rates of monomers [Fig. 16.15(b)].

Figure 16.15(c) shows representative morphologies of the Janus droplets with volume ratios of monomer phases of 2/1, 1/1, 1/2, and 1/4 [corresponding to the points A, B, C, and Fig. 16.15(b)]. The morphologies of Janus particles were "frozen" by fast photopolymerization of the constituent monomers. Figures 16.15(e) and (f) show the bright-field and fluorescence microscopy images of the resulting Janus particles that were composed of hydrophilic and hydrophobic polymers.

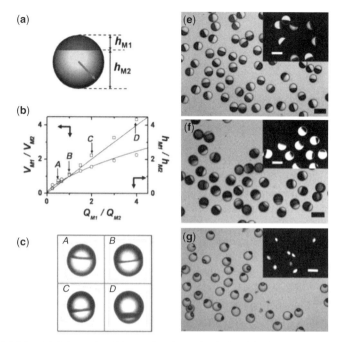

FIGURE 16.15   (a) Schematic of Janus droplet and relationship between the ratio of monomer flow rates (Q_{M1}/Q_{M2}) and the ratio of volumes (V_{M1}/V_{M2}) and the ratio of heights (h_{M1}/h_{M2}) of the constituent parts in Janus droplets. (b) Variation in volume ratio V_{M1}/V_{M2} (□) and height ratio h_{M1}/h_{M2} (○) plotted vs. the ratio of flow rates of $M1$ and $M2$. The solid lines represent calculated theoretical calculations; the empty symbols correspond to experimental results. (c) Optical microscopy images of droplets obtained under conditions indicated in (b) as A, B, C, and D, respectively. The top part of droplets is formed by $M1$. Optical microscopy images of JPs (e–g) and particles with ternary structures. (d) Bright and dark phases are polymers of $M1$ and $M2$, respectively. Insets in (e–g) show fluorescence microscopy images of the corresponding particles (bright phase is a polymer of $M2$ mixed with NBDMA dye). (Adapted, with permission, from [42].)

Ternary polymer particles containing three distinct polymer phases were obtained in a similar manner by photopolymerizing three-phase monomer droplets.

16.3.5   Synthesis of Polymer Particles with Different Compositions

Polymer colloids with different compositions are synthesized by using (1) microfluidic emulsification of a mixture of monomers with organic or inorganic additives and (2) polymerization in situ of the resulting droplets. This strategy is applicable to the production of particles with different shapes and morphologies.

Synthesis of Copolymer Particles   The synthesis of copolymer particles includes emulsification of a mixture of two or more monomers, which is accompanied by photoinitiated free-radical polymerization of the monomer mixture. Copolymer particles of poly[(triproplylene glycole diacrylate-comethyl acrylate] [14], poly[(4-hydroxybuthyl acrylate)-coacrylic acid] [15], and poly[(tripropylene glycole diacrylate)-coacrylic acid] [17] were successfully synthesized in continuous microfluidic reactors. When the constituent comonomers had a different polarity, the formation of droplets strongly depended on the composition of the monomer mixture. For example, when droplets were generated from tripropylene glycole diacrylate mixed with acrylic acid (AA), the size of droplets decreased with the increasing concentration of AA, mostly due to the lower interfacial tension between the dispersant liquid and continuous phase [17]. Moreover, when the concentration of AA in the mixture increased to 15 wt%, the formation of droplets was suppressed. The wetting properties of the monomer mixture changed as well, leading to the adherence of the comonomer thread to the walls of the microfluidic device. Therefore, for the successful production of copolymer particles, the composition of the comonomer mixture should be optimized to produce droplets with a predetermined size and narrow size distribution.

Copolymer particles synthesized by microfluidic copolymerization have potential applications in the separation, detection, and immobilization of biomolecules. For example, poly[(triproplylene glycole diacrylate)-coacrylic acid] beads were bioconjugated with bovine serum albumin that was covalently attached to fluorescein isothiocynate (FITC-BSA) (Fig. 16.16) [17].

Synthesis of Polymer Microspheres Carrying Low-Molecular-Weight Functionalities

Mixtures of liquid monomers or prepolymers with various functional species either in the form of molecules or small particles can be emulsified in microfluidic droplet generators and then polymerized to produce hybrid microbeads. The requirements of mixing include control of viscosity and interfacial tension

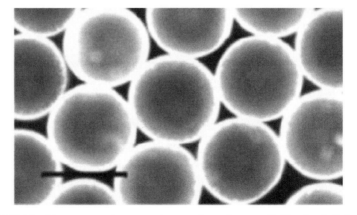

FIGURE 16.16 Fluorescent microscopy image of copolymer colloids of poly[(triproplylene glycole diacrylate)-coacrylic acid] after bioconjugation with FITC-BSA. Scale bar is 100 μm. (Adapted, with permission, from [17].)

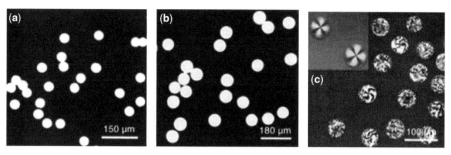

FIGURE 16.17   Optical fluorescent microscopy images of polymer particles carrying (a) an NBD dye ($\lambda_{ex} = 488$ nm), (b) CdSe quantum dots ($\lambda_{ex} = 502$ nm), and (c) polarization microscopy image of 4-cyano-4′-pentylbiphenyl embedded in polyTPGDA microspheres. Inset shows the structure of particles obtained at slow polymerization of TPGDA. (Adapted, with permission, from [14].)

between the droplet phase and continuous phase, so that these two variables do not suppress microfluidic emulsification, and absence of aggregation of functional species in the droplets.

Figure 16.17 shows examples of particles synthesized using microfluidic synthesis: dye-labeled particles, quantum dot-loaded microspheres, and polymer–liquid crystal (LC) microbeads [14]. The rate of photopolymerization had a critical effect on particle morphology. For example, when polymerization was fast, the LC was uniformly distributed through the polymeric microspheres [Fig. 16.17(c)], whereas at slow polymerization the LC segregated in the core of the microspheres [Fig. 16.17(c), inset].

Other examples of hybrid microspheres obtained by microfluidic synthesis include magnetic particles [32], microbeads carrying organic fluorochromes [46], and particles loaded with biocatalysts such as glucose oxidase (GCO) and horseradish peroxidase (HRP) [15]. Figure 16.18 shows the sensing function of the latter particle

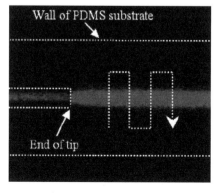

FIGURE 16.18   Fluorescence microscopy image of biocatalyst-encoded microbeads mixed with glucose and Amplex Red. (Adapted, with permission, from [15].)

carrying enzymes. After immobilizing these microbeads in the micropipette, a mixture of glucose and nonfluorescent Amplex Red in a PBS buffer was introduced into the microfluidic device. Two reactions occurred on a chip. First, the glucose converted to glucose acid and H_2O_2 was generated via the GCO-catalyzed reaction. Second, H_2O_2 reacted with Amplex Red to produce fluorescent resorufin, catalyzed by HRP. The micrograph in Fig. 16.18 shows the GCO- and HRP-catalyzed reactions in the microchannel.

16.4   CONCLUSIONS

This chapter reviews recent progress in a rapidly developing field: polymerization is microfluidic reactors. It describes the advantages of polymerization conducted in continuous flow and in droplets. The first approach produces polymers with a narrow molecular mass distribution. The second approach provides a route to the synthesis of polymer colloids with an extremely narrow size distribution and unique control over their shapes and morphologies, in addition to a relatively easy synthesis of copolymer and hybrid particles. Microfluidic synthesis is applicable to free-radical polymerization, polycondensation, ATRP, and anionic polymerization.

Thus so far, the microfluidic synthesis of polymers has not been used for the rapid throughput screening of polymerization conditions such as pH, temperature, ionic strength, compositions, catalysts, cosolvents, and concentrations of reactants. This application is extensively used in other fields of microfluidics, and it will certainly bring new insights to our understanding of polymerization mechanisms and kinetics.

BIBLIOGRAPHY

1. Jensen, K. F. (2001). Microreaction engineering—Is small better? *Chem. Eng. Sci.*, 56: 293–303.
2. Ehrfeld, W., Golbig, K., Hessel, V., Löewe, H., and Richter, T. (1999). Characterization of mixing in micromixers by a test reaction: Single mixing units and mixer arrays. *Ind. Eng. Chem. Res.*, 38: 1075–1082.
3. Lerou, J. J., Harold, M. P., Ryley, J., Ashmead, J., O'Brien, T. C., Johnson, M., Perrotto, J., Blaisdell, C. T., Rensi, T. A., Nyquist, J. (1996). Microfabricated minichemical systems. Technical feasibility. In *Microsystem Technology for Chemical and Biological Microreactors*, DECHEMA, New York, Vol. 132, pp. 51–69.
4. Chambers, R. D., and Spink, R. C. H. (1999). Microreactors for elemental fluorine. *Chem. Comm.*, 10: 883–884.
5. Borman, S. (2000). Combinatorial chemistry—Redefining the scientific method. *Chem. Eng. News*, 78: 53–53.
6. Fortt, R., Wootton, C. R., and de Mello, A. J. (2003). Continuous-flow generation of anhydrous diazonium species: Monolithic microfluidic reactors for the chemistry of unstable intermediates. *Org. Process Res. Dev.*, 7: 762–768.
7. Chan, E. M., Mathies, R. A., and Alivisatos, A. P. (2003). Size-controlled growth of CdSe nanocrystals in microfluidic reactors. *Nano Lett.*, 3: 199–201.

8. Peterson, D. S., Rohr, T., Svec, F., and Fréchet, J. M. J. (2002). Enzymatic microreactor-on-a-chip: Protein mapping using trypsin immobilized on porous polymer monoliths molded in channels of microfluidic devices. *Anal. Chem.*, 74: 4081–4088.

9. Kopp, M. U., de Mello, A. J., and Manz, A. (1998). Chemical amplificaiton: Continuous-flow PCR on a chip. *Science*, 280: 1046–1048.

10. Cheng, J., Shoffner, M. A., Mitchelson, K. R., Kricha, L. J., and Wilding, P. (1996). Analysis of ligase chain reaction products amplified in a silicon-glass chip using capillary electrophoresis. *J. Chromatogr.*, A732: 151–158.

11. Kobayashi, J., Mori, Y., Okamoto, K., Akiyama, R., Ueno, M., Kitamori, T., et al. (2004). A microfluidic device for conducting gas-liquid-solid hydrogenation reactions. *Science*, 304: 1305–1308.

12. Cabral, J. T., Hudson, S. D., Wu, T., Beers, K. L., Douglas, J. F., Karim, A., et al. (2004). Microfluidic combinatorial polymer research. *Polym. Mater.: Science & Eng.*, 90: 337–338.

13. Wu, T., Mei, Y., Cabral, J. T., Xu, C., and Beers, K. L. (2004). A new synthetic method for controlled polymerization using a microfluidic system. *J. Am. Chem. Soc.*, 126: 9880–9881.

14. Xu, S. Q., Nie, Z. H., Seo, M. S., et al. (2005). Generation of monodisperse particles by using microfluidics: Control over size, shape, and composition. *Angew. Chemie Intnl. Ed.*, 44: 724–728.

15. Jeong, W. J., Kim, J. Y., Choo, J. B., Lee, E. K., Han, C. S., Beebe, D. J., et al. (2005). Continuous fabrication of biocatalyst immobilized microparticles using photopolymerization and immiscible liquids in microfluidic systems. *Langmuir*, 21: 3738–3741.

16. Nie, Z. H., Xu, S. Q., Seo, M., Lewis, P. C., and Kumacheva, E. (2005). Polymer particles with various shapes and morphologies produced in continuous microfluidic reactors. *J. Am. Chem. Soc.*, 127: 8058–8063.

17. Lewis, P. C., Graham, R. R., Nie, Z. H., Xu, S. Q., Seo, M. S., and Kumacheva, E. (2005). Continuous synthesis of copolymer particles in microfluidic reactors. *Macromolecules*, 38: 4536–4538.

18. Seo, M. S., Nie, Z. H., Xu, S. Q., Lewis, P. C., and Kumacheva, E. (2005). Microfluidics: From dynamic lattices to periodic arrays of polymer disks. *Langmuir*, 21: 4773–4775.

19. Nisisako, T., Torii, T., and Higuchi, T. (2004). Novel microreactors for functional polymer beads. *Chem. Eng. J.*, 101: 23–29.

20. Nagaki, A., Kawamura, K., Suga, S., Ando, T., Sawamoto, A., and Yoshida, J. (2004). Cation pool-initiated controlled/living polymerization using microsystems. *J. Am. Chem. Sci.*, 126: 14702–14703.

21. Iwasaki, T., and Yoshida, J. (2005). Free radical polymerization in microreactors. Significant improvement in molecular weight distribution control. *Macromolecules*, 38: 1159–1163.

22. Honda, T., Miyazaki, M., Nakamura, H., and Maeda, H. (2005). Controlled polymerization of N-carboxy anhydrides in a microreaction system. *Lab Chip*, 5: 812–818.

23. Serra, C., Sary, N., Schlatter, G., Hadziioannou, G., and Hessel, V. (2005). Numerical simulation of polymerization in interdigital multilamination micromixers. *Lab Chip*, 5: 966–973.

24. de Mello, A. J. (2006). Control and detection of chemical reactions in microfluidic systems. *Nature*, 442: 394–402.

25. Rosenfeld, C., Serra, C., Brochon, C., and Hadziioannou, G. (2006). Use of micromixers to control the molecular weight distribution in continuous two-stage nitroxide-mediated copolymerizations. Paper presented at International Conference on Microreaction Technology, Potsdam, Germany, 2006.

26. Thorsen, T., Roberts, R. W., Arnold, F. H., and Quake, S. R. (2001). Dynamic pattern formation in a vesicle-generating microfluidic device. *Phys. Rev. Lett.*, 18: 4163–4166.

27. Anna, S. L., Bontoux, N., and Stone, H. A. (2003). Formation of dispersions using "flow focusing" in microchannels. *App. Phys. Lett.*, 82: 364–366.

28. Sugiura, S., Nakajima, M., and Seki, M. (2002). Effect of channel structure on microchannel emulsification. *Langmuir*, 18: 5708–5712.

29. Garstecki, P., Fuerstman, M. J., Stone, H. A., and Whitesides, G. M. (2006). Formation of droplets and bubbles in a microfluidic T-junction–scaling and mechanism of break-up. *Lab Chip*, 6: 437–446.

30. Seo, M. S., Nie, Z. H., Xu, S. Q., Mok, M., Lewis, P. C., Graham, R., et al. (2005). Continuous microfluidic reactors for polymer particles. *Langmuir*, 21: 11614–11622.

31. Lorenceau, E., Utada, A. S., Link, D. R., Cristobal, G., Joanicot, M., and Weitz, D. A. (2005). Generation of polymerosomes from double-emulsions. *Langmuir*, 21: 9183–9186.

32. Takeuchi, S., Garstecki, P., Weibel, D. B., and Whitesides, G. M. (2005). An axisymmetric flow-focusing microfluidic device. *Adv. Mat.*, 17: 1067–1072.

33. Quevedo, E., Steinbacher, J., and McQuade, D. T. (2005). Interfacial polymerization within a simplified microlfuidic devices: Capturing capsules. *J. Am. Chem. Soc.*, 127: 10498–10499.

34. Dendukuri, D., Tsoi, K., Hatton, T. A., and Doyle, P. S. (2005). Controlled synthesis of nonspherical microparticles using microfluidics. *Langmuir*, 21: 2113–2116.

35. Jeong, W., Kim, J. Y., Kim, S. J., Lee, S. H., Mensing, G., and Beebe, D. J. (2004). Hydrodynamic microfabrication via "on the fly" photopolymerization of microscale fibers and tubes. *Lab Chip*, 4: 576–580.

36. Kim, S. R., Oh, H. J., Baek, J. Y., Kim, H. H., Kim, W. S., and Lee, S. H. (2005). Hydrodynamic fabrication of polymeric barcoded strips as components for parallel bioanalysis and programmable microactuation. *Lab Chip*, 5: 1168–1172.

37. Dendukuri, D., Pregibon, D. C., Collins, J., Hatton, T. A., and Doyle, P. S. (2006). Continuous-flow lithography for high-throughput microparticle synthesis. *Nat. Mater.*, 5: 365–369.

38. Utada, A. S., Lorenceau, E., Link, D. R., Kaplan, P. D., Stone, H. A., and Weitz, D. A. (2005). Monodisperse double emulsions generated from a microcapillary device. *Science*, 308: 537–541.

39. Martin-Banderas, L., Flores-Mosquera, M., Riesco-Chueca, P., Rodriguez-Gil, A., Cebolla, A., Chavez, S., et al. (2005). Flow focusing: A versatile technology to produce size-controlled and specific-morphology microparticles. *Small*, 1: 688–692.

40. Roh, K. H., Martin, D. C., and Lahann, J. (2005). Biphasic Janus particles with nanoscale anisotrop. *Nat. Mater.*, 4: 759–763.

41. Roh, K. H., Martin, D. C., and Lahann, J. (2006). Triphasic nanocolloids. *J. Am. Chem. Soc.*, 128: 6796–6797.

42. Nie, Z. H., Li, W., Seo, M., Xu, S. Q., and Kumacheva, E. (2006). Janus and ternary particles generated by microfluidic synthesis: Design, synthesis, and self-assembly. *J. Am. Chem. Soc.*, 128: 9408–9412.

43. Steinbacher, J. L., Moy, R. W. Y., Price, K. E., Cummings, M. A., Roychowdhury, C., Buffy, J. J., et al. (2006). Rapid self-assembly of core-shell organosilicon microcapsules within a microfluidic device. *J. Am. Chem. Soc.*, 128: 9442–9447.

44. Perro, A., Reculusa, S., Ravaine, S., Bourgeat-Lami, E. B., and Duguet, E. (2005). Design and synthesis of Janus micro- and nanoparticles. *J. Mater. Chem.*, 15: 3745–3760.

45. Nisisako, T., Torii, T., Takahashi, T., and Takizawa, Y. (2006). Synthesis of monodisperse bicolored janus particles with electrical anisotropy using a microfluidic co-flow system. *Adv. Mater.*, 18: 1152–1156.

46. Martin-Banderas, L., Rodriguez-Gil, A., Cebolla, A., Chavez, S., Berdun-Alvarez, T., Garcia, J. M. F., et al. (2006). Towards high-throughput production of uniformly encoded microparticles. *Adv. Mater.*, 18: 559–564.

CHAPTER 17

PHOTOREACTIONS

TEIJIRO ICHIMURA, YOSHIHISA MATSUSHITA, KOSAKU SAKEDA, and TADASHI SUZUKI

17.1 THEORY

17.1.1 Photochemical and Photophysical Processes

"Photochemistry" is a science concerned with the description of the physical and chemical processes triggered by the absorption of photons [1]. It serves as a useful basis for future technologies and industries with a microreactor. Photochemical reactions differ considerably from conventional thermal reactions in the following important respects. Because the reaction is initiated with the irradiation of light and absorption of the photon by an organic molecule, the electronic structure and nuclear configuration of the excited molecule will be different from those in the ground state. The excited molecule possesses high internal energy and will give rise to the formation of a photoproduct, which is unobtainable through a thermal reaction in the ground state.

Absorption of a photon results in the excitation from a lower to a higher electronic state. The excited molecule relaxes into the stable state through photophysical or photochemical processes (Fig. 17.1): radiative and/or nonradiative processes. The fluorescence spectrum is usually independent of the excitation wavelength in the condensed phase (Kasha–Vavilov's law). Generally, the lifetime of the T_1 state ($\mu s \sim ms$), τ_T, is longer than that of the S_1 state (ns), τ_S.

A unimolecular or bimolecular chemical reaction will take place in the excited state (the S_1 or T_1 state) or higher excited state (the S_n or T_n state) such as bond fission (dissociation and decomposition), isomerization, inter- and intra-molecular hydrogen atom abstraction, charge and proton transfer, and cyclo-addition [2]. When an intense laser is used as the excitation light source, multiphoton absorption easily occurs,

Microchemical Engineering in Practice. Edited by Thomas R. Dietrich
Copyright © 2009 John Wiley & Sons, Inc.

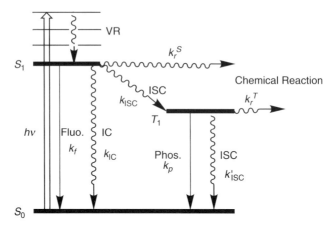

FIGURE 17.1 A schematic energy diagram of a molecule in the excited and ground states, and relaxation processes. The solid and wavy lines indicate radiative and nonradiative processes. The k value denotes a rate constant of the corresponding process. VR, vibrational relaxation; Fluo., fluorescence; IC, internal conversion; ISC, intersystem crossing; and Phos., phosphorescence.

followed by excitation to the higher excited state or ionization to form cation. The reaction quantum yield in the S_1 or T_1 state is defined as the ratio of the number of reaction products to the number of absorbed photons, equal to the rate constants k of the reaction and relaxation processes as

$$\phi_r^S = (\text{number of reaction products})/(\text{number of absorbed photons})$$

$$= k_r^S/k_S = k_r^S \tau_S \tag{17.1}$$

$$k_S = 1/\tau_S = k_f + k_{IC} + k_{ISC} + k_r^S$$

$$\phi_r^T = (\text{number of reaction products})/(\text{number of absorbed photons})$$

$$= k_r^T/k_T = k_r^T \tau_S \tag{17.2}$$

$$k_T = 1/\tau_T = k_p + k'_{ISC} + k_r^T$$

Hence, the estimation of the reaction quantum yield and lifetime of the excited state can lead to the reaction rate constant.

17.1.2 Excitation Light Source

As conventional light sources [3] for photochemistry in the UV-visible wavelength region, you can use a mercury lamp, xenon arc lamp, and lasers. The laser has become widespread in photochemistry and spectroscopy since its discovery in 1960. The laser beam, in principle, can be focused to the size of the wavelength.

17.1.3 Analysis and Detection Methods

In a photochemical reaction, product analysis is quite important to determine the quantum yield of a reaction and to investigate the reaction mechanism. Chromatography is a powerful technique for the separation of the reaction mixture and can identify each component. The electronic absorption and emission spectra measurements also provide information on the photoproducts. Direct investigation of the photochemical reaction mechanism in a microreactor is especially important because photochemistry in a bulk system may be different from that in a microreactor. Nanosecond laser flash photolysis will provide temporal information such as data on the lifetime and concentration of reaction intermediates. However, it is not applicable to a microreactor because the detection sensitivity is quite low due to the ultra-short light path length. Therefore, fluorescence, Raman, or thermal lensing spectroscopy coupled with laser irradiation and/or a microscope [4–7] can be applied to a photochemical study in a microreactor.

17.2 PHOTOREACTIONS IN MICROREACTORS

In the last decade, a microreaction system has developed using the features unique to microspace such as short molecular diffusion distance, excellent heat-transfer characteristics, laminar flow, and large surface-to-volume ratio [8–12]. Although microreaction systems have been examined successfully in a wide range of applications of analytical and organic chemistry, there are only several reports on photoreactions in microreactors, as described in the following section [13–18]. We can expect microreactors to exhibit higher spatial illumination homogeneity and better light penetration throughout the entire reactor depth in comparison to large-scale reactors. Thus, we continue to investigate the applications of microreactors in organic photoreactions. In this section, we will describe our results on asymmetric photoreactions and photocatalytic reactions in microreactors.

17.2.1 Experimental Section

Figure 17.2 shows our typical experimental set-up. Sample solution and/or gas were fed into a microreactor with a syringe pump. We used lasers and also lamps and even

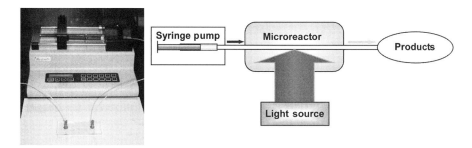

FIGURE 17.2   Typical experimental set-up.

UV-LEDs, and irradiated the sample solution from the top of the microchannel. Reaction products were analyzed by gas and high-pressure liquid chromatography (HPLC).

Microreactor Chips   We employed microreactor chips made of quartz, which is transparent to UV light. The microreactor chips have a straight microchannel 200 to 500 μm in width, 10 to 500 μm in depth, and 40 mm in length. A TiO_2 layer was immobilized in microchannels of 500-μm width for photocatalytic reactions. It has been widely recognized that the illuminated specific surface area of the photocatalyst within a reactor is the most important design parameter of photocatalytic reactors. The illuminated specific surface areas per unit of liquid of the microreactor with a micro-channel depth of 100, 300, and 500 μm were calculated to be 1.4×10^4, 7.3×10^3, and $6.0 \times 10^3 \, m^2/m^3$, respectively, without taking into account the roughness of the photocatalyst surface. Thus, the microreactors with an immobilized photocatalyst have much larger values of illuminated specific surface area of photocatalyst than typical conventional batch reactors [19].

Excitation Light Source   The reactants in the microreactor were irradiated by a KrF excimer laser (248 nm, with a pulse duration of 20 ns and repetition rate of 20 Hz), a tunable OPO laser excited with a Nd^+ : YAG Laser, or a Xe or Hg lamp. To make the most of the advantages of a miniaturized reaction vessel, a light source of minimal space and lower photon cost is suitable for microreaction systems. Therefore, in addition to lamps and lasers, we employed UV-light-emitting diodes (UV-LEDs; 365, 375, and 385 nm) for the excitation light source of photo-catalytic reactions.

17.2.2 Asymmetric Photosensitized Reactions in Microreactors

The chirality of molecules is known to play a crucial role in biological systems. Numerous efforts have been made to develop various methodologies for asymmetric chemical synthesis. Recently, the effective control and enhancement of the stereo-selectivity of asymmetric photoreactions in a batch system have been reported [20–23], which is likely a new field of vital importance in synthetic chemistry. We reexamined the two asymmetric photosensitized reactions in a microreactor [24].

Photosensitized Isomerization of (Z)-cyclo-octene   The Z-E photoisomeri-zation [21, 22] of (Z)-cyclo-octene sensitized by chiral aromatic ester gives (R)-(−)- and (S)-(+)-(E)-cyclo-octene (Scheme 17.1). The product's enantiomeric excess (ee) value is defined as follows:

$$ee = ([S] - [R])/([S] + [R]) \tag{17.3}$$

A solution of (Z)-cyclo-octene (25 mM) and the optically active saccharide esters of benzenetetracarboxylic acid (5 mM) as a sensitizer in ethyl ether was introduced into a microreactor. By varying the residence time under 248-nm laser irradiation, the E/Z ratio and product's ee value were examined and shown in Fig. 17.3. The

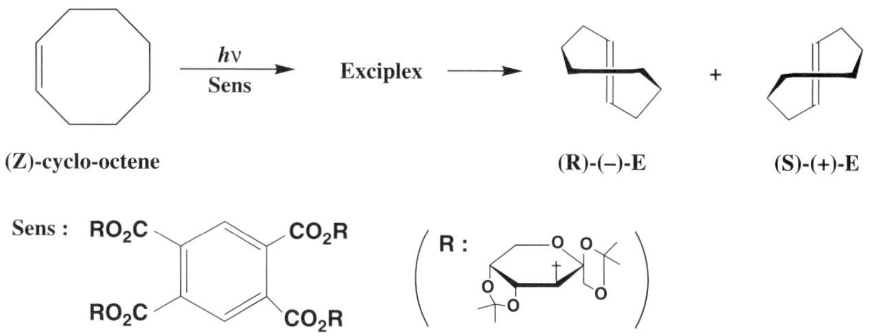

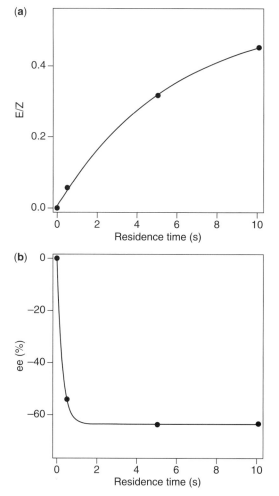

SCHEME 17.1   Photosensitized isomerization of (Z)-cyclo-octene.

FIGURE 17.3   The plots of (a) E/Z ratio and (b) ee value against residence time in the microreactor (with a width of 200 μm and depth of 20 μm).

higher the laser power becomes, the more quickly the E/Z ratio reaches its highest value. The E/Z ratio increased to 0.45 at the residence time of 10 s in the microreactor, while it was 0.28 at the residence time of 30 min in a batch system [25]. A very high ee value (-63% ee) was successfully obtained with an irradiation period less than 200 laser shots of $\sim$20-ns pulse duration, as shown in Fig. 17.3(b).

Photosensitized Addition of (R)-(+)-Limonene to Methanol The photoreaction (Scheme 17.2) gives three major products [23], that is, *cis*- and *trans*-4-isopropenyl-1-methoxy-1-methylcyclohexane (*cis* and *trans*) and exocyclic isomer (*exo*). The diastereomeric excess (de) value is defined as follows:

$$de = ([trans] - [cis])/([trans] + [cis]) \qquad (17.4)$$

A solution of (R)-(+)-limonene (25 mM) and toluene (10 mM) as a sensitizer in methanol was fed into a microreactor. For comparison, batch reactions were performed in a quartz cell (with an optical length of 3 mm) containing a 1-mL sample solution. Figure 17.4 indicates the yields of photoproducts and conversion of (R)-(+)-limonene with a Hg lamp. In the batch system, the yield of the *cis* and *trans* isomers linearly increased in 20 min and reached its plateau value, whereas the yield in the microreactor more quickly increased in linear relation to the observation time (135 s).

In order to compare the formation rate of the photoproducts in the batch and those in the microreactor systems, the observed values of quantum yield were evaluated by Eq. (17.1) and are summarized in Table 17.1. The steady-state approximation of the concentration of triplet toluene led to the formation rate constants of the *cis* (k_{cis}) and *trans* (k_{trans}) isomers, with information on the quenching rate constant of triplet toluene by (R)-(+)-limonene determined by the transient absorption measurement. The apparent formation rate constants in microreactors were successfully evaluated and are listed in Table 17.2. As the size of the microchannel became smaller, the quantum yield significantly increased. In this case, the k_{trans}/k_{cis} ratio in the microreactors became slightly higher ($\sim$5%) than that in the batch system. The de values in the batch and microreactor systems were examined against irradiation time. The de value was up to 28.5% de (15 min) in the batch system, whereas the value in the microreactor reached 30.6% de (36 s), because the microfluidic system could suppress the side reactions.

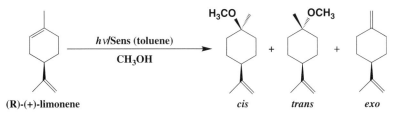

SCHEME 17.2 Photosensitized addition of (R)-(+)-limonene to methanol.

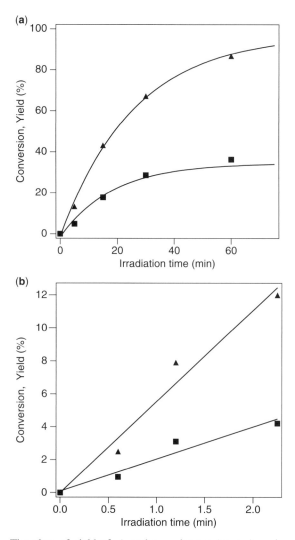

FIGURE 17.4 The plots of yield of *cis* and *trans* isomer (square), and conversion of (R)-(+)-limonene(triangle) in the batch (a) and in the microreactor (b) (with a width of 500 μm and depth of 300 μm) with a 40-W low-pressure Hg lamp.

TABLE 17.1 Quantum Yield and Formation Rate Constant for the *cis* (k_{cis}) and *trans* (k_{trans}) Isomers in the Batch and Microreactors Systems

	Quantum Yield[a]	k_{trans} $(10^8 \, \text{M}^{-1}\,\text{s}^{-1})$	k_{cis} $(10^8 \, \text{M}^{-1}\,\text{s}^{-1})$
Batch	0.008	0.37	0.21
Microreactors			
500 μm × 300 μm	0.045	2.3	1.2
200 μm × 20 μm	0.087	4.4	2.4

[a]Summation of the yield for the *cis* and *trans* isomers.

TABLE 17.2 Phocatalytic *N*-ethlylation of Benzylamine (1.0×10^{-3} M) in Microreactors Excited with 365-nm UV-LEDs and in a Batch Reactor

Reactor (Depth, μm)	Photocatalyst	Irradiation Time	Yield (%)	
			N-ethyl-benzylamine	*N,N*-diethyl-benzylamine
Batch[a]	TiO$_2$/Pt	5 h	84.4	2.4
Batch[a]	TiO$_2$/Pt	10 h	6.8	74.1
Batch[a]	TiO$_2$	—	0	0
Microreactor (500)	TiO$_2$/Pt	150 s	85	0
Microreactor (300)	TiO$_2$	90 s	98	0
Microreactor (500)	TiO$_2$	90 s	84	0
Microreactor (1,000)	TiO$_2$	90 s	70	0

[a]In suspended solution excited with a 400-W Hg lamp [31].

In conclusion, the experimental results clearly proved that a microreactor should enhance reaction efficiency due to high spatial illumination homogeneity and excellent light penetration throughout the reactor. The stereoselectivity of the photoreactions in a microreactor can be superior to that in a batch system.

17.2.3 Photocatalytic Reactions in Microreactors

The study of light-induced electron-transfer reactions in a semiconductor catalyst has become one of the most attractive research areas in photochemistry. Wide varieties of organic reactions have been successfully examined by using a semiconductor photocatalyst. A photocatalytic reaction can take place on an irradiated surface. Therefore, most research on the reaction is carried out using dispersed powders with conventional batch reactors. A separation step using powders is required after the reaction takes place. Although systems with an immobilized catalyst can avoid this step, they tend to have low interfacial surface areas.

A microreactor with an immobilized photocatalyst that has a large surface-to-volume ratio may prove advantageous in a photocatalytic reaction. Thus, we investigated the photocatalytic oxidation and reduction of organic compounds [26], and a process involving the *N*-alkylation of amines [27] in microreactors.

Photocatalytic Oxidation and Reduction Photoexcited TiO$_2$ oxidizes a reactant that donates an electron to TiO$_2$ while it reduces a reactant that receives an electron. Photoxidations of organic compounds using TiO$_2$ as a photocatalyst have been fruitfully investigated, and there are some reports on the photoreduction process as it relates to the TiO$_2$ surface [28, 29]. First, we examined the photodegradation

$$2\ ClC_6H_4OH + 13\ O_2 \xrightarrow{\ TiO_2,\ h\nu\ } 12\ CO_2 + 4\ H_2O + 2\ HCl$$

SCHEME 17.3   Photocatalytic degradation of chlorophenol.

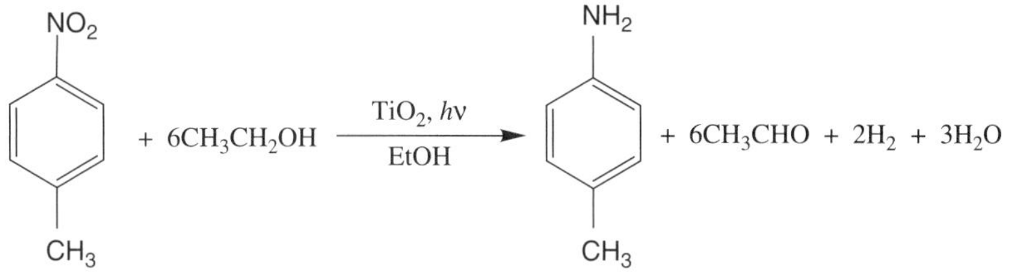

SCHEME 17.4   Photocatalytic reduction of *p*-nitrotoluene.

(photooxidation, Scheme 17.3) and photoreduction (Scheme 17.4) of organic compounds in a microreactor with an immobilized photocatalytic TiO_2 layer. As illustrated in Fig. 17.5, the reactions take place quite quickly in comparison to conventional batch reactors.

Figure 17.6 shows an action spectrum of degradation of dimethylformamide in the photocatalytic microreactor obtained by using a tunable OPO laser. It indicates that the reaction efficiency should be very sensitive to the excitation wavelength. Considering the band gap energy and action spectrum, we can expect higher reaction efficiencies with a light source of higher photon energy. Therefore, an array of 365-nm UV-LEDs (Nichia NSHU590B, with an optical power output of 10 mW) were employed for the photocatalytic reactions described in the following sections.

Multiphase Photocatalytic Oxidation   We further investigated the oxidation process of *p*-chlorophenol by using a gas–liquid–solid multiphase microreaction system. Although numerous attempts have been made to study multiphase catalytic reactions, there are still difficulties in evaluating the conduction of the reactions as a result of the low reaction yield arising from the very low efficiency of interaction and mass transfer between different phases [30]. To produce high interfacial area between the phases, we introduced an aqueous solution of *p*-chlorophenol and oxygen gas into a microchannel. The yield of phtodegradation is shown in Fig. 17.7. The horizontal axis indicates the gas injection rate, while the injection rate of the sample solution is kept at a constant value of 10 μL/min. At a lower gas injection rate, microbabbles were formed and so-called slug flow was observed. At a gas injection rate higher than 500 μL/min, a pipe flow, gas flowed through the center of the microchannel, while liquid flowing close to the photocatalyst surface was formed as schematically illustrated in Fig. 17.8. The reaction yield increased to 43% at the residence time of 14 s in the pipe flow, whereas it was 10% at the residence time of 75 s without the injection of oxygen gas. Under the pipe-flow condition, the liquid phase is always saturated with oxygen even in the final part of the microchannel.

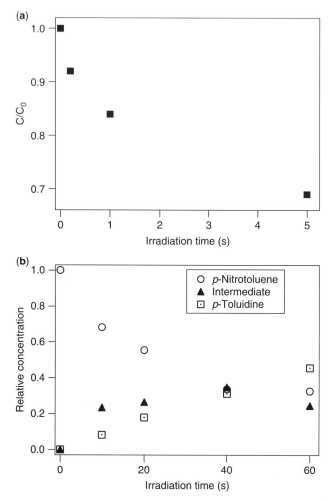

FIGURE 17.5 Photocatalytic degradation of *m*-chlorophenol $(1.1 \times 10^{-4} \text{M})$ (a) and reduction of *p*-nitrotoluene $(1.0 \times 10^{-4} \text{M})$ (b) in a microreactor of 100-μm depth and 500-μm width excited with 385-nm UV-LEDs.

In addition, as the gas injection rate increases, the thickness of the liquid phase decreases. At the gas injection rate of 750 μL/min, the thickness of the liquid phase is estimated to be 25 μm. These facts may increase the reaction yield.

Photcatalytic Alkylation It has been recognized that depositing Pt on TiO_2 enhances photocatalytic activity by serving as an electron sink and consequently slowing charge recombination. Ohtani et al. [31] studied the photocatalytic preparation of asymmetrical secondary and tertiary amines by Pt-loaded TiO_2 (TiO_2/Pt) particles suspended in a variety of alcohols as solvents by using conventional batch reactors.

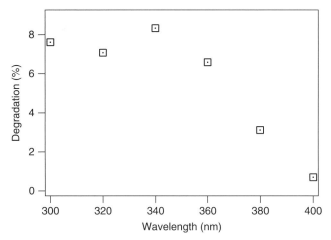

FIGURE 17.6 Action spectrum of photodegradation of dimethylformamide (1.0×10^{-4} M) in a photocatalytic microreactor excited with an OPO laser.

They reported that the *N*-alkylation of benzylamine occurred with a yield up to 84.4% after 4 h of irradiation under a 400-W high-pressure mercury lamp, while the *N*-alkylation of amines could not be observed by the irradiation of Pt-free TiO$_2$ (Scheme 17.5).

The photoirradiation of benzylamine in ethanol introduced into a microreactor with immobilized TiO$_2$/Pt led to *N*-ethylation. The reaction proceeded within only 150-s UV irradiation to yield 85% of *N*-ethylbenzylamine. The reaction mechanism can be interpreted as follows. The dehydrogenation of ethanol occurs on the surface of TiO$_2$/Pt to form acetoaldehyde and H$_2$. Substrate benzylamine is *N*-ethylated by

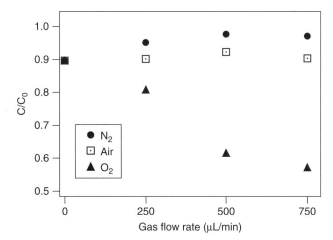

FIGURE 17.7 Phocatalytic degradation of *p*-chlorophenol (1.1×10^{-3} M) in a gas–liquid–solid multiphase microreactor of 500-μm depth and width.

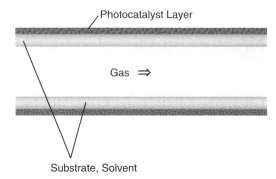

FIGURE 17.8 Schematic view of a multiphase microreactor.

the condensation of photoproduct carbonyl with benzylamine. The reduction of the resulting intermediate by H_2 occurs, yielding *N*-ethylbenzylamine.

In contrast to the result in a batch reactor, we successfully observed the *N*-alkylation reaction of benzylamine by using the microreactor with immobilized Pt-free TiO_2 as well as TiO_2/Pt. The ethylation of benzylamine proceeds very rapidly and reaction efficiency increased as the depth of the microchannel decreased (Table 17.2). The reaction proceeded within only 90 s to yield 98% of *N*-ethylbenzyl-amine with a microreactor of 300-μm depth.

The reaction efficiencies are influenced by a series of processes, including electron transfer from the conduction band of TiO_2 to the substrate and oxidation of the solvent by an electron hole. Since the electron-hole recombination within the photo-catalyst is in competition with the reaction process, the reaction efficiencies must be strongly affected by the surface-to-volume ratio of a photocatalytic reactor. Thus, the photoreaction can proceed rapidly in the microreactor, which has a remark-ably large surface-to-volume ratio as compared with conventional batch reactors. For the above reason, the *N*-alkylation of benzylamine may be observed even in the microreactor without a Pt cocatalyst. Ohtani et al. [31] reported that the UV irradiation of TiO_2/Pt suspended in ethanol led to *N*-alkylation and *N*, *N*-dialkylation. In contrast, the irradiation of benzylamine in the photocatalytic microreactor with Pt-free TiO_2 as well as TiO_2/Pt did not yield any detectable *N*, *N*-dialkylated products (Table 17.2). The absence of *N*, *N*-dialkylated products can be attributed to the nature of a continuous-flow microreaction system. In the microflow system, the residence time of the substrate is very short and the reaction vessel does not retain the reaction

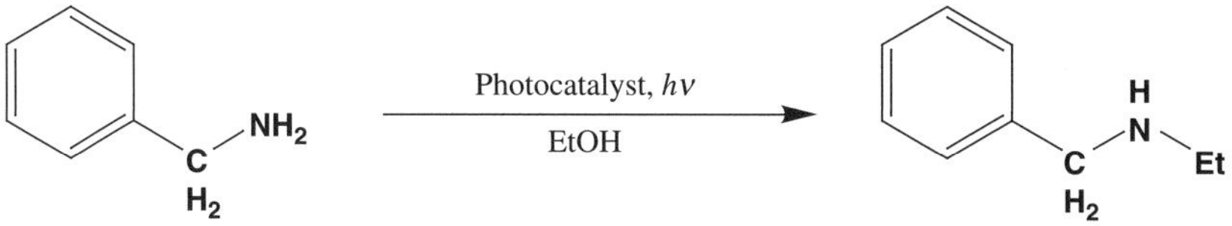

SCHEME 17.5 Ethylation of benzylamine on photocatalyst surface.

intermediates. These facts may inhibit or prevent the consecutive *N*-alkylation process of *N*-ethylbenzylamine.

Summary and Future Prospects We have investigated the application of microreactors to asymmetric photosensitized reactions concerned with the reaction efficiency and stereoselectivity. The reaction efficiency in a microreactor was significantly improved. The E/Z ratio obtained in the microreactor was about 1.6 times higher than that in the batch system. The quantum yield of the reaction of (R)-limonene in the microreactor was significantly larger than that in the batch system. Even in a very short irradiation time, high ee and de values were obtained in microreactors due to the efficient utilization of photon energy there.

We have developed a photocatalytic microreactor and examined the processes of reduction and oxidation of organic compounds, and amine *N*-alkylation reaction in microspace. These model reactions proceeded very rapidly, and the yield increased as the surface-to-volume ratio increased. In contrast to the result in a batch reactor, we successfully observed the *N*-alkylation reaction of benzylamine by using the microreactor with immobilized Pt-free TiO_2 as well as TiO_2/Pt. The use of a continuous-flow micoreactor inhibited the formation of *N*, *N*-dialkylation products.

These results suggest the applicability of microreaction systems to organic photoreactions. We must further investigate other model photoreactions to prove the advantages, especially high efficiencies and reaction selectivity, that might be introduced by such systems. The optimization of excitation wavelength and photon density, design of the microreactor, flow rate, and irradiation time are under study for the establishment of a photochemical microreaction system.

17.3 TYPICAL EXAMPLES OF OTHER RESEARCH

F. Jensen's group [13] at MIT in Cambridge, Massachusetta, carried out pioneering work in the research field of photochemical reactions in microfabricated reactors and detectors. Two different microfabricated reactors were designed for the integration of the reaction and detection modules. It also succeeded in bonding quartz substrates to micropatterned silicon devices at low temperature using a per-fluorinated polymer-CYTOP^{TM}, where quartz substrates allowed reaction and detection with UV light of shorter wavelengths (higher energies) than ones made of Pyrex glass permit. The photoreaction of benzophenone in isopropanol to form benzopinacol and acetone was investigated as a model reaction. Crystallization of the product (benzopinacol) in the microreactors was avoided with a continuous-flow system by controlling the residence time and, consequently, the extent of the reaction. Off-chip analysis of the photoproduct using HPLC confirmed the results obtained from online UV spectroscopy to observe the absorbance of the reaction mixture at different flow rates. The estimated reaction quantum yield revealed that the reactor design should improve the overall reaction efficiency, which was defined by the ratio of moles of reaction (the conversion of the reactant benzophenone) and the amount of light from the miniaturized UV lamp.

A. J. de Mello's group [12] at the Imperial College London, London, UK has studied a microfabricated nanoreactor for the safe, continuous generation and use of singlet oxygen. The term "nanoreactor" refers to reactors with an instantaneous reaction volume most conveniently measured in nanoliters. Because the efficiency of mixing and separation can be increased in nanoreactors in combination with high rates of thermal and mass transfer, nanoreactors may be ideal for processing valuable or hazardous reaction components and in many cases for improving reaction selectivities. In this study, singlet oxygen was effectively and safely generated in nanoscale reactor and used for the synthesis of ascaridole. The technique allowed for the generation of singlet oxygen without the inherent dangers of large quantities of oxygenated solvents. In addition, the low Reynolds number encountered within most nanofluidic devices could control processes taking place in continuous-flow systems. More impotant, the choice of a continuous-flow system instead of traditional batch processes can facilitate the use of a multiparallel (or scale-out) approach. Thus, this group demonstrated the applicability of nanoreactor technology to the safe, efficient, and continuous-flow synthesis of ascaridole from α-terpinene.

N. Kitamura's group [14] at Hokkaido University, Sapporo, Japan has investigated the photocyanation of pyrene(PyH) across an oil–water interface. It used two types of polymer microchannel chips in its study. The chips (with a depth of 20 μm and width of 100 μm) were fabricated with photolithography and an imprinting method. An aqueous NaCN solution and propylene carbonate solution of PyH and 1,4-dicyanobenzene were injected separately into a Y-structured chip with the same flow velocity, and the irradiation of a high-pressure Hg lamp onto the chip resulted in the formation of 1-cyanopyrene (PyCN). The results proved that the interfacial photochemical reaction of PyH proceeded very efficiently. Under optimum conditions by using a three-layer channel chip, an absolute PyCN yield of 73% was obtained with a reaction time of 210 s.

T. Kitamori's group [33] at the University of Tokyo published a review focusing on the integration of chemical and biochemical analysis systems into glass microchips for general use. By combining multiphase laminar flow driven by pressure and micro unit operations, such as mixing, reaction, extraction, and separation, continuous-flow chemical processing systems can be realized in the microchip format, whereas the application of electrophoresis-based chip technology is limited. The performance of several analysis systems was greatly improved by microchip integration because of some characteristics of microspace described elsewhere. This same group also demonstrated that several different analysis systems, such as wet analysis of cobalt ion, multi-ion, multi-ion sensor, immunoassay, and cellular analysis, could be successfully integrated on a microchip, and concluded that these microchip technologies should be promising for meeting the future demands of high-throughput chemical processing.

S. J. Haswell's group [34] at the University of Hull, Hull, UK has studied the chemical reactions in microreactors using an inverted Raman microscopic spectrometer. Raman spectroscopy has the advantage of being applicable to monitoring nonradiative (nonfluorescent) species, which means that spectroscopy, in principle, may detect any kind of reaction species with reasonable sensitivity. In this study, an inverted Raman microscope spectrometer was used to obtain information on the spatial evolution of reactant and product concentrations for a chemical reaction in a hydrodynamic flow-controlled microreactor. The Raman spectrometer was equipped with a laser (780-nm) source, confocal optics, holographic grating, and charge-coupled device (CCD) detector. The details of Raman microscopy are discussed in "Raman Analysis in Microreactor Channels," HORIBA JOBIN YVON Raman Application Note. The microreactor consisted of a T-shaped channel network of a 0.5-mm-thick glass bottom plate with a 0.5-mm-thick glass top plate. The ends of the channel network were connected to reagent reservoirs that were linked to a syringe pump to inject the solutions by a hydrodynamic flow into the channels. The microchannels were 221 μm wide and 73 μm deep. The synthesis of ethyl acetate from ethanol and acetic acid in the microreactor was investigated as a model system as Raman scattering bands for each reactant and product species were clearly resolved. It was proven that the signal intensities of each band obtained by Raman spectroscopy are propotional to the concentration for each species. Accordingly, all concentrations could be quantitatively measured after calibration. By scanning specific Raman bands within a selected area in the microchannel network at given steps in the *X-Y* plane, spatially resolved concentration profiles were obtained under steady-state flow conditions. Under the flow conditions used, different positions within the concentration profile correspond to different times after contact and mixing of the reagents, thereby enabling one to observe the time dependence of the product formation. In conclusion, Raman microscopy could provide a useful complementary technique for UV/VIS absorbance and fluorescence methods for the in situ monitoring and analysis of chemical reactions within channel networks, and could be used to optimize reactions in microreactors.

E. Verpoorte [35] at the University of Groningen, Groningen, The Netherlands reviewed micro-optics for lab-on-a-chip devices in 2003, and in her article she wrote that "the examples of microfluidic devices incorporating microfabricated light sources are still few and far between. One notable exception is the work. . . . " This statement seems to be still true at present. A small number of papers dealing with photoreactions has been published thus far, but that number is increasing remarkably.

R. Gorge's group [15] at Friedrich Schiller University, Jena, Germany has studied photocatalysis in microreactors. This paper appears to be the first report concerned with photocatalitic reaction in microreactors. A photocatalytic microreactor with immobilized titanium dioxide(TiO_2) as a photocatalyst and illuminated by UV-A LED light (385 nm) was constructed and tested for the degradation of the model substance 4-chlorophenol. The microreactor consisted of 19 channels with a cross section of approximately 200 μm × 300 μm. The intrinsic kinetic parameters of the reaction could be determined and mass-transfer limitations for the operating conditions employed could be excluded by calculating appropriate Damköhler numbers. Photonic efficiencies for the degradation of 4-chlorophenol were given. The quantum yield of a photocatalic reaction is very difficult to determine because the amount of absorbed photons by the catalyst is hard to estimate. Therefore, photonic efficiency is practically defined by the ratio of rate of reaction to incident monochromatic light intensity. This group concluded that the illuminated specific surface of the microstructured reactor should surpass that of conventional photocatalytic reactors by a factor of 4 to 400.

T. Kitamori's group [18] at the University of Tokyo has studied photocatalytic redox-combined synthesis in a microchannel chip. In this work, a titania-modified microchannel chip (TMC) was fabricated to implement an efficient photocatalytic synthesis of L-pipecolinic acid from L-lysine. The in-chip conversion rate was found to be 70 times larger than that in a batch system (cuvette) by using nm-sized titania particles with almost the same selectivity and enantiomeric excess. The experimental results have proven that TMC does have satisfactory potential to allow its application to photocatalytic synthesis as well as photodegradation.

H. Maeda's group [16] at the National Institute of Advanced Industrial Science and Technology (AIST), Tosu, Japan applied a simple method using the self-assembly of colloidal particles to modify a microcapillary inner surface and investigated photocatalytic and enzyme reactions. It arranged nano-particles on the capillary inner wall and controlled particle layer thickness and layer patterning by choosing adequate combinations of the solvent and drying temperature. In addition, this group utilized SiO_2 composite particles coated by TiO_2 (anatase type) for the purpose of the particle arrangement process in the microreactor to carry an anatase-type TiO_2 catalyst. Such a process was likely to be a simple catalyst-carrying method onto the microreactor inner wall. A similar process was also applied to a few catalytic reaction systems, including enzyme reaction and photocatalytic reaction. The experimental results showed that an enzyme reaction could be enhanced, probably due to the increased reactor surface area, and reasonable enhancement of a photocatalytic reaction was also observed for a reaction in a microreactor.

I. Ryu's group [36] at Osaka Prefecture University has investigated a photochemical $[2 + 2]$ cyclo-addition reaction in a microflow system using glass-made microchannels (with a width of 100 μm and depth of 500 μm). The reaction of cyclohexenones with vinyl acetates in a microflow system under irradiation (300 W, a Hg lamp) gave $[2 + 2]$ cyclo-addition products in good yield with a residence time of 2 h, which is a remarkable shortened reaction time compared with a batch system using the same light source.

K. Mizuno's group [17] at Osaka Prefecture University has studied the intramolecular $(2\pi + 2\pi)$ photocyclo-addition of a 1-cyanonaphthalene derivative in microreactors made of poly(dimethylsiloxane) (PDMS). By using the microreactors and flow system, both the efficiency and regioselectivity increased compared with those obtained under batch conditions.

K. Jähnisch's group [37] at the Institut für Angewandte Chemie Berlin-Adlershof, Berlin, Germany applied a falling-film microreactor for a photochemical gas–liquid reaction and demonstrated the selective photochlorination of toluene-2, 4-diisocyanate (TDI). Photochlorination of TDI with chlorine gas forms 1-chloromethyl-2, 4-diisocyanate (1Cl-TDI) by side-chain chlorination together with the ring-chain chlorinated product of 5Cl-TDI. The selectivity to form 1Cl-TDI was significantly higher in a microreactor than in a conventional batch reactor. The space-time yield in the microreaction also was orders of magnitude higher compared to a batch system. This same group also demonstrated that a falling-film microreactor should be applicable to the photooxygenation of cyclopentadiene by singlet oxygen [38]. The intermediate of explosive endoperoxide was successfully reduced to yield the final product of cyclopentendiol.

BIBLIOGRAPHY AND OTHER SOURCES

1. Wayne, C. E., and Wayne, R. P. (1996). *Photochemistry.* New York: Oxford University Press.

2. Turro, N. J. (1991). *Modern Molecular Photochemistry.* Sausalito, CA: University Science Books.

3. Moor, J. H., Davis, C. C., and Coplan, M. A. (2003). *Building Scientific Apparatus, 3rd ed. A Practical Guide to Design and Construction.* Boulder, CO: Westview Press.

4. Ichimura, T. (1999). Photodissociation dynamics of chlorinated benzene derivatives. In J. Laane, H. Takahashi, and D. A. Bandrauk (eds.), *Structure and Dynamics of Electronic Excited States*, pp. 233–262. Berlin Heidelberg: Springer-Verlag.

5. Ichimura, T., and Suzuki, T. (2000). Photophysics and photochemical dynamics of methylanisole molecules in a supersonic jet. *J. Photochem. Photobiol. C: Photochem. Rev.*, 1: 79–107.

6. Suzuki, T., Omori, T., and Ichimura, T. (2000). Excited state reaction of short-lived 2-methylbenzophenone enols studied by two-color time-resolved thermal lensing technique. *J. Phys. Chem.*, A104: 11671–11676.

7. Nagano, M., Suzuki, T., Ichimura, T., et al. (2005). Production and excited state dynamics of photo-rearranged isomer of benzyl chloride and its methyl derivatives studied by stepwise two-color laser excitation transient absorption and time-resolved thermal lensing techniques. *J. Phys. Chem.*, A109: 5825–5831.

8. Ehrfeld, W., Hessel, V., and Lowe, H. (2000). In *Microreactors.* Weinheim: Wiley-VCH.

9. Hessel, V., Hardt, S., and Lowe, H. (2004). In *Chemical Micro Process Engineering.* Weinheim: Wiley-VCH.

10. Haswell, S. J., O'Sullivana, B., and Styring, P. (2001). Kumada–Corriu reactions in a pressure-driven microflow reactor. *Lab Chip*, 1: 164–166.

11. Aoki, N., Hasebe, S., and Mae, K. (2004). Mixing in microreactors: Effectiveness of lamination segments as a form of feed on product distribution for multiple reactions. *Chem. Eng. J.*, 101: 323.

12. Suga, S., Nagaki, A., and Yoshida, J. (2003). Highly selective friedel–crafts show mono-alkylation using micromixing. *Chem. Commun.*, 3: 354–355.

13. Lu, H., Schmidt, M. A., and Jensen, K. F. (2001). Photochemical reactions and on-line UV detection in microfabricated reactors. *Lab Chip*, 1: 22–28.

14. Ueno, K., Kitagawa, F., and Kitamura, N. (2002). Photocyanation of pyrene across an oil/water interface in a polymer microchannel chip. *Lab Chip*, 2: 231–234.

15. Gorges, R., Meyer, S., and Kreisel, G. (2004). Photocatalysis in microreactors. *J. Photochem. Photobiol.*, A167: 95–99.

16. Nakamura, H., Li, X., Wang, H., Uehara, M., Miyazaki, M., Shimizu, H., et al. (2004). A simple method of self-assembled nano-particles deposition on the micro-capillary inner walls and the reactor application for photo-catalytic and enzyme reactions. *Chem. Eng. J.*, 101: 261–268.

17. Maeda, H., Mukae, H., and Mizuno, K. (2005). Enhanced efficiency and regioselectivity of intramolecular $(2\pi + 2\pi)$ photocycloaddition of 1-cyanonaphthalene derivative using microreactors. *Chem. Lett.*, 34: 66–67.

18. Takei, G., Kitamori, T., and Kim, H.-B. (2005). Photocatalytic redox-combined synthesis of L-pipecolinic acid with a titania-modified microchannel chip. *Cat. Commun.*, 6: 357–360.

19. Ray, A. K., and Beenackers, A. A. C. M. (1998). Novel photocatalytic reactor for water purification. *AIChE J.*, 44: 477–483.

20. Rau, H. (1983). Asymmetric photochemistry in solution. *Chem. Rev.*, 83: 535–547.

21. Everitt, S. R. L., and Inoue, Y. (1999). Asymmetric photochemical reactions in solution. In: *Organic Molecular Photochemistry*, V. Ramamurthy, and K. Schanze (eds.), pp. 71–130, New York: Marcel Dekker.

22. Inoue, Y., Wada, T., Asaoka, S., Sato, H., and Pete, J.-P. (2000). Photochirogenesis: Multidimensional control of asymmetric photochemistry. *Chem. Commun.*, 251–259.

23. Shim, S. C., Kim, D. S., Yoo, D. J., Wada, T., and Inoue, Y. (2002). Diastereoselectivity control in photosensitized addition of methanol to (*R*)-(+)-limonene. *J. Org. Chem.*, 67: 5718–5726.

24. Sakeda, K., Wakabayashi, K., Matsushita, Y., Suzuki, T., Ichimura, T., Wada, T., and Inoue, Y. (2007). Asymmetric photosensitized addition of methanol to (R)-(+)-(Z)-limonene in a microreactor. *J. Photochem. Photobiol. A: Chemistry*, 192: 166–171.

25. Wada, T., and Inoue, Y. (private communication).

26. Matsushita, Y., Kumada, S., Wakabayashi, K., Sakeda, K., Suzuki, T., and Ichimura, T., (2006). Photocatalytic reduction in microreactors. *Chem. Lett.*, 4: 410–411.

27. Matsushita, Y., Ohba, N., Suzuki, T., and Ichimura, T. (2008). N-Alkylation of amines by photocatalytic reaction in a microreaction system. *Catalysis Today*, 132: 153–158.

28. Joyce-Pruden, C., Pross, S., Kreisel, J. K., and Li, Y. (1992). Photoinduced reduction on titanium dioxide. *J. Org. Chem.*, 57: 5087–5091.

29. Mahadavi, F., Bruton, T. C., and Li, Y. (1993). Photoinduced reduction of nitro compounds on semiconductor particles, *J. Org. Chem.*, 58: 744–746.

30. Kobayashi, J., Mori, Y., Okamoto, K., Akiyama, R., Ueno, M., Kitamori, T., et al. (2005). A Microfluidic device for conducting gas-liquid-solid hydrogenation reactions. *Science*, 304: 1305–1308.

31. Ohtani, B., Osaki, H., Nishimoto, S., and Kagiya, T. (1986). A novel photocatalytic process of amine N-alkylation by platinized semiconductor particles suspended in alcohols. *J. Am. Chem. Soc.*, 108: 308–310.

32. Wootton, R.C. R., Fortt, R., and de Mello, A. J. (2002). A microfabricated nanoreactor for safe, continuous generation and use of singlet oxygen. *Org. Pro. Res. & Dev.*, 6: 187–189.

33. Sato, K., Hibara, A., Tokeshi, M., Hisamoto, H., and Kitamori, T. (2003). Integration of chemical and biological analysis systems into a glass microchip. *Analy. Sci.*, 19: 15–22.

34. Fletcher, P. D. I., Haswell, S. J., and Zhang, X. (2003). Monitoring of chemical reactions within microreactors using an inverted Raman microscopic spectrometer. *Electrophoresis*, 24: 3239–3245.

35. Verpoorte, E., (2003). Chip vision-optics for microchips. *Lab Chip*, 3: 42N–52N.

36. Fukuyama, T., Hino, Y., Kamata, N., and Ryu, I. (2004). Quick execution of [2 + 2] type photochemical cycloaddition reaction by continuous flow system using a glass-made microreactor. *Chem. Lett.*, 33: 1430–1431.

37. Ehrich, H., Linde, D., Morgenschweis, B. M., and Jähnisch, K. (2002). Application of microstructured reactor technology for the photochemical chlorination of alkylaromatics. *Chimia*, 56: 647–653.

38. Jähnisch, K., and Dingerdissen, U. (2005). Photochemical generation and [4 + 2]-cyclo-addition of singlet oxygen in a falling-film microreactor. *Chem. Eng. Technol.*, 28: 426–427.

CHAPTER 18

INTENSIFICATION OF CATALYTIC PROCESS BY MICRO-STRUCTURED REACTORS

LIOUBOV KIWI-MINSKER and ALBERT RENKEN

18.1 INTRODUCTION

Process intensification (PI) is the term that describes an innovative design approach in chemical engineering which aims at the miniaturization of chemical reactors and plants. This would decrease running costs and make the process more efficient, safer, and less polluting than existing ones. This new approach was pioneered in the mid-1980s by Colin Ramshaw at the University of Newcastle, Newcastle upon Tyne [1–3], who led the development of innovative intensified chemical plants during the last decade. PI is no longer considered only of academic interest: It has also become industrially feasible. Its use is often quantified by the significant increase, by at least an order of magnitude, of the ratio of equipment volume to the product yield and decreasing energy consumption. It also lowers the amount of waste and leads to better utilization of raw materials.

Chemical microstructured reactors (MSR) are devices containing open paths for fluids with dimensions in the submillimeter range. MSR mostly have multiple parallel channels with diameters between ten and several hundred micrometers in which chemical transformations occur. This gives a high specific surface area in the range of 10,000 to 50,000 m^2/m^3 and allows effective mass and heat transfer compared to traditional chemical reactors usually having $\sim 100\,m^2/m^3$. Typical examples of MSR are shown in Figs. 18.1 and 18.2.

Another important feature of MSR is that the heat exchange and reaction are often performed in the same gadget. MSR are operated under laminar flow, with the heat

Microchemical Engineering in Practice. Edited by Thomas R. Dietrich
Copyright © 2009 John Wiley & Sons, Inc.

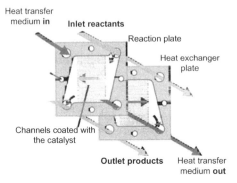

FIGURE 18.1 General view (right) and scheme of the fluid flows (left) of a microstructured reactor (Institut für Mikrotechnik Mainz, IMM). Channel length: $L = 20$ mm; width: $b = 300$ μm; height: $e = 240$ μm.

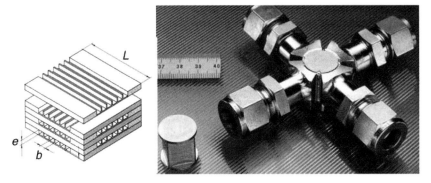

FIGURE 18.2 Schematic representation (left) and photo (right) of a microstructured heat exchanger/reactor (Forschungszentrum Karlsruhe). Channel length: $L = 14$ mm; width: $b = 100$ μm; height: $e = 78$ μm.

transfer coefficient for liquids being about $10 \, \text{kW}/(\text{m}^2 \cdot \text{K})$. This is one order of magnitude higher than in traditional heat exchangers, allowing users to avoid hot-spot formation, to attain higher reaction temperatures, and to reduce reaction volumes. This, in turn, improves energy efficiency and reduces operational costs. Integrated heat exchange makes the key difference between MSR and other structured reactors such as honeycombs. The intensification of heterogeneous catalytic processes involves, in addition to the innovative engineering of MSR, the proper design of the catalyst. This requires the simultaneous development of the catalyst and reactor. The catalyst design should be closely integrated with that of the reactor, taking into consideration the reaction mechanism, mass and heat transfer, and the energy supply addressing the high selectivity and yield of the target product.

In this chapter, the quantitative criteria for the rational use of MSR will be introduced, based on the fundamentals of chemical reaction engineering. We will discuss the major characteristics of MSR to quantify the potential gain in reactor performance by structuring the fluid flow into parallel channels. This will be followed by a discussion of the principle design of MSR with integrated catalytic phase and the range of operational conditions necessary to achieve the benefits of reactor miniaturization.

18.2 CHARACTERISTICS OF MICROSTRUCTURED REACTORS

18.2.1 Timescale of Physical and Chemical Processes

The characteristic time of chemical reactions t_r, which is defined by intrinsic reaction kinetics, can range from hours for slow organic or biological reactions to milliseconds for high-temperature (partial) oxidation reactions (Fig. 18.3). When the reaction is carried out in a reactor, heat and mass transfer interferes with the reaction kinetics.

The characteristic time of physical processes (heat and mass transfer) in conventional reactors can range from 10 to 10^{-3} s. This means that relatively slow reactions ($t_r \gg 10$ s) are carried out in the kinetic regime and the global performance of the reactor is controlled by the intrinsic reaction kinetics. The chemical reactor is designed and dimensioned to obtain the required product yield and conversion of the raw material. The attainable reactant conversion in the kinetic regime depends on the ratio of the mean residence time in the reactor, τ to the characteristic reaction time. This ratio is known as the first Damköhler number DaI:

$$\text{DaI} = \frac{\tau}{t_r} \tag{18.1}$$

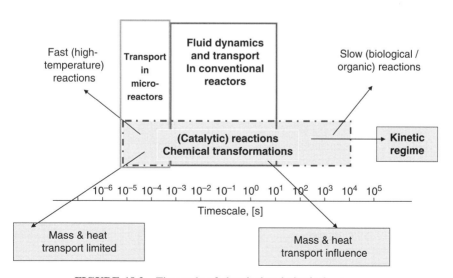

FIGURE 18.3   Timescale of chemical and physical processes.

Depending on the kinetics and type of reactor, the residence time should be several times higher than the characteristic reaction time to get conversions on the order of 90% [4, 5].

For fast chemical reactions, the characteristic reaction time is the same order of magnitude as the characteristic time of the physical processes (Fig. 18.3). The performance of a conventional reactor is influenced by mass and/or heat transfer. For very fast reactions, the global transformation rate may be completely limited by transfer phenomena. As a result, the reactor performance is diminished as compared to the maximal performance attainable in the kinetic regime and the product yield is very often reduced.

To eliminate mass transfer resistances in practice, the characteristic transfer time should be roughly one order of magnitude lower compared to the characteristic reaction time. As the mass and heat transfer performance in MSR is up to two orders of magnitude higher compared to conventional tubular reactors, the reactor performance can be considerably increased, leading to the desired intensification of the process. In addition, consecutive reactions taking place on the catalytic surface can be efficiently suppressed in MSR. Therefore, fast reactions carried out in MSR show higher product selectivity and yield.

18.2.2 Pressure Drop in MSR

The pressure drop that must be overcome during the passage of fluid through any reactor is an important characteristic related to energy demand for process optimization. Therefore, it will be considered here for multichannel microreactors, assuming noncompressible fluids. Gas properties will be used only up to approx. 600 K and at least atmospheric pressure ($p \geq 10^5$ Pa). Under these conditions, continuum mechanics are valid [6]. In addition, fluid velocities lower than 10 m/s will be considered in channels with hydraulic diameters smaller than 1 mm. Therefore, the fluid flow is laminar and compressibility effects can be neglected. The pressure drop in smooth tubes under laminar flow conditions is given by the following relation [7]:

$$\frac{\Delta p}{L_t} = 32 \, \phi \cdot \frac{\eta \cdot u}{d_t^2} \tag{18.2}$$

L_t and d_t refer to the channel length and equivalent diameter, respectively; η is the dynamic viscosity; u is the superficial fluid velocity; and ϕ is a geometric factor, which is $\phi = 1.0$ for circular tubes. In practice, channels in MSR often have a rectangular form. For rectangular channels, ϕ depends on the ratio of channel height to width, e/b (see Fig. 18.2). The correction factor becomes $\phi = 0.89$ for quadratic channels and $\phi \rightarrow 1.5$ when the ratio $e/b \rightarrow 0$ [7]. An empirical relationship is given below in Eq. (18.3):

$$\phi = 0.8735 + 0.6265 \exp\left(-3.636^* e/b\right) \tag{18.3}$$

In Fig. 18.4, the Reynolds number and pressure drop in MSR with different channel geometries are presented. Compared to MSR, the pressure drop in packed-bed

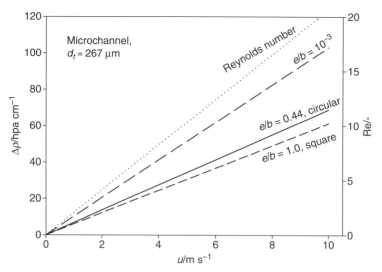

FIGURE 18.4   Reynolds number and pressure drop in microstructured multichannel reactors as function of the linear velocity of air in the channels ($T = 293$K, $p = 0.1$ MPa).

reactors with spherical particles and the same specific surface and residence time is roughly two to three times higher [8].

18.2.3 Residence Time Distribution and Mixing in MSR

One of the most important characteristics of continuous-flow reactors is flow pattern [9]. In the ideal mixed reactor, it is assumed that the concentrations and temperature within the reactor volume are uniform. Complete mixing can be achieved in a continuously operated stirred tank reactor (CSTR) or in a loop reactor applying a high recycling ratio [4, 5]. In contrast to the ideal CSTR, an ideal tubular plug-flow reactor (PFR) is characterized by a piston-type pattern of the fluid, corresponding to uniform radial concentrations and temperature. In an ideal PFR, all volume elements ΔV leave the reactor after exactly the same residence time. The reactants are continuously consumed as they flow along the reactor length to the outlet.

In practice, however, the volume elements entering the tubular reactor together will remain for different times inside the reactor, giving a distribution of residence times around the mean value. The deviation from the ideal plug-flow pattern can be caused by channeling of fluid, back-mixing, or the creation of stagnant zones. The distribution of residence times can significantly lower reactor performance, product selectivity, and product yield. Therefore, the residence time distribution (RTD) is one of the key parameters of the reactor.

The axial mixing in a tubular reactor can be described by the dispersion model. This model suggests that the RTD may be considered the result of piston flow with a superposition of longitudinal dispersion. The latter is taken into account by means of a constant effective axial dispersion coefficient D_{ax}, which has the same dimension as the molecular diffusion coefficient D_m. Usually, D_{ax} is much larger than the molecular diffusion coefficient because it incorporates all effects that cause deviation from plug flow, such as, the differences in radial velocity, eddies, and vortices.

The RTD as described by the dispersion model can be derived from the mass balance of a nonreacting species (tracer) over a volume element $\Delta V = S\Delta z$, where S is the section of the tube and z the axial coordinate. For constant fluid density and superficial velocity u, we obtain

$$S\Delta z \frac{\partial c}{\partial t} = u(c_z - c_{z+\Delta z})S + \left(-D_{ax}\frac{\partial c}{\partial z}\bigg|_z + -D_{ax}\frac{\partial c}{\partial z}\bigg|_{z+\Delta z}\right)S$$

and with $\Delta z \to 0$,
$$\frac{\partial c}{\partial t} = -u\frac{\partial c}{\partial z} + D_{ax}\frac{\partial^2 c}{\partial z^2} \tag{18.4}$$

In dimensionless form, Eq. (18.4) becomes

$$\frac{\partial C}{\partial \theta} = -\frac{\partial C}{\partial Z} + \frac{1}{\text{Bo}}\frac{\partial^2 C}{\partial Z^2} \tag{18.5}$$

with $\theta = \dfrac{t}{\tau}$, $\tau = \dfrac{L}{u}$, $Z = \dfrac{z}{L}$, $C = \dfrac{c}{\bar{c}_0}$, $\bar{c}_0 = \dfrac{n_{\text{inj}}}{V_R}$, and $\text{Bo} = \dfrac{u \cdot L}{D_{ax}}$.

c is the tracer concentration measured at the reactor outlet. The total amount of a nonreactive tracer injected as a Dirac pulse at the reactor entrance is given by n_{inj}. The Bodenstein number Bo is defined as the ratio between the axial dispersion time $t_{ax} = L^2/D_{ax}$ and mean residence time $\bar{t} = \tau = L/u$, which is identical to the space-time for reaction mixtures with constant density. For $\text{Bo} \to 0$, the axial dispersion time is short compared to the mean residence time, resulting in complete back-mixing in the reactor. For $\text{Bo} \to \infty$, no dispersion occurs. In practice, axial dispersion can be neglected for $\text{Bo} \geq 100$. If a system is open for dispersion at the reactor inlet and outlet [9], the response to a tracer pulse can be predicted with Eq. (18.6) and is shown in Fig. 18.5 (left):

$$C(\theta) = \frac{1}{2}\sqrt{\frac{\text{Bo}}{\pi \cdot \theta}}\exp\left(\frac{-(1-\theta)^2\text{Bo}}{4\theta}\right) \tag{18.6}$$

The cumulative RTD, $F(\theta)$, can be obtained by integration of Eq. (18.6) and is shown in Fig. 18.5 (right).

For reactions with a positive order, a broad RTD diminishes the performance of chemical reactors. The loss in performance can become considerable at high conversions for reactions with a high reaction order. Even more sensitive to back-mixing are the selectivity and yield of the target product, if it is an intermediate in a complex reaction network.

The dispersion in tubular reactors depends on the flow regime, characterized by the Reynolds number $\text{Re} = u \times d_t/\nu$, and the physical properties of the fluid, characterized by the Schmidt number $\text{Sc} = \nu/D_m$, where D_m is the molecular diffusion coefficient and ν the kinematic viscosity of the fluid. The flow in microchannels

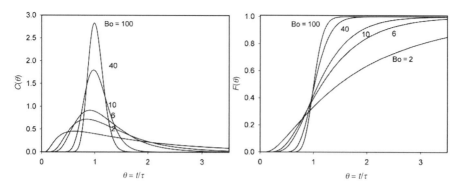

FIGURE 18.5 Residence time distribution (left) and cumulative distribution (right) predicted by the dispersion model. Parameter: Bodenstein number Bo.

with diameters between 10 and 500 μm is mostly laminar and has a parabolic velocity profile. Therefore, the molecular diffusion in axial and radial directions plays an important role in RTD. The diffusion in the radial direction tends to diminish the spreading effect of the parabolic velocity profile, whereas in the axial direction the molecular diffusion increases the dispersion. The axial dispersion coefficient depends on the molecular diffusion, flow velocity, and channel diameter [10, 11]. The relationship is given in Eq. (18.7):

$$D_{ax} = D_m + \chi \frac{u^2 d_t^2}{D_m}, \quad \text{for} \quad \frac{L}{d_t} > 0.04 \frac{u \cdot d_t}{D_m} \tag{18.7}$$

The parameter χ depends on the channel shape (1/192 for circular tubes).

The Bodenstein number in microchannels can be estimated with Eq. (18.8):

$$\frac{1}{Bo} = \frac{D_{ax}}{u \cdot L} = \frac{D_m}{L^2} \frac{L}{u} + \frac{1}{192} \frac{d_t^2}{D_m} \frac{u}{L} = \frac{D_m}{L^2} \frac{L}{u} + \frac{1}{192} \frac{4 \cdot R_t^2}{D_m} \frac{u}{L}$$

$$\frac{1}{Bo} = \frac{\tau}{t_{D,ax}} + \frac{1}{48} \frac{t_{D,rad}}{\tau}, \quad \text{with} \quad t_{D,ax} = \frac{L^2}{D_m}, t_{D,rad} = \frac{R_t^2}{D_m} \tag{18.8}$$

The first term in Eq. (18.8) corresponds to the ratio between space-time and the characteristic axial diffusion time $t_{D,ax}$. The diffusion coefficient lies in the order of $10^{-5}\,\text{m}^2\text{s}^{-1}$ and $10^{-9}\,\text{m}^2\text{s}^{-1}$ for gases and liquids, respectively. Typical lengths of MSR are several centimeters and space-time falls in the range of seconds. Therefore, the axial dispersion in microchannels is mainly determined by the ratio of space-time to radial diffusion time $t_{D,rad}$ [the second term in Eq. (18.8)]. The Bodenstein number can be estimated with Eq. (18.9):

$$Bo \cong 48 \cdot \frac{\tau}{t_{D,rad}} \cong 50 \cdot \tau \cdot \frac{D_m}{R_t^2} \tag{18.9}$$

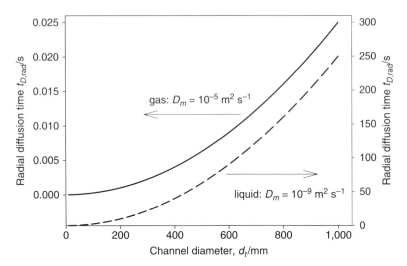

FIGURE 18.6   Characteristic radial diffusion time (micro mixing time) as function of channel diameter in MSR.

It follows that axial dispersion can be neglected (Bo $\geq$ 100), if the space-time is at least twice the radial diffusion time. In Fig. 18.6, the characteristic radial diffusion time for different channel diameters are summarized. Accordingly, the axial dispersion of gases in microchannels can be neglected, if their diameters are less than 1,000 μm and space-time is longer than 0.1 s. To the contrary, plug-flow behavior in liquid systems is much more difficult to obtain: The channel diameter should be less than 40 μm for a residence time of $\tau = 1$ s.

Due to the small volume of a single channel, many channels have to be used in parallel to obtain the sufficient performance of an MSR. A uniform distribution of the reaction mixture over thousands of microchannels is required to obtain a suitable performance of the MSR [12]. Flow maldistribution will enlarge the RTD in the multitubular MSR and lead to reduced reactor performance along with reduced product yield and selectivity [13, 14]. Several works in the field have presented design studies of flow distribution manifolds [6, 15, 16]. Maldistribution may also be caused by small deviations of the channel diameter introduced during the manufacturing process, leading to an enlargement of the RTD as shown schematically in Fig. 18.7.

The small deviations of the channel diameter may result from the nonuniform coating of the channel walls with catalytic layers. If the number of parallel channels is large ($N > 30$), a normal distribution of the channel diameters with standard deviation σ can be assumed. The relative standard deviation $\hat{\sigma}_d = \sigma_d / \bar{d}$ influences the pressure drop over the microreactor [17]:

$$\Delta p = \frac{128 \cdot \eta \cdot \dot{V}_{tot} \cdot L}{\pi \cdot N \cdot \bar{d}_t^4 \cdot (1 + 6\hat{\sigma}_d^2)} \qquad (18.10)$$

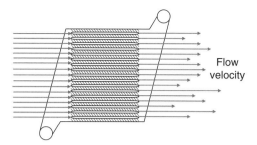

FIGURE 18.7   Flow maldistribution in multichannel MSR.

The relation (18.10) shows that a variation in the channel diameter leads to a decrease of the pressure drop at a constant overall volumetric flow. As the pressure drop for each channel is identical, the variation of the diameter results in a variation of the individual flow rates $\dot{V}_i$ and residence time $\tau_i = V_i/\dot{V}_i$.

Supposing there is plug flow in each channel ($Bo_i \rightarrow \infty$), the overall dispersion is inversely proportional to the relative standard deviation and can be estimated by Eq. (18.11) [17]:

$$Bo_{reactor} \cong \frac{d_t^2}{2\sigma_d^2} \tag{18.11}$$

As a consequence, the plug-flow behavior in a multichannel microreactor ($Bo_{reactor} \geq 100$) can be assumed for the relative standard deviation of the channel diameters $\sigma_d/d_t \leq 0.07$.

Residence time distribution in the MSR shown in Fig. 18.8 was determined experimentally by Rouge et al. [13, 18]. The entire device consisted of 340 parallel

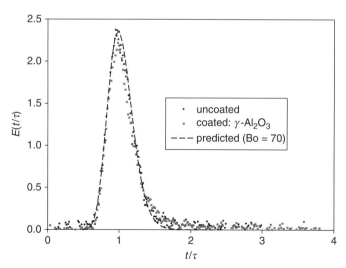

FIGURE 18.8   Residence time distribution in MSR. 340 channels, width: 300 μm, height: 240 μm, flow rate: 0.33 cm^3 s^{-1}, $\tau = 2.5$ s [13].

channels with an equivalent diameter of $d_t = 267\ \mu m$ and length of $L = 20\ mm$. Under the experimental conditions, a mean residence time of $\bar{t} = \tau = 2.5\ s$ was given. Whereas plug-flow behavior can be expected for each channel (Bo $\gg$ 100), a Bodenstein number for the multitubular MSR of Bo $= 70$ was obtained (Fig. 18.8). This may be explained by flow differences between the parallel channels.

18.3 MICROSTRUCTURED REACTORS FOR HETEROGENEOUS CATALYTIC REACTIONS

For heterogeneous fluid–solid systems, the catalytic active component is immobilized on the surface of a solid support containing the reactants that are passing through the reactor. The performance of a solid catalyst is proportional to its specific surface area. Therefore, porous media with a large inner surface area are commonly used as support for the catalytic active component. The presence of different phases requires transport between these phases to allow a reaction to take place. Since the driving force for the transfer of mass or heat is provided by concentration or temperature gradients, the mass and heat production in the reactor are not only always determined by the chemical (intrinsic) kinetics, but also by the rates of transport phenomena. The possible different physical and chemical steps starting from a reactant A_1 and ending in a reaction product A_2 via a heterogeneously catalyzed reaction path are illustrated in Fig. 18.9. We will consider microstructured multichannel reactors with a porous catalytic layer immobilized on the channel walls, so-called catalytic wall reactors. The following steps are involved: (*empty active site; A_i^*: adsorbed A_i on an active site)

1. Transfer of reactant A_1 from the bulk phase in the channel to the external surface of the catalytic layer
2. Transport (diffusion) from the external surface through the pores toward the active sites on the interior surface
3. Chemisorption of A_1 on an active site * as A_1^*

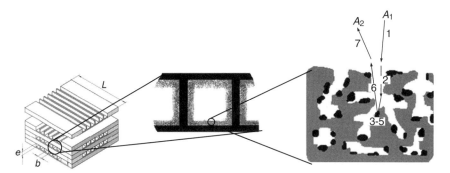

FIGURE 18.9 Chemical and physical steps involved in heterogeneously catalyzed reaction $A_1 \rightarrow A_2$.

4. Surface reaction of A_1^* to A_2^* active site *

5. Desorption of A_2 and liberating

6. Transport of A_2 through the pores toward the external surface of the catalytic layer

7. Transfer of A_2 from the external surface to the bulk fluid phase in the channel

Steps 3 to 5 are strictly chemical and consecutive to each other. The transfer steps 1 and 7 are strictly physical and in series with the steps 3 to 5: The transfer occurs separately from the chemical reaction. The transport steps 2 and 6, however, cannot be separated from the chemical steps 3 to 5 since the transport inside the pores occurs simultaneously with the chemical reaction.

Both external transfer and internal transport phenomena diminish reactor performance for fast reactions and may have a negative effect on product yield and selectivity. In the next sections, the design criteria for avoiding strong mass transfer effects during chemical transformation will be discussed.

18.3.1 Influence of Internal Mass Transport on the Reactor Performance

In general, the number of active sites on the outer surface of the catalytic layer can be neglected. The reactant has to be transported through the pores inside the layer to reach the active sites. Due to the chemical reaction occurring simultaneously with mass transport within the porous layer, a concentration gradient is established and interior surfaces are exposed to lower reaction concentrations than those near the exterior. Under isothermal conditions, the average reaction rate throughout the catalytic layer will be almost always lower than it would be if there were no mass transport influences.

A measure for the degree of internal diffusion limitations is given by the internal effectiveness factor, defined as

$$\eta_{in} = \frac{\text{reaction rate with internal diffusion limitation}}{\text{reaction rate at the external surface}} \quad (18.12)$$

Corresponding to

$$\eta_{in} = \frac{\dfrac{1}{V_{cat}} \displaystyle\int r_{V,L}(c_1)\, dV}{r_{V,cat}(c_{1,s})} \quad (18.13)$$

V_{cat} is the volume of the catalytic layer on the MSR wall, c_1 the reactant concentration, $c_{1,s}$ the reactant concentration at the external surface, and $r_{V,cat}$ the reaction rate referred to in the layer volume. The effectiveness factor η_{in} depends on the characteristic diffusion time of the reactant in the layer, $t_D = \delta^2/D_{1,eff}$, and the characteristic reaction time. For an irreversible first-order reaction, the latter is given by $t_r = 1/k_{V,cat}$, where δ corresponds to the thickness of the catalytic layer, $D_{1,eff}$ is the effective diffusion coefficient within the porous catalyst, and $k_{V,cat}$ is the volumetric first-order rate constant.

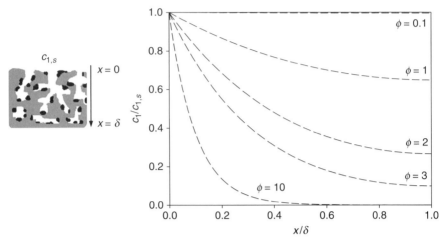

FIGURE 18.10 Reactant concentration profiles within a catalytic layer for different values of the Thiele modulus. Irreversible first-order reaction.

The concentration profile in the layer can be calculated with the material balance. For thin catalytic layers, a slab geometry of the catalyst can be assumed, resulting in the following relationship between the reactant concentration and distance from the surface given in Eq. (18.14) [19]:

$$\frac{c_1}{c_{1,s}} = \frac{\cosh\left[\phi\left(1 - \frac{x}{\delta}\right)\right]}{\cosh \phi} \tag{18.14}$$

The concentration profile is determined by the Thiele modulus ϕ defined as

$$\phi = \sqrt{\frac{t_D}{t_r}} = \delta\sqrt{\frac{k_{V,\text{cat}}}{D_{1,\text{eff}}}} \tag{18.15}$$

The influence of ϕ on the concentration profile is illustrated in Fig. 18.10.

Based on the concentration profile in the layer [Eq. (18.14)], the effectiveness factor η_{in} [Eqs. (18.12) and (18.13)] can be determined from

$$\eta_{\text{in}} = \frac{\tanh \phi}{\phi} \tag{18.16}$$

In Fig. 18.11, the effectiveness factor for a catalytic layer is shown as function of the Thiele modulus. To get a high performance of the catalytic microstructured wall reactor, the thickness of the catalytic layer should be as thick as possible, as long as the effectiveness factor does not fall under 0.95, corresponding to a Thiele modulus of $\phi < 0.4$.

In general, the intrinsic reaction rate is not known. Therefore, different criteria for the absence of significant diffusion effects based on observable effective reaction

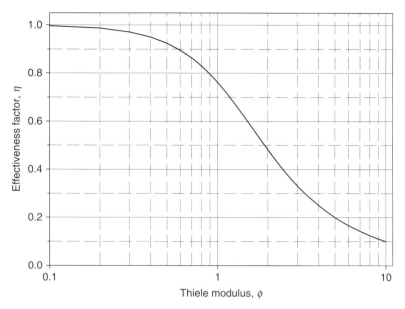

FIGURE 18.11   Effectiveness factor of a catalytic layer with slab geometry as a function of the Thiele modulus.

rates are proposed in the literature and summarized by Maers [20]. To ensure an effectiveness factor of $\eta \geq 0.95$ in an isothermal catalyst, the following criterion, modified for a catalytic layer in microchannels, results:

$$\delta_{max} \leq b \sqrt{\frac{D_{1,\text{eff}} \cdot c_{1,s}}{r_{V,\text{cat,eff}}}} \tag{18.17}$$

$D_{1,\text{eff}}$ and $r_{V,\text{cat,eff}}$ are the effective diffusion coefficient for the reactant and the observed reaction rate, respectively; $c_{1,s}$ is the reactant concentration on the outer catalyst surface. The parameter b depends on the formal reaction order m: $b = 0.8$ ($m = 0$); $b = 0.3$ ($m = 1$); $b = 0.18$ ($m = 2$).

Strong exothermal or endothermal reactions may provoke a temperature profile within the catalytic layer, influencing the reaction rate and selectivity. The importance of the temperature profile depends, in addition to the reaction rate and layer thickness, on the reaction enthalpy ΔH_R, the activation energy E, and the thermal conductivity of the porous catalyst λ_{eff}. For quasi-isothermal behavior, the observed rate $r_{V,\text{cat,eff}}$ must not differ from the rate that would prevail at constant temperature by more than about 5%. The resulting criterion, estimated for a catalytic layer in microchannels, is given in Eq. (18.18):

$$\delta_{max} \leq 0.3 \sqrt{\frac{R}{E} \frac{\lambda_{\text{eff}} \cdot T_s^2}{|\Delta H_R| r_{V,\text{cat,eff}}}} \tag{18.18}$$

T_s corresponds to the temperature on the catalyst surface and R is the gas constant. To ensure the high reactor performance of the microstructured wall reactor, the maximum acceptable thickness of the catalytic layer, $\delta_{\max}$, is desired.

18.3.2 Influence of External Mass Transfer on the Reactor Performance

In general, the thickness of the catalytic layer is kept sufficiently small to avoid the influence of internal mass transfer on kinetics. In this way, only the transfer of the reactants from the bulk to the catalytic wall must be considered as well as the reaction rate per unit of the outer surface of the catalytic layer. For an irreversible first-order reaction, the rate is given by

$$r_s = k_s c_{1,s} \tag{18.19}$$

The molar flux from the fluid phase to the surface of the layer is proportional to the concentration gradient between the bulk and surface:

$$J_1 = k_g(c_{1,b} - c_{1,s}) \tag{18.20}$$

Under stationary conditions, the molar flux from the fluid phase reaching the catalytic surface and the rate of transformation per surface unit must be identical: $J_1 = r_s$. It follows for a first-order reaction:

$$k_s c_{1,s} = k_g(c_{1,b} - c_{1,s}) \tag{18.21}$$

Solving Eq. (18.21) for the concentration of A_1 at the catalyst surface gives

$$c_{1,s} = \frac{k_g}{k_g + k_s} c_{1,b} = \frac{1}{1 + \mathrm{DaII}} c_{1,b} \tag{18.22}$$

and for the observed (effective) reaction rate:

$$r_{s,\mathrm{eff}} = k_s \frac{k_g}{k_g + k_s} c_{1,b} = k_s \frac{1}{1 + \mathrm{DaII}} c_{1,b} \tag{18.23}$$

The effective reaction rate is determined by the ratio of the characteristic transfer time t_D and characteristic reaction time t_r, the second Damköhler number:

$$\mathrm{DaII} = \frac{t_D}{t_r} = \frac{k_s}{k_g} \tag{18.24}$$

Low values of DaII ($t_D \ll t_r$) correspond to a situation in which the effect of the mass transfer can be neglected. The observed reaction rate agrees with the intrinsic rate:

$$r_{s,\mathrm{eff}} = k_s \frac{1}{1 + \mathrm{DaII}} c_{1,b} \cong k_s c_{1,b} = r_s \tag{18.25}$$

At high values of DaII, the rate is completely limited by mass transfer from the fluid phase to the surface; the surface concentration is nearly zero ($c_{1,s} \cong 0$):

$$r_{s,\text{eff}} = k_s \frac{1}{1 + \text{DaII}} c_{1,b} \cong k_g c_{1,b} = J_1 \tag{18.26}$$

The existence of a significant difference between the concentration in the bulk of the fluid phase and the surface of the catalytic layer leads to lower reaction rates for positive reaction orders. This can be expressed by the external effectiveness factor η_{ex}:

$$\eta_{\text{ex}} = \frac{r_s(c_{1,s})}{r_s(c_{1,b})} = \frac{r_{s,\text{eff}}}{r_s(c_{1,b})} \tag{18.27}$$

For an intrinsic first-order reaction, Eq. (18.27) becomes

$$\eta_{\text{ex}} = \frac{1}{1 + \text{DaII}} \tag{18.28}$$

Due to the small channel diameters in MSR, laminar flow can be considered. The radial velocity profile in a single channel develops from the entrance to the position where a complete Poiseuille profile is established. The length of the entrance zone depends on the Re number and can be estimated from the following empirical relation [21, 22]:

$$L_e \leq 0.06 \cdot \text{Re} \cdot d_t \tag{18.29}$$

Within the entrance zone, the mass transfer coefficient diminishes, reaching a constant value. The dependency can be described with Eq. (18.30) in terms of Sherwood numbers, $\text{Sh} = k_g d_t / D_m$ [23, 24]:

$$\text{Sh} = B \left(1 + 0.095 \frac{d_t}{L} \text{Re} \cdot \text{Sc} \right)^{0.45} \tag{18.30}$$

The constant B in Eq. (18.30) corresponds to the asymptotic Sh number for constant concentration at the wall, which is identical to the asymptotic Nusselt number Nu, characterizing the heat transfer in laminar flow at constant wall temperature. The constant B depends on the geometry of the channel, as summarized in Table 18.1.

If the entrance zone in the tube can be neglected, the mass transfer is constant and given by B. As for the heat transfer, for a circular tube-shaped reactor:

$$\text{Sh}_\infty = 3.66 \quad \text{for } L \geq 0.05 \text{Re} \cdot \text{Sc} \cdot d_t \quad \text{(constant wall concentration)} \tag{18.31}$$

If the mass transfer is accompanied by a chemical reaction at the catalyst surface taking place on the reactor wall, the mass transfer depends on the reaction kinetics [25]. For a zero-order reaction, the rate is independent of the concentration and the mass flow from the bulk to the wall is constant, whereas the reactant concentration at the catalytic wall varies along the reactor length. For this situation, the asymptotic Sh number in circular tube reactors becomes $\text{Sh}'_\infty = 4.36$ [25]. The same value is

TABLE 18.1 Mass and Heat Transfer Characteristics for Different Channel Geometries [23]

Geometry	Constant B in Eq. (18.30).
Circular	3.66
Ellipse; width : height $= 2$	3.74
Parallel plates	7.54
Rectangle; width : height $= b/e = 4$	4.44
Rectangle; width : height $= b/e = 2$	3.39
Square	2.98
Equilateral triangle	2.47
Sinusoidal	2.47
Hexagonal	3.66

obtained when reaction rates are low compared to the rate of mass transfer. If the reaction rate is high (very fast reactions), the concentration at the reactor wall can be approximated to zero within the entire reactor and the final value for Sh is $Sh_\infty = 3.66$. As a consequence, the Sh number in the reacting system depends on the ratio of the reaction rate to the rate of mass transfer characterized by the second Damköhler number defined in Eq. (18.24).

Villermaux [25] proposed a simple relation to estimate the asymptotic Sh number as a function of DaII (Fig. 18.12):

$$\frac{1}{Sh''_\infty} = \frac{1}{Sh'_\infty} + \frac{DaII}{DaII + 1.979}\left(\frac{1}{Sh_\infty} - \frac{1}{Sh'_\infty}\right) \qquad (18.32)$$

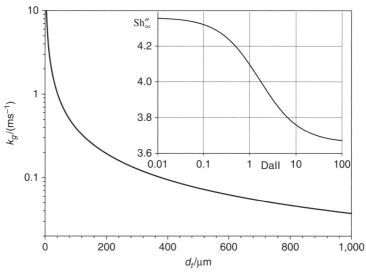

FIGURE 18.12 Mass transfer coefficient in a microchannel ($D_m = 10^{-5}\,m^2\,s^{-1}$; DaII > 100) and variation of the asymptotic Sh number with the DaII (first-order reaction).

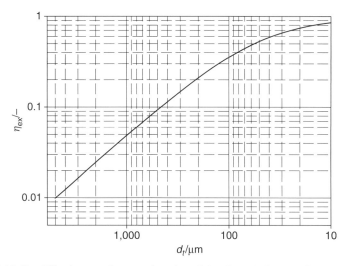

FIGURE 18.13   Effectiveness factor as function of the channel diameter for a fast gas-phase reaction (DaII = 100 for d_t = 5 mm).

Thus, the mass transfer coefficient in multitubular MSR depends, in addition to the molecular diffusion coefficient, on the channel diameter d_t and second Damköhler number, DaII:

$$k_g = \frac{Sh''_\infty D_m}{d_t} \tag{18.33}$$

This is shown for a gas-phase reaction in Fig. 18.12.

As the mass transfer coefficient increases with decreasing channel diameter, the effective reaction in the MSR is considerably augmented. This is shown, as an example, in Fig. 18.13. The decrease of the channel diameter from 5 mm to 50 μm leads to approx. 100 times higher productivity per catalyst mass and an increase in the performance per reactor volume of four orders of magnitude.

18.4 MAIN DESIGN PARAMETERS OF CATALYTIC MSR

18.4.1 Introduction of Catalytic Phase in MSR

One of the main problems in using microstructured reactors for heterogeneously catalyzed gas-phase reactions is the introduction of the catalyst in the reaction zone. Therefore, the MSR are classified here by type of catalytic bed and according to their design criteria.

18.4.2 Randomly Packed Channels

The straightforward way to introduce the catalyst into MSR is to fill the microchannels with catalyst powder. The packed-bed MSR are mostly used for catalyst screening [26–28], but there are also examples of their utilization in the distributed production of chemicals. The advantage of packed-bed MSR stems from the fact that the catalyst developed for traditional reactors can be applied [29].

The drawback of micro-packed-bed reactors is the high pressure drop. In addition, each channel must be packed identically to avoid maldistribution, which is known to lead to broad residence time distribution in the reactor system. This is difficult to obtain in practice.

18.4.3 Catalytic Wall Reactors

To avoid high pressure drop in randomly packed MSR, multichannel reactors with catalytically active walls have been used. Typical channel diameters in MSR fall in the range of 50 to 1,000 μm, with a length between 20 and 100 mm. Up to 10,000 channels are assembled in one reactor unit (Figs. 18.1 and 18.2). However, the geometric surface of the microchannels is not sufficient to carry out catalytic reactions with high performance. Therefore, the specific surface area has to be increased by chemical treatment of the channel walls or by coating them with a porous layer. The porous layer serves as a catalyst or as support for a catalytic phase. Different techniques have been developed and tested for this purpose in recent years [30] and will be summarized below.

The anodic oxidation of aluminium to create a porous layer of alumina in microchannels was pioneered by Hönicke and coworkers [31–33]. The layer of α-alumina obtained has a regular pore structure oriented perpendicularly to the flow direction. In general, this porous layer forms the support for the catalytically active components. Electrolyte concentration, temperature, and electric potential influence oxidation efficiency and pore density. An example is shown in Fig. 18.14. In addition, subsequent hydrothermal-thermal treatment controls the morphology of the porous oxide layer, leading to an increase in the surface area [34].

A thin porous alumina layer on a metal surface can be created by the high-temperature treatment of Al containing steel (e.g., DIN 1.4767, FeCrAlloy). This alloy is often used for metallic monoliths of automotive exhaust converters.

200 μm

FIGURE 18.14 SEM image of a diagonal cross section of a diffusion-welded and subsequently anodically oxidized microstructured aluminium reactor [30].

By heating the alloy for 5 h at $1,000°C$, a thin alumina film of approx. 5 μm is formed on the surface [35]. This Al_2O_3 film can be used as support for catalytically active metals, as explored by Aartun et al. [36], or for the adherence of subsequently deposited thicker layers [37, 38].

A novel approach was reported by Reuse et al. [38, 39]. These authors developed a method of wash-coating using commercial catalysts for the steam reforming of methanol. The copper-based catalyst was micromilled to nano-particles and deposited on the microchannel walls from acetone suspensions. The activity of the catalytic layer exceeded the values of the granulated catalyst. This was explained by a mechanical activation by milling [40, 41]. The method proposed allows the tedious and time-consuming development of an active catalytic layer to be shortened. Similar procedures were reported by Bravo et al. [42], Park et al. [43, 44], and Pfeifer et al. [45] for the catalytic steam reforming in microchannels.

The electrophoretic deposition of Al_2O_3, ZnO, and CeO_2 nano-particles from colloidal solutions is another method for the formation of oxide layers. The layer properties are controlled by the composition and pH of the solution, as well as the applied voltage and current density [46].

Another promising method to obtain oxide layers on the walls of microchannels is a sol-gel technique. It has the advantage of producing a wide variety of compositions with tailored porosities and surface textures, and is also widely used for the preparation of particulate porous catalytic supports [34, 47, 48].

Zeolite-coated microchannel reactors with a 1- to 2-μm layer demonstrate higher productivity per mass of catalyst as compared to conventional zeolite-packed beds [49]. This results from the large surface-to-volume ratio of the zeolite film, which makes the entire catalyst surface available for the reaction. A stainless steel MSR coated by Fe-ZSM-5 was tested and found effective during the hydroxylation of benzene with N_2O [50].

An elegant way to prepare catalytic wall MSR is by means of the direct formation of zeolite crystals on the metallic structure [51–53]. An MSR coated with ZSM-5 proved to be effective in the selective catalytic reduction (SCR) of NO with ammonia without mass transfer limitations. In this case, a higher SCR reaction rate was observed, as compared to granulated zeolites.

Chemical vapor deposition (CVD) is a valuable tool for obtaining porous ceramic coatings on the inner walls of microchannels. Jänicke et al. [54] deposited an alumina layer, increasing the specific surface by a factor of ~ 100. Thin nano-structured films of Mo_2C were grown on Si(100) by metal-organic CVD by Chen et al. [55] using $MO(CO)_6$ as a precursor. The flame combustion synthesis is widely used for catalyst manufacture based on carbon black, fumed silica, and titania [56–58]. Thybo et al. [59] investigated a flame spray technique for the one-step deposition of a porous layer in to an MSR using a shadow mask to cover everything except the channels during exposition.

The coating techniques presented lead to metal oxide coatings. Techniques to produce carbon-based coatings in microstructures have been less investigated in spite of the fact that carbon is a common support in a variety of applications. Schimpf et al. [60] studied carbon-coated microstructures for heterogeneously

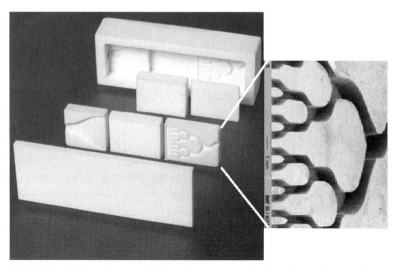

FIGURE 18.15 Ceramic MSR with exchangeable functional elements [61].

catalyzed hydrogenations. Carbon-coated MSR have been prepared via the carbonization of polymers. The amount of carbon deposited depends on the composition of the precursor monomer solution and on the presence of a template. The activity and selectivity of the carbon after loading with ruthenium in the hydrogenation of acrolein were affected by the composition of the polymers, the time of carbonization due to the carbon surface functional groups, and also the porosity of the carbonaceous layer.

Microstructured reactors made of metals or polymers are not suitable for chemical reactions at high temperatures and/or with corrosive reactants involved.

Ceramic materials are a valuable alternative for this kind of applications. Knitter and Liauw [61] developed a modular MSR made of alumina and tested for the oxidative coupling of methane and isoprene selective oxidation to citraconic anhydride (Fig. 18.15). High thermal and chemical resistance at temperatures up to 1,000°C was demonstrated.

18.4.4 New Developments [62]

The drawback of randomly packed microchannels by a catalyst is the high-pressure drop. In addition, each channel must be packed identically to avoid maldistribution, which is known to lead to broad residence time distribution (RTD) during the passage of reactants through the reactor. To avoid problems related to the randomly packed beds, the use of structured catalytic beds was recently proposed [63–67]. The novel concept was applied for MSR containing a structured catalytic bed arranged with parallel filaments or wires. This micro "string reactor" yields flow hydrodynamics similar to those of multichannel microreactors. The channels for gas flow between the filaments have an equivalent hydraulic diameter in the range of a few microns, ensuring laminar flow, and short diffusion times in the radial direction, leading to a narrow RTD. This novel MSR was applied for the oxidative steam reforming of methanol (OSRM) [66, 67]. OSRM is based on the combination of exothermic oxidation and endothermic reforming in the same reactor. At 300°C, a formally autothermal reaction occurs for the following composition of reactant feed:

$$4CH_3OH + 3H_2O + 0.5O_2 \longrightarrow 4CO_2 + 11H_2$$

$$\Delta_r H^{573} \cong 0 \, kJmol^{-1}$$

(18.34)

Reactors in this case operate autothermally, that is, do not require any external heating or cooling once having reached the reaction temperature. The main difficulty in carrying out OSRM is due to the much faster methanol oxidation compared to the reforming rate. As a consequence, heat is generated mostly at the reactor entrance, whereas the heat consumption occurs in the middle and rear of the reactor. In conventional reactors, with randomly packed beds and low axial and radial heat conductivity, a pronounced axial temperature profile is developed [68]. They are characterized by a hot spot at the reactor entrance and "cold spot" in the second part of the reactor [69]. The high temperature may damage the catalyst and the low temperature diminishes the rate of the reforming reaction, leading to poor reactor performance. Thus, temperature control is crucial for reactor performance.

To avoid axial temperature profiles, the catalyst in the form of thin metallic wires (with diameters in the millimeter range) was introduced into a "macro" tubular reactor (Fig. 18.16). This design provides laminar flow with narrow RTD and low-pressure drop throughout the catalytic bed [65]. Brass wires with high heat conductivity (120 W/(m · K)) are chosen for the microstructured string reactor as they contain Cu and Zn catalyzing the reforming/oxidation of methanol. To obtain metal wires with high specific surface area, a thin metal/aluminium alloy is formed on the wire outer surface [71]. The aluminium is leached out either by an acid or a base boiling solution, resulting in a thin porous layer with a morphology similar to that of Raney metals. In OSRM, the microstructured string reactor showed high selectivity for CO_2 and H_2 ($\geq 98\%$). Hot and cold spots in the reactor were efficiently reduced down to $\Delta T < 1.5 \, K$ at methanol conversions $X > 50\%$ [72].

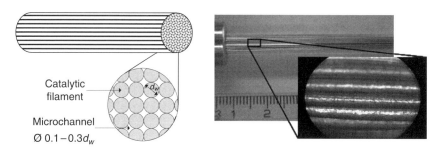

FIGURE 18.16 Schematic presentation and photo of the microstructured string reactor [70].

The metallic grids used in the studies reported above were made of wires with a diameter of approx. 0.1 mm with meshes in the same order of magnitude. The openings can be significantly reduced by using metallic fiber filters. Sintered metal fiber filters (SMF) have uniform micron-sized filaments sintered into a homogeneous three-dimensional structure. SMF present porosities of 80 to 90% and in high permeability. Fibers made of alloys (stainless steel, Inconel, and Fecralloy) exhibit high mechanical strength, and chemical and thermal stability. The high thermal conductivity of the metal fiber matrix provides an efficient radial heat transfer and can act also as a static micromixer. These are advantageous for their use as building units in microstructured reactors. Yuranov et al. produced a thin uniform zeolite film with controlled thickness coating the metal microfibers [53]. The coating consisted of highly interconnected crystals about 1 μm in size with prismatic MFI morphology. The zeolite/SMF elements can be assembled in the form of disks, presenting a three-level catalyst structure: (1) nano-structure of the zeolite film, (2) microstructure of the porous three-dimensional media of sintered metal fibers, and (3) macrostructure of the layered catalytic bed formed from the composite elements.

Novel catalyst structures with microchannels able to reduce the diffusion resistance in fast heterogeneous gas-phase reactions were developed by Bae et al. [73]. They fabricated a structured microchannel catalyst by means of a modified ceramic tape casting process.

Small-size microchannel catalysts (with a diameter of ~1 cm) prepared for microreactor tests are shown in Fig. 18.17. The microchannel was loaded inside a stainless steel microreactor tube and heat-treated to burn away the fugitive layer and sinter the catalyst powder as a self-supporting form. After burning the fugitive layer, well-defined channels were created. The microchannel reforming catalysts were tested with natural gas and gasoline-type fuels at space velocities of up to 250,000 h^{-1}. The catalysts have also been used in engineering-scale reactors (10 kW$_e$ with a diameter of 7 cm) with similar product qualities. Compared with pellet catalysts, the microchannel catalysts offer a nearly 5-fold reduction in catalyst weight and volume.

2.5 cm

FIGURE 18.17   Microchannel catalyst for autothermal reforming of hydrocarbon fuels [73].

The small volume and weight of microreactors lead to short transient periods after the start-up or after a change in the reaction conditions. Microreactors allow fast heating and cooling of the reaction mixture within a fraction of a second [74, 75]. Therefore, reactions can be carried out under defined short reaction times, avoiding parallel and consecutive reactions. An example is the catalytic dehydrogenation of methanol to water-free formaldehyde. The reaction takes place in the range of 1,000 to 1,200 K, attaining the complete conversion of methanol within milliseconds. Formaldehyde is unstable under the reaction conditions and decomposes to carbon monoxide and hydrogen. This is why the reaction mixture is quenched rapidly at the reactor outlet. When using a microstructured heat exchanger, temperature gradients of up to 6,400 K/s were attained for freezing the reaction mixture, leading to formaldehyde yields of more than 80% at the nearly complete conversion of methanol [76].

The excellent dynamic behavior of microreactors can be exploited for running reactions under non-steady-state conditions by changing periodically the concentration and/or temperature. The periodic operation of catalytic chemical reaction may lead to increased reactor performance, improved selectivity, and reduced catalyst deactivation compared to the reaction performed under steady-state conditions [77]. The advantages of fast periodic concentration oscillations in microreactors were reported by Rouge et al. [13, 14]. Recently, microstructure devices for "fast temperature cycling" also have been described [58, 78]. The devices obtained a periodic temperature change of 100 K within less than a second. The oxidation of carbon monoxide was chosen as a model reaction. Under fast temperature cycling, a considerably higher CO_2 yield compared to the steady-state value was obtained.

18.5 CONCLUSIONS

Microstructured reactors (MSR) have been recognized more and more in recent years as a novel tool for the intensification of industrial chemical processes. MSR are particularly suited for highly exothermic and fast reactions since they allow users to avoid hot-spot formation, to attain higher reaction temperatures, and to reduce reaction volumes. This improves energy efficiency and reduces operational costs. Moreover, broader reaction conditions, including those up to the explosion regime, can be afforded. The small diameters of the reactor channels ensure a short radial diffusion time, leading to a narrow residence time distribution. It also precludes strong mass transfer resistance. This is advantageous for consecutive processes since high selectivity for the intermediate is desired. In addition, the small accumulations of reactants and products lead to inherent safety during the reactor operation. Based on the research carried out as of 2008, a clear picture of the advantages and limits to the use of microreactors for catalytic reactions has emerged. For process intensification, highly active catalysts are required. It is expected that industry will soon participate in this fascinating area of research and contribute to new developments.

BIBLIOGRAPHY

1. Cross, W. T., and Ramshaw, C. (1986). Process intensification: Laminar flow heat transfer. *Chemical Engineering Research and Design*, 64(4): 293–301.

2. Ramshaw, C. (1985). Process intensification: A game for n players. *Chemical Engineer*, 416.

3. Ramshaw, C., and Arkley, K. (1983). Process intensification by miniature mass transfer. *Process Engineering*, 64(1): 29–30.

4. Baerns, M., and Renken, A. (2004). Chemische Reaktionstechnik. In *Winnacker-Küchler: Chemische Technik: Prozesse und Produkte*, pp. 453–643. Weinheim: Wiley-VCH.

5. Levenspiel, O. (1999). *Chemical Reaction Engineering*, 3rd ed., New York: John Wiley & Sons.

6. Commenge, J. M., Falk, L., Corriou, J. P., and Matlosz, M. (2002). Optimal design for flow uniformity in microchannel reactors. *AIChE Journal*, 48(2): 345–358.

7. *VDI-Wärmeatlas*, 9th ed. (2002). Berlin: Verein Deutscher Ingenieure, New York: Springer.

8. Renken, A., and Kiwi-Minsker, L. (2006). Chemical reactions in continuous flow microstructured reactors. In *Micro Process Engineering*, N. Kockmann, ed., pp. 173–201. Weinheim: Wiley-VCH.

9. Baerns, M., Hofmann, H., and Renken, A. (1999). *Chemische Reaktionstechnik*, 3rd ed., Weinheim: Wiley-VCH.

10. Taylor, G. (1953). Dispersion of soluble matter in solvent flowing slowly through a tube. *Proceedings of the Royal Society of London*, Series A(A219): 186–203.

11. Aris, R. (1956). On the dispersion of a solute in a fluid flowing through a tube. *Proceedings of the Royal Society of London Series*, Series A(A235): 67–77.

12. Wörz, O., Jäckel, K.-P., Richter, T., and Wolf, A. (2001). Microreactors—A new efficient tool for reactor development. *Chemical Engineering Technology*, 24: 138–142.

13. Rouge, A., Spoetzl, B., Gebauer, K., Schenk, R., and Renken, A. (2001). Microchannel reactors for fast periodic operation: The catalytic dehydrogenation of isopropanol. *Chemical Engineering Science*, 56: 1419–1427.

14. Rouge, A., and Renken, A. (2001). Performance enhancement of a microchannel reactor under periodic operation. *Studies in Surface Science and Catalysis*, 133: 239–246.

15. Commenge, J. M., Rouge, A., Renken, A., Corriou, J. P., and Matlosz, M. (2001). Dispersion dans un microréacteur multitubulaire: Etude expérimentale et modélisation. *Récents Progrés en Génie des Procédés*, 15: 329–336.

16. Delsman, E. R., De Croon, M. H. J. M., Pierik, A., Kramer, G. J., Cobden, P. D., and Hofmann, C., et al. (2004). Design and operation of a preferential oxidation microdevice for a portable fuel processor. *Chemical Engineering Science*, 59(22–23): 4795–4802.

17. Delsman, E. R., Croon, M. H. J. M. D., Elzinga, G. D., Cobden, P. D., Kramer, G. J., and Schouten, J. C. (2005). The influence of differences between microchannels on micro reactor performance. *Chemical Engineering & Technology*, 28(3): 367–375.

18. Rouge, A. (2001). Periodic operation of a micro-reactor for heterogeneously catalysed reactions: The dehydration of isopropanol Ph.D. dissertation, EPF-Lausanne.

19. Satterfield, C. N. (1970). *Mass Tranfer in Heterogeneous Catalysis*. Cambridge, MA: MIT Press.

20. Mears, D. E. (1971). Tests for transport limitations in experimental catalytic reactors. *Industrial & Engineering Chemistry Process Design and Development*, 10: 541–547.

21. Sherony, D. F., and Solbrig, C. W. (1970). Analytical investigation of heat or mass transfer and friction factors in a corrugated duct heat or mass exchanger. *International Journal of Heat and Mass Transfer*, 13(1): 145–146.

22. Hoebink, J. H. B. J., and Marin, G. B. (1998). Modeling of monolihic reactors for automotive exhaust gas treatment. In *Structured Catalysts and Reactors*, Cybulski, J. and Moulijn, J. A. (ed.), New York: Marcel Dekker, pp. 209–237.

23. Cybulski, A., and Moulijn, J. A. (1994). Monoliths in heterogeneous catalysis. *Catalytic Reviews—Science and Engineering*, 36(2): 179–270.

24. Hayes, R. E., and Kolaczkowski, S. T. (1994). Mass and heat transfer effects in catalytic monolith reactors. *Chemical Engineering Science*, 49(21): 3587–3599.

25. Villermaux, J. (1971). Diffusion dans un reacteur cylindrique. *International Journal of Heat and Mass Transfer*, 14(12): 1963–1981.

26. Rodemerck, U., Ignaszewski, P., Lucas, M., Claus, P., and Baerns, M. (2000). Parallel synthesis and testing of heterogeneous catalysts. In *Proceedings of 3rd International Conference on Microreaction Technology, IMRET 3*. Frankfurt: Springer, pp. 287–293.

27. Jensen, K. F. (2001). Microreaction engineering—is small better? *Chemical Engineering Science*, 56(2): 293–303.

28. Ajmera, S. K., Delattre, C., Schmidt, M. A., and Jensen, K. F. (2002). Microfabricated differential reactor for heterogeneous gas phase catalyst testing. *Journal of Catalysis*, 209(2): 401–412.

29. Losey, M. W., Isogai, S., Schmidt, M. A., and Jensen, K. F. (2000). Microfabricatad devices for multiphase catalytic processes. In *Proceedings of 4th International Conference on Microreaction Technology, IMRET 4*. Atlanta, GA: AIChE; pp. 416–422.

30. Haas-Santo, K., Gorke, O., Pfeifer, P., and Schubert, K. (2002). Catalyst coatings for microstructure reactors. *Chimia*, 56(11): 605–610.

31. Hönicke, D., and Wiessmeier, G. (1996). Heterogeneously catalyzed reactions in a micro-reactor. DECHEMA Monographs. *Microsystem Technology for Chemical and Biological Microreactors*, 132: 93–107.

32. Wiessmeier, G., and Hönicke, D. (1996). Microfabricated components for hetero-geneously catalysed reactions. *Journal of Micromechanics and Microengineering*, 6: 285–289.

33. Wiessmeier, G., Schubert, K., and Hönicke, D. (1997). Monolithic microstructure reactors possessing regular mesopore systems for the successful performance of heterogeneously catalyzed reactions. In *Proceedings of 1st International Conference on Microreaction Technology, IMRET 1*, W. Ehrfeld (ed.), pp. 20–26. Berlin: Springer.

34. Ganley, J. C., Riechmann, K. L., Seebauer, E. G., and Masel, R. I. (2004). Porous anodic alumina optimized as a catalyst support for microreactors. *Journal of Catalysis*, 227(1): 26–32.

35. Reuse, P. (2003). Production d'hydrogène dans un réacteur microstructuré. Couplage themique entre le steam reforming et l'oxydation totale du méthanol. PhD dissertation No. 2830, EPF-Lausanne.

36. Aartun, I., Gjervan, T., Venvik, H., Gorke, O., Pfeifer, P., and Fathi, M., et al. (2004). Catalytic conversion of propane to hydrogen in microstructured reactors. *Chemical Engineering Journal*, 101(1–3): 93–99.

37. Wang, Y., Chin, Y. H., Rozmiarek, R. T., Johnson, B. R., Gao, Y., and Watson, J., et al. (2004). Highly active and stable $Rh/MgO-Al_2O_3$ catalysts for methane steam reforming. *Catalysis Today*, 98(4): 575–581.

38. Reuse, P., Renken, A., Haas-Santo, K., Gorke, O., and Schubert, K. (2004). Hydrogen production for fuel cell application in an autothermal micro-channel reactor. *Chemical Engineering Journal*, 101(1–3): 133–141.

39. Reuse, P., Tribolet, P., Kiwi-Minsker, L., and Renken, A. (2001). Catalyst coating in microreators for methanol steam reforming: Kinetics. In *Proceedings of 5th International Conference on Microreaction Engineering*, *IMRET 5*, Ehrfeld, W., ed., pp. 322–331. Strasbourg: Springer.

40. Mitchenko, S. A., Khomutov, E. V., Shubin, A. A., and Shul'ga, Y. M. (2004). Catalytic hydrochlorination of acetylene by gaseous HCl on the surface of mechanically pre-activated K_2PtCl_6 salt. *Journal of Molecular Catalysis A: Chemical*, 212(1–2): 345–352.

41. Isupova, L. A., Tsybulya, S. V., Kryukova, G. N., Alikina, G. M., Boldyreva, N. N., and Vlasov, A. A., et al. (2002). Physicochemical and catalytic properties of $La1-xCaxFeO_3-0.5x$ Perovskites prepared using mechanochemical activation. *Kinetics and Catalysis*, 43(1): 129–138.

42. Bravo, J., Karim, A., Conant, T., Lopez, G. P., and Datye, A. (2004). Wall coating of a $CuO/ZnO/Al_2O_3$ methanol steam reforming catalyst for micro-channel reformers. *Chemical Engineering Journal*, 101(1–3): 113–121.

43. Seo, D. J., Yoon, W.-L., Yoon, Y.-G., Park, S.-H., Park, G.-G., and Kim, C.-S. (2004). Development of a micro fuel processor for PEMFCs. *Electrochimica Acta*, 50(2–3): 715–719.

44. Park, G.-G., Seo, D. J., Park, S.-H., Yoon, Y.-G., Kim, C.-S., and Yoon, W.-L. (2004). Development of microchannel methanol steam reformer. *Chemical Engineering Journal*, 101(1–3): 87–92.

45. Pfeifer, P., Schubert, K., and Emig, G. (2005). Preparation of copper catalyst washcoats for methanol steam reforming in microchannels based on nanoparticles. *Applied Catalysis A: General*, 286(2): 175–185.

46. Pfeifer, P., Görke, O., and Schubert, K. (2002). Washcoat and electrophoresis with coated and uncoated nanoparticles on microstructured foils and microstructured reactors. *Proceedings of 6th International Conference on Microreaction Technology*, *IMRET 6*. New Orleans, LA, pp. 125–130.

47. Brinker, C. J., and Scherer, G. W. (1990). *Sol-Gel Sciences*. Orlands, FL: Academic Press.

48. Gonzalez, R. D., Lopez, T., and Gomez, R. (1997). Sol-Gel preparation of supported metal catalysts. *Catalysis Today*, 35(3): 293–317.

49. Coronas, J., and Santamaria, J. (2004). The use of zeolite films in small-scale and micro-scale applications. *Chemical Engineering Science*, 59(22–23): 4879–4885.

50. Hiemer, U., Klemm, E., Scheffler, F., Selvam, T., Schwieger, W., and Emig, G. (2004). Microreaction engineering studies of the hydroxylation of benzene with nitrous oxide. *Chemical Engineering Journal*, 101(1–3): 17–22.

51. Rebrov, E. V., Seijger, G. B. F., Calis, H. P. A., de Croon, M. H. J. M., van den Bleek, C. M., and Schouten, J. C. (2001). The preparation of highly ordered single layer

ZSM-5 coating on prefabricated stainless steel microchannels. *Applied Catalysis A: General*, 206(1): 125–143.

52. Yuranov, I., Dunand, N., Kiwi-Minsker, L., and Renken, A. (2002). Metal grids with high-porous surface as structured catalysts: Preparation, characterisation and activity in propane total oxidation. *Applied Catalysis B: Environmental*, 36: 183–191.

53. Yuranov, I., Renken, A., and Kiwi-Minsker, L. (2005). Zeolite/sintered metal fibers composites as effective multi-structured catalysts. *Applied Catalysis*, 281(1–2): 55–60.

54. Jañicke, M. T., Kestenbaum, H., Hagendorf, U., Schüth, F., Fichtner, M., Schubert, K. (2000). The controlled oxidation of hydrogen from an explosive mixture of gases using a microstructured reactor/heat exchanger and Pt/Al_2O_3 catalyst. *Journal of Catalysis*, 191(2): 282–293.

55. Chen, H. Y., Chen, L., Lu, Y., Hong, Q., Chua, H. C., Tang, S. B., et al. (2004). Synthesis, characterization and application of nano-structured Mo_2C thin films. *Catalysis Today*, 96(3): 161–164.

56. Stark, W. J., Wegner, K., Pratsinis, S. E., and Baiker, A. (2001). Flame Aerosol synthesis of vanadia-titania nanoparticles: structural and catalytic properties in the selective catalytic reduction of NO by NH_3. *Journal of Catalysis*, 197(1): 182–191.

57. Johannessen, T., and Koutsopoulos, S. (2002). One-step flame synthesis of an active Pt/TiO_2 catalyst for SO_2 oxidation—A possible alternative to traditional methods for parallel screening. *Journal of Catalysis*, 205(2): 404–408.

58. Jensen, S., Olsen, J. L., Hansen, H., and Quaade, U. J. (2005). Reaction rate enhancement of catalytic CO oxidation under forced thermal oscillations in microreactors with real time gas detection. In *Paper 134h presented at International Conference on Microreaction Technology, IMRET 8*. Atlanta, GA, April 10–14.

59. Thybo, S., Jensen, S., Johansen, J., Johannessen, T., Hansen, O., and Quaade, U. J. (2004). Flame spray deposition of porous catalysts on surfaces and in microsystems. *Journal of Catalysis*, 223(2): 271–277.

60. Schimpf, S., Bron, M., and Claus, P. (2004). Carbon-coated microstructured reactors for heterogeneously catalyzed gas phase reactions: Influence of coating procedure on catalytic activity and selectivity. *Chemical Engineering Journal*, 101(1–3): 11–16.

61. Knitter, R., and Liauw, M. A. (2004). Ceramic microreactors for heterogeneously catalysed gas-phase reactions. *Lab on a Chip*, 4(4): 378–383.

62. Kiwi-Minsker, L., and Renken, A. (2005). Microstructured reactors for catalytic reactions. *Catalysis Today*, 110(1–2): 2–14.

63. Wolfrath, O., Kiwi-Minsker, L., Reuse, P., and Renken, A. (2001). Novel membrane reactor with filamentous catalytic bed for propane dehydrogenation. *Industrial & Engineering Chemistry Research*, 40: 5234–5239.

64. Wolfrath, O., Kiwi-Minsker, L., and Renken, A. (2001). Filamenteous catalytic beds for the design of membrane microreactor: Propane dehydrogenation as a case study. In *Proceedings of 5th International Conference on Microreaction Engineering, IMRET 5*, W. Ehrfeld, ed., pp. 191–201, Strasbourg: Springer.

65. Kiwi-Minsker, L., Wolfrath, O., and Renken, A. (2002). Membranereactor microstructured by filamentous catalyst. *Chemical Engineering Science*, 57: 4947–4953.

66. Horny, C., Kiwi-Minsker, L., and Renken, A. (2003). Autothermal micro-structured reactors based on filamentous catalysts. *Paper presented at 3rd International Symposium on Multifunctional Reactors, ISMR3*, Bath, UK, August 27–29.

67. Horny, C., Kiwi-Minsker, L., and Renken, A. (2004). Microstructured string-reactor for autothermal production of hydrogen. *Chemical Engineering Journal*, 101(1–3): 3–9.

68. Jenkins, J., and Shutt, E. (1989). The hot spot reactor. *Platinum Metals Review*, 33: 118–127.

69. Geissler, K. (2002). Wasserstoffgewinnung aus Methanol für PEM-Brennstoffzellen-Anwendung. PhD dissertation No. 2442, EPF-Lausanne.

70. Horny, C., Kiwi-Minsker, L., and Renken, A. (2004). Microstructured string reactor for autothermal production of hydrogen. *Chemical Engineering Journal*, 101(1–3): 3–9.

71. Kiwi-Minsker, L. (2002). Novel structured materials for structured catalytic reactors. *Chimia*, 56(4): 143–147.

72. Horny, C. (2005). Développement d'un réacteur microstructuré basé sur des filaments métalliques catalytiques. Production autotherme d'hydrogène par steam-reforming oxydatif du méthanol. PhD dissertation No. 3271, EPF-Lausanne.

73. Bae, J.-M., Ahmed, S., Kumar, R., and Doss, E. (2005). Microchennel development for autothermal reforming of hydrocarbon fuels. *Journal of Power Sources*, 139(1–2): 91–95.

74. Alépée, C., Paratte, L., Renaud, P., Maurer, R., and Renken, A. (2000). Fast heating and cooling for high temperature chemical microreactors. In *Proceedings of 4th International Conference on Microreaction Engineering*, W. Ehrfeld (ed.), pp. 514–525, Berlin: Springer.

75. Alepée, C., Vulpescu, L., Cousseau, P., Renaud, P., Maurer, R., and Renken, A. (2000). Microsystem for high-temperature gas phase reactions. *Measurement & Control*, 33(9): 265–268.

76. Maurer, R., and Renken, A. (2003). Dehydrogenation of methanol to anhydrous formaldehyde in a microstructured reactor system. *Transactions IChemE*, Part A, 81(7): 730–734.

77. Silveston, P. L., Hudgins, R. R., and Renken, A. (1995). Periodic operation of catalytic reactors—Overview. *Catalysis Today*, 25: 91–112.

78. Brandner, J. J., Emig, G., Liauw, M. A., and Schubert, K. (2004). Fast temperature cycling in microstructure devices. *Chemical Engineering Journal*, 101(1–3): 217–224.

CHAPTER 19

MICROSTRUCTURED IMMOBILIZED ENZYME REACTORS FOR BIOCATALYSIS

MALENE S. THOMSEN and BERND NIDETZKY

19.1 INTRODUCTION

Biocatalysis has become a key technology in the synthesis of fine chemicals and pharmaceuticals [1–6]. Its widespread use in industry is driven by the need for optically pure molecules [3–6] and the increased requirement for environmentally clean ("green") processes [7–10]. Biocatalysts possess several advantages over chemical counterpart catalysts that facilitate the achievement of these important features of process quality. They typically show excellent stereo- and regio-selectivity and catalyze their reactions with high atom efficiency (because side reactions are normally prevented effectively) [1–6, 9]. They operate under mild conditions, usually in an aqueous solvent, and can be produced at competitive costs using recombinant DNA technology and microbial fermentation. Their properties (e.g., stability and dependence of activity on environmental factors such as pH) can be made process-compatible using tailoring by protein engineering [1, 2]. Several hundred processes featuring a biocatalytic step have been implemented for the manufacture of kilogram quantities of product in industry [1, 2, 11, 12]. Whole cells or isolated enzymes, very often in the form of an insoluble, immobilized protein preparation, are used [1, 12, 13].

A typical bioconversion process scheme is shown in Fig. 19.1. It involves the production and optionally the purification and immobilization of the relevant enzyme, biocatalytic transformation, and isolation of the product. For each unit operation, there are various options on how to improve the individual process step. For example, recombinant DNA technology may be employed to engineer microorganisms for optimal enzyme production [14], and enzyme immobilization onto a solid carrier may be useful to stabilize the activity under operational conditions and enable continuous processing [15]. Multiple interacting variables between individual

Microchemical Engineering in Practice. Edited by Thomas R. Dietrich
Copyright © 2009 John Wiley & Sons, Inc.

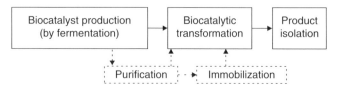

FIGURE 19.1   Unit operations of a typical bioconversion process.

unit operations increase the level of complexity and make optimization of the performance of the bioprocess as a whole a difficult task [16–18]. The integration of enzyme production and biocatalyst preparation by immobilization [15, 19–21] or enzymatic transformation and product recovery [22, 23] are interesting examples of current research and development.

The implementation of a new bioprocess in industry is therefore often slow because it requires consideration of competing options for the biocatalyst and the overall process design. A delay in the transfer of processes established in the laboratory to manufacturing scale is costly because it has a strong impact on the time of product release to the market. If one were able to collect bioprocess information early and at the microliter scale, the time required for development could be decreased significantly. Automated microscale processing techniques represent a possible solution [18, 24, 25].

19.1.1   Microscale (Bio)processing Methods

Lye and colleagues have summarized the advantages of automated microliter scale techniques for bioprocess development [18]. It is possible to screen large libraries of biocatalysts in shorter periods of time. The requirement of expensive reagents and solvents is reduced. Design data for process operation can be obtained in a time-efficient manner, and the translation of processes from discovery to actual production can be faster. Systems for high-throughput (HTP) experimentation, developed in microwell format, are widely used in analytical and screening applications [26–28]. More recently, microfluidic devices have also been introduced as novel tools offering HTP capability [29]. Their use is characterized by continuous-flow experimentation and therefore complements rather than substitutes the studies in microwells.

An important point of consideration is that the vast majority of HTP screening studies were not intended to provide information for bioprocess design and did not take into account relevant engineering principles. Several groups have now focused on establishing systems and protocols with which it will be possible to obtain a sound engineering basis for microscale experimentation and eventually generate design data for each step of the bioprocess (see [18] and the references listed there).

19.1.2   Microwell Formats and Other Miniaturized Vessel-Type Systems

Microwell systems are compatible with batch and fed-batch processes. The scale of operation ranges from a few tens of microliters to a few milliliters. The microwell plates in which the experiments are performed have a standardized footprint of typically 86×128 mm. The geometry and volume of the individual microwell can vary. The generation of quantitative data for the process design requires that mixing and mass transfer in the microwells are well understood [18, 25, 30, 31]. The problem of liquid handling (addition and removal from individual wells) must also be solved [32].

Several groups have measured the volumetric oxygen mass transfer coefficient ($k_L a$, h^{-1}) and the resulting dependence on the design of the well (standard round well and deep square well) and methods of agitation (shaking and mechanical stirring) and aeration (microspargers and surface aeration) [32–34]. Lye and co-workers have investigated jet mixing in microwells and showed that rapid mixing can be achieved in less than 1 s [35]. They argue that the addition of small volumes to microwells in simple shaken systems imparts insufficient energy into the liquid such that mixing times on the order of minutes result. Inhomogeneities are a probable cause of variations in HTP assays. Heinzle and coworkers determined that mixing times in round-bottomed microwells fall in the range of 5 to 500 s, depending on well geometry and shaking speed [30]. Oxygen transfer rates can be as high as 40 mmol/(L h), as shown by several authors (for a review, see [18]; [36–40]). Wells with a square cross section show higher oxygen transfer rates than wells with the standard round format [37]. The groups of Heinzle [30], Büchs [33], and Lye [18, 31, 34] have investigated the minimum shaking speed that is required to promote rapid mixing and oxygen transfer. Hermann et al. determined the critical shaking speed to be the point at which the centrifugal force owing to shaking first exceeds the surface tension of the liquid [33]. Fluid motion in shaken microwell plates was observed using high-speed video photography. It was shown that increases in shaking speed beyond the critical point promote increasingly turbulent fluid motion and faster mixing. At a low shaking speed, the mixing times were on the order of minutes and the formation of dead zones was clearly seen [18, 30]. Doig et al. carried out a dimensional analysis in which they established for the first time a correlation between $k_L a$ and gas–liquid interfacial area (which, in turn, depends on fluid properties, well geometry, and shaking conditions) [34].

Microwell systems have already been applied in various fields of bioprocess research, including microbial fermentation (e.g., [25, 31, 36, 39]), mammalian cell culture [41], metabolite [42] and enzyme [25] production, bioconversion [31, 38], and enzyme kinetics studies [43]. Downstream processing in the microwell format has only recently attracted attention, and the results of studies on microfiltration [44] and chromatography [45, 46] have been reported. Notably, there has been recent work on the integration of microwell systems and microfluidic devices (to be described below).

Miniaturized (milliliter scale) bioreactors equipped with a stirrer, air sparger, and instrumentation for the measurement of pH and dissolved oxygen have been described and used in practice, for example, for microbial cultures [32, 47–50]. Weuster-Botz and coworkers [32, 50] developed a milliliter scale device for HTP bioprocess design. Their miniaturized bioreactor is equipped with a magnetically driven gas-inducing impeller and features oxygen transfer coefficients as high as in laboratory and industrial stirred tank reactors. They established a bioreaction block with 48 10-mL reactors that is integrated within a widely used Tecan platform, a liquid-handling robot that can be used for pH control, fed-batch operation, and automated sample removal. This fully equipped microbioreactor system was used for the automated and parallelized cultivations of *Escherichia coli* using at-line measurements of pH and optical density (cell concentration) in microtiter plates. The parallel in situ monitoring of dissolved oxygen using measurements of the fluorescence lifetime of fluorophors immobilized inside the milliliter scale reactors is currently under development.

19.1.3 Microfluidic Formats

Microfluidic systems have attracted much attention in recent years as a result of their use in bioanalytics as well as cellular and developmental studies [51]. For example, samples containing a single cell have been examined and manipulated [52]. Protein crystallization is another emerging area of application of microfluidics in the biosciences [53]. Microfluidic technologies have been applied to examine cell–microenvironment interactions [54] and for tissue engineering [55]. The reader is referred to recent reviews and several milestone papers in the field [51, 54, 56–60].

From a bioengineering point of view, microfluidic formats represent a further reduction in both the experimental scale (several tens of nanoliters to several hundreds of microliters) and footprint of the device, compared with the microwell formats discussed above. The characteristics of fluids in microchannels, such as laminar flow, represent a defining feature for most microfluidic systems. Solute transfer occurs solely (or mainly) by diffusion. The mass transfer (as well as the heat transfer) can still be rapid (and not rate-determining) because the flow path is reduced to dimensions approaching the continuous-phase boundary layer (for discussion and review, see [56]). Experimentation using devices with a microfluidic format is typically carried out under conditions of continuous flow. If we consider that large-scale production is often performed in flow-through processes, it is not completely understandable why the HTP developments on the laboratory scale are almost exclusively restricted to working in batches [61]. Because of the constant reaction parameters used, experimentation in continuous-flow reactors facilitates automation and provides reproducibility as well as process reliability [59–63]. Microfluidic systems therefore offer various advantages for bioprocess development, including, for example, efficient biocatalyst screening and selection, and rapid optimization of conditions for a particular unit operation. The opportunity to perform multistep bioconversions under conditions in which the individual steps are separated in space and time and no intermediate purification is carried out represents an important fundamental advantage not offered by microwell systems.

There is, however, the question of how process conditions in a microfluidic device relate to those prevailing in a large-scale bioreactor. The translation of results obtained in a continuous microstructured reactor into a much larger conventional stirred tank is not straightforward. Furthermore, it remains to be demonstrated for different bioprocesses that numbering-up is an economically viable option to enhance production capacity to pilot and manufacturing scales. If we consider the unit operations in Fig. 19.1, it would seem that microstructured reactors hold particularly great promise for carrying out biocatalytic transformations with immobilized enzymes (see below). Microfluidic reactors for studies of microbial and mammalian cell culture in batch and chemostat operation have been described [64–67]. The reactors incorporate online optical density, pH, and dissolved oxygen sensors. The supply of oxygen in amounts sufficient to support the optimal growth of bacteria can, however, be a problem in the reported microfluidic reactors. The development of new gas–liquid–solid microstructured reactors may represent a potential solution [68].

A substantial amount of work has been devoted to the study of enzymatic microfluidic reactors for chemical analysis, kinetic studies, and biocatalytic organic synthesis [69–71]. The remainder of this chapter will focus on microstructured reactors containing an immobilized enzyme as the catalytically active component. (The use of a soluble enzyme is not common in continuous reaction systems because unless recycled by a suitable method such as ultrafiltration, the catalyst is lost to the effluent.)

19.2 STEPS TOWARD A MICROSTRUCTURED IMMOBILIZED ENZYME REACTOR

Two types of bioreactors are commonly used in industry to perform enzyme-catalyzed transformations: the stirred tank reactor, and the packed or fluidized bed reactor [1]. The substitution of these well-established reactors with a system based on microfluidic technology will happen only if there are clear and compelling advantages in so doing. It is therefore necessary to consider, in a rigorous and case-specific manner, the potential assets of microstructured reactors for biocatalytic synthesis other than serving as a tool during initial process development in the laboratory (see above). No clear and generally valid answers can be given at the present time, although the following benefits may be anticipated [59]: intensification of processing, especially for multiphase reactions; easier integration of unit operations such as biocatalytic reaction and product removal by liquid–liquid extraction, for example, ability to carry out multistep organic syntheses in the absence of intermediate purification; easier implementation of kinetically controlled reactions; change of product properties; faster transfer of research results into production and therefore earlier start of production at lower costs; and finally easier scale-up of production capacity.

Two possible designs exist for the microstructured immobilized enzyme reactor. One option is to attach the enzyme directly to the surface of the microchannel(s). Another is to immobilize the enzyme on (typically spherical) microparticles that are then filled in the microchannel to generate a microbed. (The pressure difference required to operate a fixed microbed reactor under continuous-flow conditions could be quite high, however.)

19.2.1 Fabrication of Microstructure

The materials used are essentially those seen in other microfluidic applications. In bioanalytical applications, the elastomer polydimethylsiloxane (PDMS) is most often employed [72]. Thermoplastics such as polymethylmethacrylate (PMMA), polycarbonate, polystyrene, and polyvinylchloride have also been utilized [73]. For biocatalytic synthesis in microreactors showing production capacity, stainless steel may be the ultimate material of choice [59, 60], but we are not aware of reports describing its use with enzymes. Glass and fused silica would present other useful possibilities, but are also not widely used [59]. The known methods of microfabrication are employed to generate the desired functional microstructures [74, 75]. The relevant techniques are covered exhaustively in detailed reviews and in this book.

A key point that will dictate the selection of material for microstructure fabrication is the chemical functionalization of the surface. Obviously, not all materials are equally well suited for enzyme immobilization. Glass and fused silica are often treated with amino-siloxanes to generate a layer of reactive amino groups on the surface (e.g., [1]). Stainless steel requires surface modification, for example, coating with a layer of aluminium oxide, to facilitate chemistry with enzymes [59, 62]. Several methods exist to make PDMS reactive for the covalent attachment of enzymes [76]. Roman et al. describe the sol-gel modification of a microfluidic device made from PDMS [77].

Different strategies for the coating of PDMS have been proposed. Marie et al., for example, describe the attachment of a biotin-labeled derivative of PEG that could be used for the directed attachment of appropriately tagged proteins [78]. Oxidation with air plasma to generate surface silanol groups is a well-known procedure to "activate" the surface of PDMS for chemical derivatization [76]. The self-assembly of polyethyleneimine and poly(acrylic acid) [79], poly(oxyethylene) [80], and poly(vinyl alcohol) [81] on PDMS surfaces has also been described. Likewise, methods for surface modifications of PMMA have been proposed [82]. The modification of PDMS with polyacryamide was suggested as another option [83]. If microparticles are used for enzyme immobilization and later assembled in the microchannels, the choice of the material for microstructure fabrication will obviously be less important.

19.2.2 Biocatalyst Immobilization

The majority of commercialized processes, which operate on an industrial scale of 1,000 or greater annual tons of product, utilize immobilized cells or enzymes [1]. The first reports on the immobilization of proteins go back in time to the beginning of the twentieth century. Immobilization is therefore clearly not a "new" technology. However, the importance of immobilized enzymes in industrial practice and the requirement of integrating immobilized enzymes into new functional devices such as biosensors are strong drivers of current research, with the goal of finding improved methods of immobilization [13]. Generally, the immobilization of an enzyme involves chemical or physical attachment to a soluble or insoluble binding partner, or entrapment. Its major aim is retention of the enzyme under conditions in which the soluble protein would be lost due to diffusion or convection. Immobilization thus facilitates continuous processing. It is also performed to enhance the operational stability of the biocatalyst. The methods used can be classified broadly according to whether the final preparation of the biocatalyst is soluble or insoluble.

Working with homogeneous or heterogeneous catalytic systems obviously has distinct advantages and disadvantages, and which of the two possibilities is preferable must be evaluated in a case-specific manner. The literature indicates that biocatalytic particles are often the favored choice (for reviews, see [13, 19, 20]). The immobilization of an enzyme onto the walls of a microchannel is a special case of attachment to an insoluble carrier, as discussed below (e.g., [84]). Immobilization onto microparticles will utilize conventional procedures (i.e., binding of protein in batch followed by filling the channels with functionalized beads).

Immobilization Methods These can be classified according to whether binding to the surface occurs through physical adsorption, covalent bonds, or (pseudo)specific protein–ligand interactions (for a review, see [13]). Another useful distinction is whether the immobilization is random regarding the protein site(s) undergoing reaction or oriented (i.e., site-selective) (Fig. 19.2). Adsorption alone is rarely used because elution of the enzyme during continuous processing would be a problem. Adsorption can be combined with protein cross-linking in a postadsorption step. Cross-linking with bifunctional reagents such as glutardialdehyde is a well-established procedure to immobilize proteins to surfaces bearing free amino

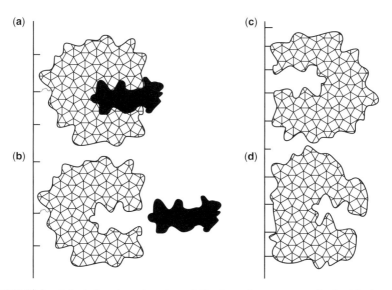

FIGURE 19.2 Oriented and random immobilization of an enzyme. In the ideal case, as shown in (a) and (b), the structure of the enzyme is unchanged and the substrate can easily access the active site. Immobilization via a protein "tag" that confers affinity to a ligand attached to the carrier often positions the enzyme in a suitable orientation. Panels (c) and (d) show two situations that can occur by random immobilization. In (c), the active site is turned toward the carrier and access to the active site is hindered. In (d), the shape of the enzyme has been changed, thereby distorting the active site of the enzyme. Multipoint attachment of the enzyme can cause distortions of the native conformation. However, it is thought to be often responsible for the extra stability conferred to enzymes by immobilization.

groups. Reduction of the formed Schiff base with borohydride is sometimes used to make the attachment permanent, in consideration of the fact that the initially formed linkages are not completely resistant to hydrolysis.

The literature on enzyme immobilization is a rich source of potentially useful alternative methods for covalent binding as well as entrapment of enzymes. The reader is referred to a recent authoritative review of the field [13]. It should be noted, however, that practical use is normally an important criterion in choosing a protocol for biocatalyst immobilization. The use of sol-gel procedures to coat surfaces and entrap enzymes at the same time holds considerable promise for application [70].

The methods described above attach enzymes to the surface in a randomly oriented manner. Because not all binding modes of the enzyme will be compatible with retention of activity and preserve the catalytic properties of the soluble enzyme, it is often desirable to obtain more control over the site(s) with which the protein is linked to the surface (Fig. 19.2). The use of chimeric proteins showing high affinity for binding ligands that have been assembled on the surface of the microstructure presents a possible solution. Immobilized metal ions and biotin are examples of often used ligands. The enzyme of interest is engineered genetically to harbor the relevant "tag" (i.e., a peptide conferring ligand-binding affinity) typically on the *N*- or *C*-terminus of the protein.

Immobilization is achieved by simply recirculating enzyme solution through the microstructure until the loading of protein is complete (Fig. 19.3). There are clear advantages to using biological affinity for immobilization. First of all, the noncovalent interactions stabilizing the enzyme–ligand complex are normally strong, preventing the leakage of catalyst under conditions of processing. Second, the enzyme of interest does not need to be purified prior to immobilization because proteins lacking the tag do not bind or do so only weakly. Third, the microstructure can be regenerated using competitive elution of the bound protein. Inactivated enzyme can thus be removed and new catalyst loaded.

Homogeneous Immobilization vs. Patterning   At face value, the simplest form of enzyme immobilization into microchannels is to cover the surface with a homogeneous monolayer of protein. Using oriented attachment via enzyme–ligand

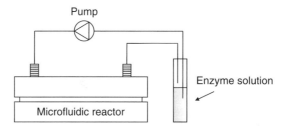

FIGURE 19.3   Enzyme immobilization in a microfluidic reactor. The enzyme solution is recirculated over the microstructured plate until protein binding is exhaustive.

interactions, it is usually possible to obtain a relatively homogeneous surface coverage. However, using random methods of attachment, this is anything else than easy to achieve. Cross-linking with glutardialdehyde normally leads to the formation of clusters of aggregated enzyme on the surface of a carrier, as shown in Fig. 19.4. Methods have been described that avoid cross-linking during the covalent attachment of proteins onto surfaces. However, the binding efficiency (expressed as units of active enzyme immobilized per unit of surface area) is often best when cross-linking is employed, and therefore, in spite of its lack of control over the final structure of the immobilizate, this method is widely used for biocatalyst preparation. Honda et al., for example, report enzyme immobilization on microchannel surfaces by glutardialdehyde-mediated cross-linking and aggregation [84].

There are various applications of microfluidic reactors in which it would be desirable to place enzymes at an appropriate position in the microstructure. Consider a biotransformation in two steps where the product of the first reaction is the substrate of the second enzyme, converting it to the relevant product. The methods of immobilization described above exercise very little control over positioning of the proteins. Ideally, the patterning of enzymes would be possible in situ using flow conditions. Holden et al. described an interesting approach by photoattachment chemistry [85]. However, generally applicable and most of all, practical strategies for patterned biocatalyst immobilization in microchannels still need to be developed. Delamarche has reviewed microcontact printing as a tool to pattern proteins on surfaces with good spatial control [86]. Applications of the described techniques will probably occur in bioanalytics rather than biocatalysis. If one used microbeads for immobilization, the "patterning" of carrier-bound enzymes in a microchannel could simply be achieved by assembling the respective biocatalytic particles in appropriate order. A useful compilation of methods employed for the immobilization of enzymes on microchannel surfaces is given by Miyazaki and Maeda [70].

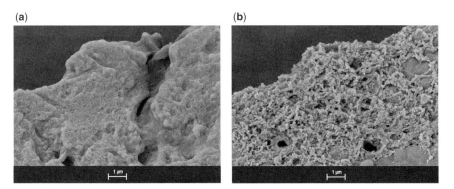

FIGURE 19.4 Electron micrographs of PDMS (containing pyrogenic silicic acid) before (a) and after (b) the immobilization of an enzyme [116]. Immobilization involved the silanization of the microchannel surface, activation of the surface amino groups with glutardialdehyde, and protein binding.

19.3 SELECTED EXAMPLES FROM THE LITERATURE

Three recent reviews provide systematic coverage of the enzymatic microreactors used in chemical analysis [69–71]. Given that the focus of this book is chemical engineering, we do not consider the analytical applications and the reader is referred to the cited literature at this chapter's end (and the references given there).

19.3.1 Enzyme Characterization and Kinetic Studies

Kerby et al. measured the kinetic parameters of immobilized alkaline phosphatase in a microfluidic reactor [87]. They attached the biotinylated enzyme to streptavidin-coated silica microbeads and used a glass microfluidic chip equipped with an in-channel weir for the capture of particles. The authors analyzed the flow rate dependence of the kinetic parameters to determine the influence of mass-transfer resistances. They also examined reactor performance at high levels of substrate conversion.

Mao et al. developed microfluidic devices for the HTP determination of enzyme kinetic parameters and temperature dependence of enzyme activity [88]. They used the immobilization of streptavidin-conjugated alkaline phosphatase to microchannels coated with biotinylated phospholipid bilayers. Measurements at different substrate concentrations were performed simultaneously using laminar flow-controlled dilution on the specially designed microfluidic device. The authors also demonstrated a two-step enzymatic reaction on the same microstructure using glucose oxidase and horseradish peroxidase. Another device was described by Mao et al. with which a linear temperature gradient for parallel measurements can be generated [89]. The activation energy for phosphatase was determined.

Seong et al. investigated the reactions of immobilized horseradish peroxidase and β-galactosidase in a continuously operated packed-bed microreactor [90]. They analyzed the data by using the Lilly–Hornby equation that was originally derived by balancing the convective transport of substrate with its consumption in a Michelis–Menten type of reaction under steady-state conditions.

Koh and Pishko studied the kinetics of alkaline phosphatase and urease using a poly(ethylene glycol)-based hydrogel microstructure that contained the enzyme and was incorporated in a microfluidic device fabricated from PDMS [91]. Liu et al. performed a kinetic analysis of glucose odixation by glucose oxidase adsorbed onto PET sheets [92]. Delouise and Miller report on the kinetic parameters of gluthatione-S-transferase bound on silicon [93]. Jiang et al. applied online frontal analysis of peptides produced by trypsin immobilized on modified cellulose for the determination of kinetic parameters [94].

Gleason and Carbeck describe "microscale steady-state kinetic analysis" as a new technique that enables the rapid and parallel determination of enzyme kinetic parameters [95]. Their approach differs from others just described in that the enzyme, alkaline phosphatase, is immobilized on the wall of a macroscopic flow chamber but its reaction is analyzed in the hydrodynamic boundary layer of the catalytic surface that effectively defines the microscopic reaction volume.

19.3.2   Biocatalytic Transformations Using Continuous-Flow Conditions

Kanno et al. report on the on-chip synthesis of oligosaccharides using glycosidase-catalyzed transgalactosylation [96, 97]. The authors did not use an immobilized enzyme. However, a remarkable result of their study was that the product yield could be enhanced significantly in comparison with the conventional batch conversion. Other hydrolytic enzymes—namely, lipase [98–101] and protease [102–106]—were also tested in microchannel reactors. Different approaches of entrapment were used for their immobilization. The results show that yields of product could be improved and requirements for reactant reduced through the use of microreactors. Jones et al. designed an enzyme reactor in which urease was immobilized on the microreactor wall by incorporating the enzyme directly into the PDMS material used for microstructure fabrication [107, 108]. Belder et al. performed enantioselective catalysis, using a soluble epoxide hydrolase, and analysis on a microfluidic chip [109]. NAD(H)-dependent enzymatic redox reactions were also carried out in microfluidic systems [110, 111]. Methods for the electrochemical regeneration of NADH were described [111]. Oxidative enzymatic reactions examined in microchannel systems include the degradation of p-chlorophenol catalyzed by laccase [112] and hydroxylation of an alkaloid catalyzed by a bacterial P450 enzyme [113].

More recently, Ku et al. [114] reported on the polyketide synthesis and functionalization using two microchannel reactors in series. The first reactor contained type III polyketide synthase immobilized onto agarose microbeads via interactions of the protein "his-tag" and Ni-nitrilotriacetic acid groups displayed on the surface of the particles. The second reactor contained soybean peroxidase attached on the surface of the microchannel. The study shows the rapid evaluation of different reaction conditions for the two-step enzymatic conversion and the effect of the varied conditions on the final product structure. Microstructured multienzyme reactors may become more widely used in the future for combinatorial synthesis. Honda et al. [115] introduced the technology of cross-linked enzyme aggregates to microstructured reactors. They deposited an aggregate formed between an aminoacylase and poly-L-lysine in the presence of glutardialdehyde on the inner wall of a silica capillary. The enzyme reactor thus obtained showed good operational stability, and the immobilized enzyme activity was more resistant to thermal stress and organic solvents than that of the free enzyme. We expect that this list of examples of biocatalytic conversions in microstructured reactors will rapidly increase in the near future.

ACKNOWLEDGMENTS

Financial support was received from the Austrian Federal Ministry of Transport, Innovation, and Technology (Program "Factory of Tomorrow", Project 807947). We also thank D. Kirschneck (Microinnova), M. Koncar (VTU Engineering), and Mag. H. Reichl (Hämosan) for their support.

BIBLIOGRAPHY

1. Buchholz, K., Kasche, V., and Bornscheuer, U. T. (2005). *Biocatalysis and Enzyme Technology.* Weinheim: Wiley-VCH.

2. Bommarius, A. S., and Riebel, B. R. (2004). *Biocatalysis.* Weinheim: Wiley-VCH.

3. Shoemaker, H. E., et al. (2003). Dispelling the myths—Biocatalysis in industrial synthesis. *Science*, 299: 1694–1697.

4. Schmid, A., et al. (2001). Industrial biocatalysis today and tomorrow. *Nature*, 409: 258–268.

5. Bornscheuer, U. T., and Buchholz, K. (2005). Highlights in biocatalysis—Historical landmarks and current trends. *Eng. Life Sci.*, 5: 309–323.

6. Pollard, D. J., and Woodley, J. M. (2007). Biocatalysis for pharmaceutical intermediates: The future is now. *Trends Biotechnol.*, 25: 66–73.

7. Sheldon, R. A., Arends, I., and Hanefeld, U. (2007). *Green Chemistry and Catalysis.* Weinheim: Wiley-VCH.

8. Lancaster, M. (2002). *Green Chemistry.* London: Royal Society of Chemistry.

9. Sheldon, R. A. (2000). Atom efficiency and catalysis in organic synthesis. *Pure Appl. Chem.*, 72: 1233–1246.

10. Haswell, S. J., and Watts, P. (2003). Green chemistry: Synthesis in micro reactors. *Green Chem.*, 5: 240–249.

11. Straathof, A. J. J., et al. (2002). The production of fine chemicals by biotransformation. *Curr. Opin. Biotechnol.*, 13: 548–556.

12. Liese, A., Seelbach, K., and Wandrey, C. (eds.). (2006). *Industrial Biotransformations.* Weinheim: Wiley-VCH.

13. Cao, L. (2006). *Carrier-Bound Immobilized Enzymes.* Weinheim: Wiley-VCH.

14. Gellissen, G. (ed.). (2005). *Production of Recombinant Proteins.* Weinheim: Wiley-VCH.

15. Woodley, J. M. (2006). Choice of biocatalyst form for scalable processes. *Biochem. Soc. Trans.*, 34: 301–303.

16. Lilly, M. D., and Woodley, J. M. (1996). A structured approach to design and operation of biotransformation processes. *J. Ind. Microbiol.*, 17: 24–29.

17. Buckland, B. C., et al. (2000). Biocatalysis for pharmaceuticals—Status and prospects for a key technology. *Metab. Eng.*, 2: 42–48.

18. Micheletti, M., and Lye, G. J. (2006). Microscale bioprocess optimisation. *Curr. Opin. Biotechnol.*, 17: 611–618.

19. Cao, L. (2005). Immobilised enzymes: Science or art? *Curr. Opin. Chem. Biol.*, 9: 217–226.

20. Cao, L., Langen, L., and Sheldon, R. A. (2003). Immobilised enzymes: Carrier-bound or carrier-free? *Curr. Opin. Biotechnol.*, 14: 387–394.

21. Dib, I., Stanzer, D., and Nidetzky, B. (2007). *Trigonopsis variabilis* D-amino acid oxidase: Control of protein quality and opportunities for biocatalysis through production in *Escherichia coli. Appl. Environ. Microbiol.*, 73: 331–333.

22. Lye, G. J., and Woodley, J. M. (1999). Application of in situ product removal techniques to biocatalytic processes. *Trends Biotechnol.*, 17: 395–402.

23. Stark, D., and von Stockar, U. (2002). In situ product removal (ISPR) in whole-cell bio-technology during the last twenty years. *Adv. Biochem. Eng. Biotechnol.*, 80: 149–176.

24. Lye, G. J., et al. (2003). Accelerated design of bioconversion processes using automated microscale processing techniques. *Trends Biotechnol.*, 21: 29–37.

25. Lye, G. J., Dalby, P. A., and Woodley, J. M. (2002). Better biocatalytic processes faster: New tools for the implementation of biocatalysis in organic synthesis. *Org. Proc. Res. Dev.*, 6: 434–440.

26. Sundberg, S. A., et al. (2000). High-throughput and ultra-high-throughput screening: Solution and cell-based approaches. *Curr. Opin. Biotechnol.*, 11: 47–53.

27. Stahl, S., et al. (2000). Implementation of rapid microbial screening procedure for bio-transformation activities. *J. Biosci. Bioeng.*, 89: 367–371.

28. Yazbeck, D. R., et al. (2003). Automated enzyme screening methods for the preparation of enantiopure pharmaceutical intermediates. *Adv. Synth. Catal.*, 345: 524–532.

29. Erickson, D., and Li, D. (2004). Integrated microfluidic devices. *Anal. Chim. Acta*, 507: 11–26.

30. Weiss, S., et al. (2002). Modeling of mixing in 96-well microplates observed with fluor-escence indicators. *Biotechnol. Prog.*, 18: 821–830.

31. Micheletti, M., et al. (2006). Fluid mixing in stirred bioreactors: Implications for scale-up predictions from microlitre-scale microbial and mammalian cell cultures. *Chem. Eng. Sci.* 61: 2939–2949.

32. Puskeiler, R., Kaufman, K., and Weuster-Botz, D. (2005). Development, parallelization, and automation of a gas-inducing milliliter-scale bioreactor for high-throughput biopro-cess design (HTBD). *Biotechnol. Bioeng.*, 89: 512–523.

33. Hermann, R., Lehmann, M., and Büchs, J. (2003). Characterization of gas-liquid mass transfer phenomena in microtiter plates. *Biotechnol. Bioeng.*, 81: 178–186.

34. Doig, S. D., et al. (2005). Modelling surface aeration rates in shaken microtitre plates using dimensionless groups. *Chem. Eng. Sci.*, 60: 2741–2750.

35. Nealon, A. J., et al. (2006). Quantification and prediction of jet macro-mixing times in static microwell plates. *Chem. Eng. Sci.*, 61: 4860–4870.

36. Duetz, W. A., et al. (2000). Methods for intense aeration, growth, storage, and replication of bacterial strains in microtiter plates. *Appl. Environ. Microbiol.*, 66: 2641–2646.

37. Hermann, R., et al. (2001). Optical method for the determination of the oxygen-transfer capacity of small bioreactors based on sulfite oxidation. *Biotechnol. Bioeng.*, 74: 355–363.

38. Doig, S. G., et al. (2002). The use of microscale processing techniques for quantification of biocatalytic Baeyer-Villiger oxidation kinetics. *Biotechnol. Bioeng.*, 80: 42–49.

39. Duetz, W. A., and Witholt, B. (2001). Effectiveness of orbital shaking for the aeration of suspended bacterial cultures in square-deepwell microtiter plates. *Biochem. Eng. J.*, 7: 113–115.

40. Duetz, W. A., and Witholt, B. (2004). Oxygen transfer by orbital shaking of square vessels and deepwell microtiter plates of various dimensions. *Biochem. Eng. J.*, 17: 181–185.

41. Girard, P., et al. (2001). Small-scale bioreactor system for process development and optimization. *Biochem. Eng. J.*, 7: 117–119.

42. Minas, W., Baiely, J. E., and Duetz, W. (2000). Streptomycetes in micro-cultures: Growth, production of secondary metabolites and storage and retrieval in 96-well format. *Antonie van Leeuwenhoek*, 78: 297–305.

43. John, G. T., and Heinzle, E. (2001). Quantitative screening method for hydrolases in microplates using pH indicators: Determination of kinetic parameters by dynamic pH monitoring. *Biotechnol. Bioeng.*, 72: 620–627.

44. Jackson, N. B., Liddell, J. M., and Lyle, G. J. (2006). An automated microscale technique for the quantitative and parallel analysis of microfiltration operations. *J. Membr. Sci.*, 276: 31–41.

45. Mazza, C. B., et al. (2002). High-throughput screening and quantitative structure-efficacy relationship models of potential displacers for proteins in ion-exchange systems. *Biotechnol. Bioeng.*, 80: 60–72.

46. Rege, K., et al. (2004). Parallel screening of selective and high-affinity displacers for proteins in ion-exchange systems. *J. Chromatogr.*, A1033: 19–28.

47. Kostov, Y., et al. (2001). Low cost microbioreactor for high-throughput bioprocessing. *Biotechnol. Bioeng.*, 72: 346–352.

48. Lamping, S. R., et al. (2003). Design of a prototype miniature bioreactor for high throughput automated bioprocessing. *Chem. Eng. Sci.*, 58: 747–758.

49. Kumar, S., Wittmann, C., and Heinzle, E. (2004). Minibioreactors. *Biotechnol. Lett.*, 26, 1–10.

50. Weuter-Botz, D., et al. (2005). Methods and milliliter scale devices for high-throughput bioprocess design. *Bioproc. Biosyst. Eng.*, 28: 109–119.

51. Whitesides, G. M. (2006). The origin and future of microfludics. *Nature*, 442: 368–373.

52. Wheeler, A. R., et al. (2003). Microfluidic device for single-cell analysis. *Anal. Chem.*, 75: 3581–3586.

53. Hansen, C. L., et al. (2002). A robust and scalable microfluidic metering method that allows protein crystal growth by free interface diffusion. *Proc. Natl. Acad. Sci. USA*, 99: 16531–16536.

54. El-Ali, J., Sorger, P. K., and Jensen, K. F. (2006). Cells on chips. *Nature*, 442: 403–411.

55. Khademhosseini, A., et al. (2006). Microscale technologies for tissue engineering and biology. *Proc. Natl. Acad. Sci. USA*, 103: 2480–2487.

56. De Mello, A. (2006). Control and detection of chemical reactions in microfluidic systems. *Nature*, 442: 394–402.

57. Dittrich, P. S., Tachikawa, K., and Manz, A. (2006). Micro total analysis systems. Latest advancements and trends. *Anal. Chem.*, 78: 3887–3908.

58. Janasek, D., Franzke, J., and Manz, A. (2006). Scaling and the design of miniaturized chemical-analysis systems. *Nature*, 442: 374–380.

59. Ehrfeld, W., Hessel, V., and Löwe, H. (2000). *Microreactors.* Weinheim: Wiley-VCH.

60. Hessel, V., et al. (2005). *Chemical Micro Process Engineering. Fundamentals, Modeling And Reactions.* Weinheim: Wiley-VCH.

61. Jas, G., and Kirschning, A. (2003). Continuous flow techniques in organic synthesis. *Chem. Eur. J.*, 9: 5708–5723.

62. Jänisch, K., et al. (2004). Chemistry in microstructured reactors. *Angew. Chemie Int. Ed.*, 43: 406–446.

63. Watts, P., and Wiles, C. (2007). Recent advances in synthetic micro reaction technology. *Chem. Commun.*, 443–467.

64. Zanzotto, A., et al. (2004). Membrane-aerated microbioreactor for high-throughput bio-processing. *Biotechnol. Bioeng.*, 87: 243–254.

65. Szita, N., et al. (2005). Development of a multiplexed microbioreactor system for high-throughput bioprocessing. *Lab Chip*, 5: 819–826.

66. Zhang, Z., et al. (2006). Microchemostat-microbial continuous culture in a polymer-based, instrumented microbioreactor. *Lab Chip*, 6: 906–913.

67. Maharbiz, M., et al. (2004). Microbioreactor arrays with parametric control for high throughput experimentation. *Biotechnol. Bioeng.*, 85: 376–381.

68. Hessel, V., et al. (2005). Gas-liquid and gas-liquid-solid microstructured reactors: Contacting principles and applications. *Ind. Eng. Chem. Res.*, 44: 9750–9769.

69. Urban, P. L., Goddall, D. M., and Bruce, N. C. (2006). Enzymatic microreactors in chemical analysis and kinetic studies. *Biotechnol. Adv.*, 24: 42–57.

70. Miyazaki, M., and Maeda, H. (2006). Microchannel enzyme reactors and their applications for processing. *Trends Biotechnol.*, 24: 463–470.

71. Krenkova, J., and Foret, F. (2004). Immobilized microfluidic enzymatic reactors. *Electrophoresis*, 25: 3550–3563.

72. Sia, S., and Whitesides, G. M. (2003). Microfluidic devices fabricated in poly(dimethyl-siloxane) for biological studies. *Electrophoresis*, 24: 3563–3576.

73. Fiorini, G. S., and Chiu, D. T. (2005). Disposable microfluidic devices: Fabrication, function, and application. *BioTechniques*, 38: 429–446.

74. Becker, H., and Gärtner, C. (2000). Polymer microfabrication methods for microfluidic analytical applications. *Electrophoresis*, 21: 12–26.

75. Brand, O., et al. (eds.). (2006). *Micro Process Engineering*. Weinheim: Wiley-VCH.

76. Makamba, H., et al. (2003). Surface modification of poly(dimethylsiloxane) microchannels. *Electrophoresis*, 24: 3607–3619.

77. Roman, G. T., et al. (2005). Sol-gel modified poly(dimethylsiloxane) microfluidic devices with high electroosmotic mobilities and hydrophilic channel wall characteristics. *Anal. Chem.*, 77: 1414–1422.

78. Marie, R., et al. (2006). Use of PLL-*g*-PEG in micro-fluidic devices for localizing selective and specific protein binding. *Langmuir*, 22: 10103–10108.

79. Makamba, H., et al. (2005). Stable permanently hydrophilic protein-resistant thin-film coatings on poly(dimethylsiloxane) substrates by electrostatic self-assembly and chemical crosslinking. *Anal. Chem.*, 77: 3971–3978.

80. Hellmich, W., et al. (2005). Poly(oxyethylene) based surface coatings for poly(dimethyl-siloxane) microchannels. *Langmuir*, 21: 7551–7557.

81. Wu, D. P., et al. (2005). Multilayer poly(vinyl alcohol)-adsorbed coating on poly(di-methylsiloxane) microfluidic chips for biopolymer separation. *Electrophoresis*, 26: 211–218.

82. Liu, J., et al. (2004). Surface-modified poly(methyl methacrylate) capillary electrophoresis microchips for protein and peptide analysis. *Anal. Chem.*, 76: 6948–6955.

83. Xiao, D., et al. (2004). Surface modification of the channels of poly(dimethylsiloxane) microfluidic chips with polyacrylamide for fast electrophoretic separations of proteins. *Anal. Chem.*, 76: 2055–2061.

84. Honda, T., et al. (2005). Immobilization of enzymes on a microchannel surface through cross-linking polymerisation. *Chem. Commun.*, 5062–5064.

85. Holden, M. A., Jung, S.-Y., and Cremer, P. S. (2004). Patterning enzymes inside microfluidic channels via photoattachment chemistry. *Anal. Chem.*, 76: 1838–1843.

86. Delamarche, E. (2004). Microcontact printing of proteins. In C. M. Niemeyer and C. A. Mirkin (eds.), *Nanobiotechnology*, Weinheim: Wiley-VCH, pp. 31–52.

87. Kerby, M. B., Legge, R. S., and Tripathi, A. (2006). Measurement of kinetic parameters in a microfluidic reactor. *Anal. Chem.*, 78: 8273–8280.

88. Mao, H., Yang, T., and Cremer, P. S. (2002). Design and characterization of immobilized enzymes in microfluidic systems. *Anal. Chem.*, 74: 379–385.

89. Mao, H., Yang, T., and Cremer, P. S. (2002). A microfluidic device with linear temperature gradient for parallel and combinatorial measurements. *J. Am. Chem. Soc.*, 124: 4432–4435.

90. Seong, G. H., Heo, J., and Crooks, R. M. (2003). Measurement of enzyme kinetics using a continuous-flow microfluidic system. *Anal. Chem.*, 75: 3161–3167.

91. Koh, W.-G., and Pishko, M. (2005). Immobilization of multienzyme microreactors inside microfluidic devices. *Sens. Actuators,* B106: 335–342.

92. Liu, A. L., et al. (2006). Off-line form of the Michaelis-Menten equation for studying the reaction kinetics in a polymer microchip integrated with enzyme microreactor. *Lab Chip*, 6: 811–818.

93. DeLouise, L. A., and Miller, B. L. (2005). Enzyme immobilization in porous silicon: Quantitative analysis of the kinetic parameters for glutathione-*S*-transferases. *Anal. Chem.*, 77: 1950–1956.

94. Jiang, H. H., et al. (2000). On-line characterization of the activity and reaction kinetics of immobilized enzyme by high-performance frontal analysis. *J. Chromatogr.*, A903: 77–84.

95. Gleason, N. J., and Carbeck, J. D. (2004). Measurement of enzyme kinetics using microscale steady-state kinetic analysis. *Langmuir*, 20: 6374–6381.

96. Kanno, K., et al. (2002). Rapid enzymatic transglycosylation and oligosaccharide synthesis in a microchip reactor. *Lab Chip*, 2: 15–18.

97. Hisamoto, H., et al. (2003). Chemico-functional membrane for integrated chemical processes on microchip. *Anal. Chem.*, 75: 350–354.

98. Urban, P. L., et al. (2006). On-line low-volume transesterification-based assay for immobilized lipases. *J. Biotechnol.*, 126: 508–518.

99. Pijanowska, D. G., et al. (2001). The pH-detection of triglycerides. *Sens. Actuators*, B78: 263–266.

100. Nakamura, H., et al. (2004). A simple method of self-assembled nano-particles deposition on the micro-capillary inner walls and the reactor application for photocatalytic and enzyme reactions. *Chem. Eng. J.*, 101: 261–268.

101. Park, C. B., and Clark, D. S. (2002). Sol-gel encapsulated enzyme arrays for highthroughput screening of biocatalytic activity. *Biotechnol. Bioeng.*, 78: 229–235.

102. Miyazaki, M., et al. (2003). Simple method for preparation of nanostructure on microchannel surface and its usage for enzyme immobilisation. *Chem. Commun.*, 648–649.

103. Miyazaki, M., et al. (2004). Preparation of functionalized nanostructures on microchannel surface and their use for enzyme microreactors. *Chem. Eng. J.*, 101: 277–284.

104. Miyazaki, M., et al. (2005). Efficient immobilization of enzymes on microchannel surface through His-tag and application for microreactor. *Protein Pept. Lett.*, 12: 207–210.

105. Kawakakami, K., et al. (2005). Development and characterization of a silica monolith immobilized enzyme micro-bioreactor. *Ind. Eng. Chem. Res.*, 44: 236–240.

106. Wu, H., et al. (2004). Titania and alumina sol-gel derived microfluidic enzymatic reactors for peptide mapping: Design, characterization, and performance. *J. Proteome Res.*, 3: 1201–1209.

107. Jones, F., Lu, Z., and Elmore, B. (2002). Development of novel microscale system as immobilized enzyme bioreactor. *Appl. Biochem. Biotechnol.*, 98: 627–640.

108. Jones, F., et al. (2004). Immobilized enzyme studies in a microscale bioreactor. *Appl. Biochem. Biotechnol.*, 113: 261–272.

109. Belder, D., et al. (2006). Enantioselective catalysis and analysis on a chip. *Angew. Chemie Int. Ed.*, 45: 2463–2466.

110. Zhao, D. S., and Gomez, F. A. (1998). Double enzyme-catalyzed microreactors using capillary electrophoresis. *Electrophoresis*, 19: 420–426.

111. Yoon, S. K., et al. (2005). Laminar flow-based electrochemical microreactor for efficient regeneration of nicotinamide cofactors for biocatalysis. *J. Am. Chem. Soc.*, 30: 10466–10467.

112. Maruyama, T., et al. (2003). Enzymatic degradation of *p*-chlorophenol in a two-phase flow microchannel system. *Lab Chip*, 3: 308–312.

113. Srinivasan, A., et al. (2004). Bacterial P450-catalyzed polyketide hydroxylation on a microfluidic platform. *Biotechnol. Bioeng.*, 88: 528–535.

114. Ku, B., et al. (2006). Chip-based polyketide biosynthesis and functionalization. *Biotechnol. Prog.*, 22: 1102–1107.

115. Honda, T., et al. (2006). Facile preparation of an enzyme-immobilized microreactor using a cross-linking enzyme membrane on a microchannel surface. *Adv. Synth. Catal.*, 348: 2163–2171.

116. Thomsen, M., Pölt, P., and Nidetzky, B. (2007). Development of a microfluidic immobilised enzyme reactor. *Chem. Commun.*, 2527–2529.

CHAPTER 20

MULTIPHASE REACTIONS

J. G. E. (HAN) GARDENIERS

20.1 INTRODUCTION

Many of the chemical reactions used in industry today involve multiple phases, such as combinations of a gas and a liquid in which the gas only dissolves to a limited extend, and of two immiscible liquids. In most cases, catalysts are also involved, many of which are introduced in a solid form, for example, as a packed bed of particles, monolithic porous structure, or wall coating.

The processes that determine the yield and selectivity of multiphase reactions are, in addition to the intrinsic chemical reaction kinetics, basically the same as those discussed in Chapter 6 for separation units, that is, partitioning of the species of interest between the different phases, and the mass transfer of that species within a single phase. As was also pointed out in Chapter 6, in general, mass transfer limitations are more prominent in the liquid rather than gas phase. Just as in extractions, two principles are of interest to increase mass transfer during the reactions of species present in two immiscible phases: decreasing the distances over which mass transfer has to occur (in addition, also leading to higher concentration gradients), or increasing the contact area available for transfer. A combination of both principles is also feasible.

To understand the different options to enhance or, more generally, to control mass transfer, in the next paragraph two-phase flow diagrams will be described. Of particular interest are the conditions under which specific flow regimes are stable, in order to arrive at well-defined reaction systems. By "well-defined," it is meant that besides an enhancement of mass transfer, reaction selectivity can also be controlled, by means of the minimization of residence time distribution (RTD). After useful flow regimes are discussed, a number of reaction examples will be given to illustrate the different options.

Microchemical Engineering in Practice. Edited by Thomas R. Dietrich
Copyright © 2009 John Wiley & Sons, Inc.

20.2 TWO-PHASE FLOW REGIMES

The behavior of flowing two-phase systems is of special interest here because virtually all microreactor systems are of the continuous-flow type. More specifically, almost all microreactors involve microchannels or microstructured fluid ducts through which a gas, liquid, or combination of one or more gases and liquids flows continuously.

Gas–liquid flow in channels of many different dimensions has been studied very intensively over the years for applications in the field of heat exchange (cooling and heating). In this field, a gradual shift can be observed from the use of larger-diameter channels, on the order of 10 to 20 mm, to that of channels with diameters close to or even below 100 μm. The reason for this shift is the same as why microreactor channels have become smaller over the years: The increased surface-to-volume ratio of a smaller channel enables a higher heat flux out of or into the fluid flowing through a microchannel. Kandlikar et al. [1] give a very good overview on the topic of heat transfer in mini and microchannels.

In contrast to gas–liquid two-phase flow in microchannels used for cooling purposes, in which the fraction of gas (or rather, evaporated liquid) changes continuously due to boiling after heat uptake when the fluid passes the microchannel, in a gas–liquid two-phase flow microchannel microreactor the fraction of gas is generally constant (in fact, it decreases slightly due to gas consumption in the reaction). This situation compares best to that described in the literature as "adiabatic flow," which is generally studied by introducing a gas–liquid mixture of specific void fraction (i.e., gas volume fraction) into a microchannel, through specially designed inlets to achieve the appropriate mixing of gas and liquid. As the term "adiabatic" indicates, no heat exchange between the fluid and its surroundings is enabled, and therefore the gas and liquid mass fractions will remain constant throughout the length of the microchannel (note that the volume fractions may change slightly due to an axial pressure gradient which may compress the gas volume more or less, depending on the axial position in the channel). For microchannels with diameters somewhat below 0.5 mm, flow diagrams up to a few mm have the general qualitative appearance shown in Fig. 20.1.

At low liquid superficial velocities, an annular flow pattern is observed, in which the liquid flows as a film along the walls of the channel and the gas flows through the core. In bubbly flow, observed for low gas fractions at moderate velocities, the non-wetting gas flows as small bubbles dispersed in the continuous, wetting liquid. Slug flow, also frequently called "plug," "segmented," "bubble train," "intermittent," or "Taylor" flow [4] (see also Fig. 6.7 in Chapter 6), is a flow pattern in which gas bubbles span the cross section of the channel. The length of the liquid slugs and gas bubbles is mainly determined by the inlet conditions [4]. At the highest velocities, small satellite bubbles appear at the rear of the slug, eventually leading to aerated slugs and a rather chaotic flow pattern that is called "churn" flow, also known as "dispersed" flow [3]. For high velocities at low liquid fraction, the annular flow pattern consists of a thin wavy liquid film flowing along the wall, sometimes with a mist of gas and entrained liquid in the core.

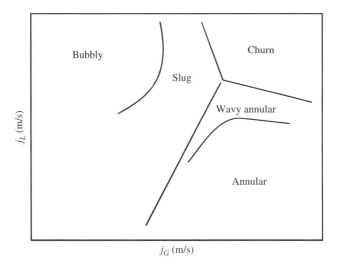

FIGURE 20.1   Typical gas–liquid two-phase flow diagram (adapted from [2, 3]). j_L and j_G are the superficial liquid and gas velocity, respectively.

For detailed descriptions of the different flow patterns, the reader is referred to recent literature; for example, in [5] Jensen et al. provide a good overview of recent knowledge on two-phase flow in microchannels. These authors also discuss in detail the very limited amount of information on microchannels with diameters 100 μm and below. Namely, in microchannels with a hydraulic diameter much smaller than the Laplace constant (which scales surface tension to gravitational forces and is on the order of 1 mm for water at atmospheric conditions and room temperature), two-phase flow characteristics become significantly different. For example, due to capillarity effects, bubbly or churn flow is not observed in such small microchannels.

In order to achieve good control over mass transfer conditions and residence time distribution in a microchannel reactor, the two most useful regimes for a gas–liquid two-phase flow system are annular and segmented flow. In the following sections, microreactor examples in which these two flow patterns have been used will be discussed in detail.

Bubbly flow, in principle, may also be useful for gas–liquid microreactors, but generally bubbles that are much smaller than the channel diameter can behave unpredictably, tend to stick to channel walls and corners under conditions where liquid wetting is not perfect, or grow and shrink due to collision and agglomeration with other bubbles. Bubble size under such conditions is generally inhomogeneous and difficult to control, and as a result reaction conditions also become uncontrollable. Nevertheless, examples do exist in which this type of flow pattern has been successfully used for chemistry, for example, in the work of Löb et al., in which a micromixer was used to create a foam with very high void fraction [6]. This example has already been discussed in Chapter 6.

The above discussion was concerned with gas–liquid two-phase flow systems. For liquid–liquid (i.e., two immiscible liquids) systems, the information available about flow patterns and their regimes with respect to liquid velocity and the liquid–liquid ratio is limited. Several studies exist on oil–water mixtures with applications in off-shore technology. Figure 20.2 is a flow diagram adapted from the study by Bensakhria et al. on a 25-mm diameter pipe [7], but nearly identical results were found for paraffin–water combinations in tubes of 7-mm diameter [8]. Similar flow patterns as in gas–liquid flow are observed. This diagram includes a regime of stratified flow, due to the fact that gravity plays a role for this large channel diameter ("stratified" in this case means that the fluids separate in two streams, with the denser fluid running on the bottom side of the channel). It can also be observed that the curve is not symmetrical, as one may have expected, but shows only the annular flow situation with a water film on the tube wall and an oil core, but not the inverse situation with an oil film and a water core. This is a happy coincidence discovered a long time ago and helps in transporting the viscous oil with a limited pressure drop along the transporting tubes, since the wall friction is mainly determined by the water.

In microreactor practice in liquid–liquid flow situations, the most utilized flow pattern is that where the two liquids run in parallel streams alongside one another, and interchange occurs at the interface that is essentially flat along the axial direction of the channel; see, for example, [9]. It has to be pointed out, though, that the empirical flow map shown in Fig. 20.2 indicates that such a flow pattern, in general, does not constitute a stable situation. In fact, the two liquid streams when entering the microchannel will either reform to annular flow or break up into droplets or slugs, depending on the conditions. Due to the interfacial tension between the two liquids, which tends to reduce interfacial area, and the viscous stresses that act to extend and drag the interface downstream, the interface is destabilized. For example, in the case

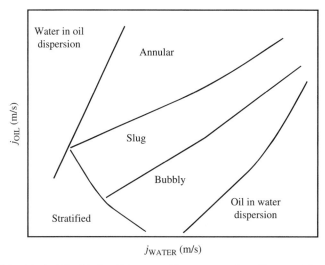

FIGURE 20.2 Typical liquid–liquid (water–oil) two-phase flow diagram (adapted from [7]).

of oil and water streams joining at a T-junction, droplets of radius $R \sim h/\text{Ca}$ (with h being the hydraulic diameter of the channel and Ca the capillary number, as defined in Chapter 6) will be formed [10]. The droplets become smaller for larger Ca, which is equivalent to higher fluid velocity and viscosity (i.e., higher shear), or lower surface tension.

20.3 EXAMPLES OF MICROREACTORS BASED ON SEGMENTED FLOW

20.3.1 Gas–Liquid Segmented Flow Where the Gas is a Reagent

The most common application of segmented (Taylor) flow in microreactors is the case in which a gas has to react with a component of the liquid. It has been shown that the hydrodynamics of this type of flow allow for good mass-transfer rates from the gas bubbles to the liquid [11, 12] and extremely high mass transfer from the gas bubbles to the wall, which contains catalytic coating [13].

The main reason for the high mass-transfer rates within the liquid slug is the presence of vortices within the slug (see Chapter 6, Fig. 6.7, and the accompanying text). These vortices enhance mixing within the liquid slug. On the other hand, the mixing of liquid content between the slugs is reduced due to the presence of a gas bubble, in addition to a very thin, stagnant liquid film on the wall at the position of the bubble that connects two consecutive slugs. This film does not allow a large flux of species between the slugs because diffusion through it is the only mechanism of mass transport between slugs. However, it allows for a high mass transfer of gas to the surface at the location where the thin layer surrounds the bubble. The thickness of this thin film is inversely proportional to fluid velocity. The best gas transfer is therefore obtained at lower velocity [4].

All this leads to the conclusion that a single Taylor flow-based microchannel can be considered close to an ideal plug flow reactor. Evidence for this was found in residence time distribution studies using tracers, which gave quite narrow tracer exit curves [14]. These curves had a sharp rise and somewhat dispersed tail, which can be understood from the fact that under the experimental conditions studied, the film may be considered a virtually stagnant zone, and the exchange of the tracer with this zone can never lead to an earlier exit of the tracer than expected, based on the mean velocity of the slug in which the tracer was injected [4]. The tail occurs by the exchange of tracer from its original slug to the stagnant film and from that to the upstream slugs. The effect of different fluidic parameters on RTD in Taylor flow microchannel reactors has recently been studied by Salman et al. [15].

Taylor flow has been used for three-phase (gas–liquid–solid catalyst) applications in monolith reactors. Monoliths are ceramic structures of parallel straight channels with a typical diameter of 1 mm and typical wall thickness of 100 μm. Due to the open structure, the pressure drop over the length of the microchannels is low, whereas with the small channels a high surface area is available, which ensures fast mass transfer if a thin layer of catalytically active material is deposited on the walls (e.g., by a washcoat process) or if a catalyst is impregnated in the wall material (which has to be porous in that case).

On an industrial scale, monolith reactors are used for the production of hydrogen peroxide by the anthraquinone process [16]. It has been studied at the laboratory scale, for example, for the oxidation of glucose [17], dehydrogenation of ethylbenzene [18], and Fischer–Tropsch process [19].

An interesting example of safe chemistry in a microreactor is that of a membrane contactor (see also Chapter 6) in combination with Taylor flow. A multiphase reactor was developed in which the problem of generating explosive mixtures of oxygen and hydrocarbon vapor was eliminated by the introduction of a porous membrane contactor that served three functions: (1) continuous oxygen feeding along the reactor length, (2) retention of homogeneous catalyst, (3) separation of oxygen and hydrocarbon streams [20]. The reaction carried out in this reactor was the selective oxidation of 1-butene to methyl-ethyl-ketone by an aqueous Pd^{2+} heteropoly-anion catalyst. The separation of oxygen and hydrocarbon streams is ensured by a liquid layer at the membrane interface. The hydrophilic carbon membrane was found to remain fully wetted by the water phase. The rate-limiting step was the gas–liquid mass transfer of 1-butene in Taylor flow that was established and carefully controlled on the tube side of the membrane. Under these conditions, the concentration of reactants in the liquid phase was negligible, which is necessary to obtain separation of the gaseous feed components.

Voloshin et al. tested the feasibility of a continuous plug-flow bioreactor composed of a tube made of Teflon through which a culture of *Pseudomonas putida* was pumped simultaneously with air to provide air bubbles that separated the tubular culture. They compared it with growth in a batch bioreactor and found that when the residence time in the plug-flow bioreactor was greater than the time needed to reach the stationary phase in batch mode, the maximum biomass density reached in both was the same and all benzoate was consumed. The drawbacks for practical application were found to be fluctuations of cell concentration in the outflow cultural liquid due to cell aggregation and adhesion to the tubing wall, and inadequate aeration [46].

A microbubble column, developed by IMM, which is based on Taylor flow, has been discussed in detail already in Chapter 6 [21]. Besides its use for CO_2 absorption from nitrogen mixtures in a NaOH solution, this system has also been utilized for the direct fluorination of aromatic compounds with fluorine gas [22]. Aromatic direct fluorinations are fast gas–liquid reactions whose rate is expected to be strongly dominated by the rate of mass transfer of reactants from the liquid phase to gas–liquid interface. It may be clear from the previous discussion of mass transfer processes in Taylor flow that the internal mixing occurring within the liquid slugs enhances these reactions.

A capillary microreactor for the selective hydrogenation of α,β-unsaturated aldehydes in an aqueous solution with a dissolved Ru(II)-catalyst was developed by Önal et al. [48]. The catalyst was thus physically separated from the reactant and product in the organic phase. Hydrogen was used as a reducing agent. Segmented flow was used, with alternating organic and aqueous slugs, with hydrogen gas bubbles in the organic phase. The overall reaction rate was found to be strongly dependent on the liquid–liquid mass transfer. By decreasing the diameter of the capillary from 1 to 0.5 mm, the specific surface and internal recirculation in the organic phase increased, leading to better volumetric mass transfer and a 3-fold increase in the global reaction rate.

20.3.2 Gas–Liquid Segmented Flow Where the Gas is Inert

As said in the previous paragraph, a single Taylor flow-based microchannel can be considered close to an ideal plug-flow reactor, with the understanding that mixing within a slug is excellent due to the presence of vortices in the slug, and no back-mixing occurs between consecutive slugs of liquid. These two properties of Taylor flow have attracted attention for its application in the synthesis and coating of nanoparticles and nanocrystals.

Nanocrystals of semiconductors or "quantum dots" (see [23] for a review) exhibit a variety of unique size- and shape-dependent physical and chemical properties, for example, a size-dependent fluorescence wavelength. To broaden their application range, semiconductor nanocrystals are often coated with several atomic layers of a different semiconductor, which reduces nonradiative recombination and results in brighter emission, or with a biological tag to make the particles suitable for life science applications.

Their synthesis poses a significant challenge, because robust methods for preparing nanoparticles of homogeneous and predictable size and shape are required. Older synthetic methods rely on a convoluted interplay of kinetic and thermodynamic factors; for example, one widely adopted method consists of rapid nucleation by the injection of a precursor into a hot bulk liquid, followed by growth at a lower temperature in the presence of stabilizing surfactants [23]. Specifying the precise conditions of such reactions is difficult, and therefore more reliable alternative methods are needed. The intrinsic advantage of microfluidic reaction systems is that temperature and concentration can be changed rapidly and reproducibly on the scale of micrometers and milliseconds, as desired for nanocrystal synthesis, but also more precise control over RTD is achieved when Taylor flow is used.

The concept of using gas-separated liquid reactant slugs has been elaborated on by Jensen and coworkers in a number of publications [24–27, 47]. These authors used liquid slugs separated by gas bubbles in order to create small containers. Figure 20.3(a) (from [27]) compares the variance of the particle size distribution (PSD) obtained for the growth of silica particles (with a typical size of ~500 nm) in a laminar flow reactor and in a slug flow (i.e., Taylor flow) reactor. Due to the parabolic flow profile, particles that nucleate and grow in regions close to the wall experience a slower velocity and therefore a longer reaction time, so that they grow to be larger. This leads to a larger variance in the PSD. However, this effect is only found for the shorter reaction times, up to 10 min. As can be seen in Fig. 20.3(b), in this time period the growth rate is highest and therefore most sensitive to changes in residence time. Another explanation (not discussed by the authors of [27]) for the observation that PSD variance in the laminar flow situation approaches that for slug flow may be the fact that particles, once they have reached a specific size, cannot

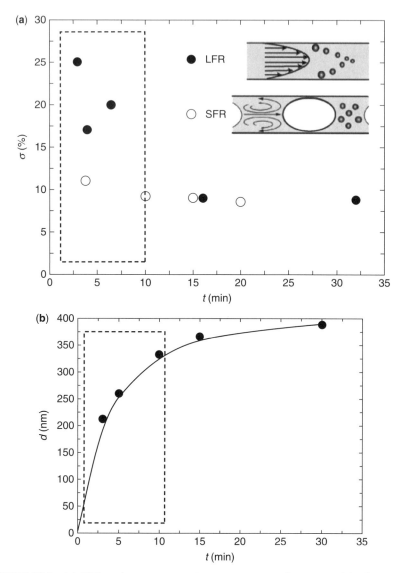

FIGURE 20.3   (a) PSD variance, expressed as a percentage of mean particle diameter, for laminar flow reactor (LFR) and segmented flow reactor (SFR). Inset shows flow characteristics for each reactor type. (b) Batch growth curve of silica particles. Dotted rectangle indicates zone where growth is most sensitive to variations in residence time (reprinted from [27]).

approach the wall so closely anymore and are transported more toward the center, an effect that has been used in so-called hydrodynamic chromatography to separate particles in microchannels [28].

An interesting example of the same concept as above, but for a different purpose, is the work by Fabel et al. [29]. These authors have used a postcolumn microfluidic microreactor device that divides the eluent of an HPLC column into fractions by breaking up the liquid stream into analyte slugs separated by air. They have done this in order to reduce the RTD for enzymatic reactions that are used for the identification of compounds with toxic potential. The toxin present in a slug inhibits enzymatic activity in that particular slug, which can be detected through the reduced conversion of substrate that is also present in the liquid. Work has also been done on a porous mullite monolith as a support for enzyme-catalyzed reactions, using gas– liquid Taylor flow [30]. Here, Taylor flow served mainly to establish a narrow RTD and good mixing within the individual liquid slugs. Lactase from *Aspergillus oryzae* and lipase from *Candida rugosa* were immobilized on the monolith, and activity and stability were compared to those of similarly prepared enzymes on cor- dierite monoliths. The use of high-porosity monoliths yielded more stable and more active bioreactors.

20.3.3 Liquid–Liquid Segmented Flow Microfluidics for Crystallization

In analogy to the previously described method of creating "nanoreactors," consisting of liquid slugs separated by gas bubbles, Alivisatos and coworkers have used liquid bubbles dispersed in an inert fluid for the preparation of fluorescent nanocrystals. Although the size distribution of nanocrystals was already small in their original work, in which they tuned crystal size by changing reaction time, temperature, and precursor concen- tration in a microfluidic chip [31], in a later publication [32] they further improved the CdSe crystal size distribution by using a perfluorinated carrier fluid in which octa- decene droplets with cadmium and selenium precursors were injected through a flow- focusing nanojet structure with a step increase in channel height. These droplets flowed through a high-temperature ($240-300°C$) glass microreactor. The authors state that isolating the reaction solution in droplets prevented particle deposition and hydro- dynamic dispersion, allowing the reproducible synthesis of nanocrystals.

Mixing and reaction, including crystallization, in aqueous slugs separated by oil have been studied by Ismagilov and coworkers in a number of publications. These authors do not use the microfluidic system for production, but for the investigation of reaction kinetics and protein crystallization conditions. They have developed microfluidic structures with inlets through which laminar streams of two aqueous sol- utions of reagents with an inert aqueous liquid separating them were continuously injected into a flow of water-immiscible and inert perfluorodecaline in the main microchannel (Fig. 20.4) [33]. In this microchannel, the laminar stream spon- taneously breaks up into a train of plugs (of ∼0.5 nL) separated by perfluorodecaline. Plugs of three or more reagents may be formed in the same manner. Also, the system allows the user to continuously change the concentrations of reagents in the plugs by controlling the relative flow rates of the reagent streams [34]. Note that the authors of [33] use the term "plug" for the liquid segment that has a meniscus such that the seg- ment approaches droplet shape, so that the liquid in the droplet does not touch the wall. This would be the case for a water segment in an oily substance, in a microchan- nel with a hydrophobic wall (e.g., in PDMS).

A plug or slug moving in a straight channel would generate a steady recirculating flow in that plug or slug. This internal circulation would lead to mixing, but with the

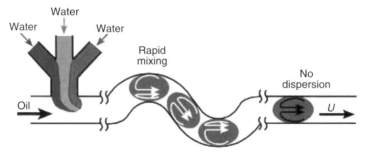

FIGURE 20.4   Schematic presentation of microfluidic device in which two aqueous reagents (dark gray) with a separating fluid (light gray) enter a serpentine channel with a flowing immiscible fluid and form droplets (slugs). Internal recirculation in the slugs is indicated by arrows (from [33]).

risk of regions with no mixing at all. To understand this, one has to transform the two-dimensional pattern in Fig. 6.7 in Chapter 6 into a three-dimensional picture, in which the flow pattern (i.e., in a tubular microchannel) would be cylindrical, with the flow going to the right along the central axis of the channel and to the left all along the wall surface. Thus, there will be a region, forming a ring that surrounds the central axis of the slug, where the liquid does not move at all, and in which therefore no mixing will also occur. The exact flow pattern depends on the shape of the microchannel; in a channel with a rectangular cross section, the circulation pattern would probably more resemble the two-dimensional image in Fig. 6.7 of Chapter 6. And for that case, it would be clear that (in projection) two halves exist within the liquid which basically do not mix, so that axial mixing is inefficient.

From a scan of recent literature, it becomes clear that the circulations within the slug are not yet completely understood, which is the reason why several authors have undertaken flow visualization studies, using particle image velocimetry (PIV) and particle tracking techniques [35, 36]. Unfortunately, these techniques were only used to obtain two-dimensional images, more specifically, for a projection of the flow pattern on a symmetry plane going through the middle of a channel. An attempt has been made recently to evaluate the three-dimensional flow patterns and mixing quality arising from them, for plugs moving through a serpentine microchannel like the one shown in Fig. 20.4 [37]. Namely, it has been established that when a plug or slug passes a serpentine, the internal flow pattern is redirected at each bend and mixing is enhanced by an order of magnitude [38]. A scaling argument for the dependence of the mixing time was derived, which reads $t \sim (aw/u)\log(\text{Pe})$, where $w(m)$ is the cross-sectional dimension of the microchannel, a is the dimensionless length of the plug measured relative to w, u (m/s) is the flow velocity, Pe is the Péclet number (Pe $= wu/D$), and D (m²/s) is the diffusion coefficient of the reagent being mixed. It was experimentally found that under favorable conditions, sub-ms mixing could be obtained, which opens a route for studying fast kinetics [33, 34]. One further finding is that one favorable condition for fast mixing by using plugs in a serpentine microchannel is the choice of two liquids that have nearly the same viscosity [37, 39].

As mentioned above, aqueous slugs separated by oil have been employed by Ismagilov and coworkers for the investigation of protein crystallization conditions. The main issue in the field of protein crystallization is finding the exact solution conditions within a large range of possible solution compositions (salts and other additives) and concentrations (e.g., protein supersaturation), temperature, and gradients of all these parameters in space and time (e.g., solvent evaporation rates and cooling rates). The window for the formation of single crystals of the exact crystal structure (one out of potentially many so-called polymorphs) is generally very narrow.

Still, the availability of protein crystals of excellent quality is crucial for developments in proteomics, as well as in pharmaceutical product development. For proteomics, the most precise technique available at the moment to determine the three-dimensional structures of proteins, in terms of bond distances and angles, is X-ray crystallography (although high-resolution NMR can give similar information, the interpretation of X-ray data is more straightforward). In the pharmaceutical industry, finding a bio-active formulation of a protein-based drug with the appropriate release properties within the human body is an important issue. The key is generally finding a suitable crystal structure for the protein. Besides drug formulation, the purification of proteins at the industrial scale is also of great importance; in that case, finding the appropriate conditions is not trivial. Very generally, it can be said that although methods exist which have large predictive value for crystallization conditions [40], in most cases a trial-and-error method is preferred, which involves the screening of a large amount of crystallization conditions per protein.

The latter is exactly what has been done by Ismagilov and coworkers: They have used their microfluidic formulators to define plugs of a specific composition (which differs from plug to plug) and stored these plugs in a microchannel to essentially wait for crystals to form. This is generally done in a two-step procedure, where the first step screens a large number of reagents to identify the appropriate combination that produces protein crystals (this is called "sparse-matrix screening"), while in the second step the concentration of each reagent in the combination is finely screened to grow crystals of diffraction quality ("gradient screening") [41–45]. X-ray diffraction is actually done in the microfluidic chip, which is composed of an amorphous material that does not interfere with these measurements.

20.3.4 Liquid–Liquid Segmented Flow Microreactors for Polymerization and Emulsification

Conventional emulsification equipment like colloid milling machines and rotor-stator systems commonly result in droplets with broad size distributions. For many applications, however, precise droplet-size control of the emulsion within a narrow range is desired, for example, for maintaining the stability of the emulsion and in order to obtain emulsions with new functional roles. Emulsification is also often used as a process step in the fabrication of polymer microspheres. First a polymer solution is emulsified, followed by removal of the solvent through evaporation, by a thermal or UV polymerization step. Occasionally, a polymer melt is emulsified and successively cooled down to solidify the droplets.

Besides the size distribution, another drawback of some of the conventional emulsification methods is that the energy input during the process is very high, which may result in a temperature rise. The latter is especially undesirable for ingredients for food and pharmaceutical emulsions. An alternative technique is membrane emulsification, which works with low energy input, which is a great advantage over conventional methods [49, 50]. In this method, the disperse phase is pressed through the membrane, forming droplets in the continuous phase at the other side of the membrane. If the membrane has a narrow pore size distribution and a controlled distance between the pores to avoid coalescence, this process generally will yield monodisperse emulsions. A very promising but somewhat costly way to manufacture such membranes is by using photolithography and etching techniques; these processes give extremely well-defined pore size and pitch [51].

Microchannel emulsification is a recently developed promising approach for preparing monodisperse droplets (and beads, see below), with an average size ranging from a few μm up to ∼100 μm [52, 53]. Basically, three main types of microchannel structures are used for emulsification (Fig. 20.5):

- Type I could be described as a one-dimensional version of membrane emulsification, that is, the dispersed phase is pressed through an array of nozzles into a microchannel with the continuous phase [52, 53]. Alternatively, a laminar mixer structure may be used to create basically the same result [6], a concept that is also useful for generating foams.

- Type II consists of a T-junction at which the dispersed phase is pressed into a continuous stream of the second phase [54]. Due to shearing at the junction, a droplet breaks off at a specific size, which depends on the balance of capillary forces and shear forces.

- Type III is based on hydrodynamic focusing of the dispersed phase between two streams of the continuous phase (or "sheath flow" [54]), which results in what may be described as a two-dimensional annular flow pattern. Either this annular pattern breaks up at some distance downstream of the point where the liquids are joined, by itself due to the shearing effect of the two outer streams on the central stream [55, 56], or the combined stream may be pressed through a nozzle to increase shear forces on the annular flow pattern so that it becomes unstable at a shorter channel distance and breaks up into uniform droplets. This method, more so than the previous other two, also allows the formation of droplet-in-droplet structures [55] (also possible with two consecutive T-junctions [59]) or two-color droplets [54].

All these emulsifier types may also be used to manufacture polymer beads, if the droplet contains the monomer precursor mix and a polymerization process is initiated and finalized downstream of the position where the droplet is generated. The latter may be done by UV illumination [53], which, if it is done directly after the point of droplet formation, may also lead to freezing any internal structure that the droplets

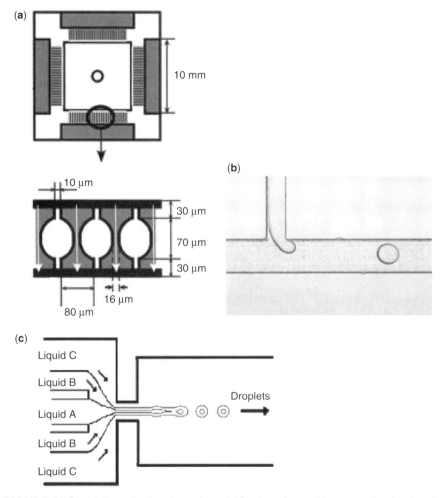

FIGURE 20.5   (a) Type I microchannel emulsification plate, with a zoom-in showing the nozzle openings of 16-μm width (from [52]). (b) Type II emulsifier, a T-junction (from [54]). (c) Type III emulsifier, which uses hydrodynamic focusing with the aid of sheath flow; the case shown here may be used to generate droplet-in-droplet structures (from [55]).

have, such as two colored halves [54] or droplet-in-droplet [55]. Similarly, if the photopolymerization is performed in a confined geometry, the particle will take on (part of) that geometry; see Fig. 20.6 [56]. Disk- and rod-shaped particles have also been demonstrated using a Type II emulsifier [57]. Finally, the Type III emulsifier has been used to make ~30 to 50-μm-wide fibers and tubes [58].

Microfluidic emulsifiers that work on the industrial scale include the interdigital micromixer developed by IMM [60] and the microchannel module developed by Velocys [61], which uses a porous plate to introduce the dispersed phase into the continuous phase, giving small uniform droplet sizes of 1 μm or less.

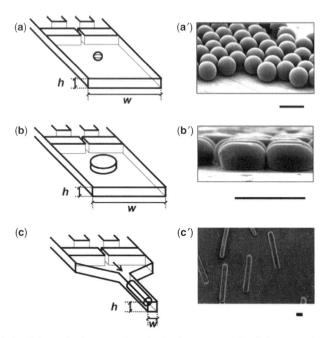

FIGURE 20.6 Schematic (a–c) and optical microscopy (a′–c′) images of poly-TPGDA, tri(propylene glycol) diacrylate, particles with different shapes: microspheres (a, a′), disks (b, b′), and rods (c, c′). The particles were obtained via photopolymerization of droplets produced at water and monomer flows of 8 and 0.1 mL/h, respectively, and a concentration of 4 wt% of photoinitiator in the monomer phase. The black scale bar is 50 μm (from [56]).

20.3.5 Liquid–Liquid Segmented Flow Microreactors for Phase-Transfer Chemistry

Many reactions in industry are highly exothermic and involve two liquid phases [62], and although the reaction generally occurs in only one of those phases, the mass transfer between the two immiscible liquids and the chemical reaction are closely linked. Microreactor technology (or more generally, process intensification) offers several advantages for performing such processes, the most important one for this case being the excellent temperature control. An example is the work done by Dummann et al. [63], who chose the highly exothermic nitration of an aryl group as a test reaction. In these nitrations, which are of industrial importance, the homogeneously catalyzed reaction takes place in the aqueous phase, which consists of a mixture of concentrated sulfuric and nitric acids. The authors used a Taylor flow capillary microreactor, which gives a constant, uniform specific surface area for mass transfer between the two phases. In the nitration reaction, besides the mononitrated product, by-products are formed via consecutive or parallel reactions; for example, oxidation to form phenolic products take place in the organic phase, while the nitrations occur in the aqueous phase. Interfacial mass transfer and residence time distribution therefore strongly influence product yield and selectivity. The capillary microreactor was shown to behave like a plug-flow reactor, permitting a high mass transfer between the phases, and also allowed for some definitive conclusions on the mechanism of by-product formation, that is, by-products were mainly formed via parallel reactions. More information about microreactor work on aromatic nitrations can be found in Section 4.3.1 of [64].

The reaction between substances that are present in different phases is often inhibited because of the insolubility of the substances in one of the two immiscible liquids. Instead of looking for an appropriate mutual solvent, which may sometimes be difficult to find, one may also use a process called "phase-transfer catalysis," a term introduced by Starks [65]. The reaction is enabled by the use of minute amounts of an agent that transfers one reactant across the interface into the other phase so that the reaction can proceed.

Starks has shown that the rates of some displacement, oxidation, and hydrolysis reactions conducted in two-phase systems are dramatically enhanced by the presence of quaternary ammonium and phosphonium salts. These salts are excellent agents for the transport of anions from the aqueous to organic phase, where the anion-tetralkylammmonium complex reacts with the target compound in the organic phase, followed by a transfer of the ammonium salt back to the aqueous phase. In other cases, the mechanism is the transfer of the target organic substance, by ammonium salt, first to the aqueous phase, where it can react to form a water-insoluble substance that falls back to the organic phase. Other mechanisms may also apply [66].

It may be evident that phase transfer processes benefit from a large interfacial area and short diffusion distances, which implies that microreactor technology is extremely suitable to perform such reactions. Ueno et al. demonstrated the use of a phase transfer catalyst in a microchannel reactor for the benzylation of ethyl-2-oxocyclopentane-carboxylate with benzyl bromide in Taylor flow with segments of CH_2Cl_2 and NaOH [67]. The reaction in the microreactor was found to be more efficient than that in a round-bottom flask with vigorous stirring. Other examples are the alkylation of malonic ester, also in Taylor flow and with the same solvent combination [68], and the Claisen–Schmidt reaction of benzaldehyde with acetone in an ethanol–water combination [69] that was studied in both Taylor flow and stratified flow (i.e., in two parallel streams of the two liquids; see below). Taylor flow was found to give the higher conversion of the two flow types, due to a larger interfacial area.

Ryu and coworkers at Osaka Prefecture University have performed in microreactors coupling reactions in a two-phase system, of which one phase was an ionic liquid. Ionic liquids are organic salts with melting points under 100°C, often even lower than room temperature. At least one ion has a delocalized charge and one component is organic, which prevents the formation of a stable crystal lattice. Ionic liquids are employed more and more frequently as substitutes for the traditional organic solvents in chemical reactions, particularly as solvents for transition metal catalysis, for which remarkable improvements have been shown if the organic solvent is replaced by an ionic liquid. The most common ionic liquids are imidazolium and pyridinium derivatives, and phosphonium or tetralkylammonium compounds. Ionic liquids have practically no vapor pressure that facilitates product separation by distillation. For detailed information about the properties and use of ionic liquids, the reader is referred to [70, 71].

Ionic liquids form biphasic systems with many organic product mixtures, which gives rise to the possibility of the convenient extraction and recovery of homogeneous catalysts. The two-phase characteristics of ionic liquid-organic solvent systems also pave the way for the use of microreactors, similarly as was discussed earlier in this chapter. Following this idea, Ryu and coworkers have carried out Sonogashira coupling in ionic liquids [72]. In a microflow system containing a micromixer commercially available from IMM, Mainz, having 2×15 interdigitated channels (with a width of 40 μm and depth of 200 μm), they performed the reaction of iodobenzene with phenylacetylene in the presence of a catalytic amount of $PdCl_2(PPh_3)_2$ in the ionic liquid 1-butyl-3-methylimidazolium hexafluorophosphate. The coupling product, diphenylacetylene, was formed in 93% yield. In another publication [73], the authors used the same micromixer system to perform the Mizoroki–Heck reaction of iodobenzene with butyl acrylate, using the low-viscosity ionic liquid 1-butyl-3-methyl-imidazolium bis(trifluoro-methyl-sulfonyl)imide and in the presence of a homogeneous Pd catalyst. For both reactions, efficient catalyst recycling from the microflow system was demonstrated.

In [72, 73] no further details about the flow structure in the microreactor are given, but in a more recent publication [74], Ryu et al. examined the applicability of a Taylor flow microfluidic system for the Pd-catalyzed carbonylation of aromatic iodides in the presence of a secondary amine, which is being extensively investigated with regard to single/double carbonylation selectivity. Since CO has limited solubility in ionic liquids, carbonylation selectivity is presumed to be controlled by the diffusion efficiency of CO, which can be enhanced by the larger surface area achieved in a Taylor flow concept. It was indeed found that product selectivity was significantly enhanced in the microflow system compared to a batch reactor.

20.4 EXAMPLES OF MICROREACTORS BASED ON OTHER FLOW TYPES

20.4.1 Gas–Liquid

Several examples exist of microreactors with a packed-bed catalyst, through which a two-phase gas–liquid mixture is flown. The basic idea behind such a configuration is to achieve a high catalyst surface area per volume of reactor, but especially in earlier work, no extra attention was paid to also obtaining a well-defined liquid–gas interface. The Taylor flow concept will most probably not apply to packed beds, but under well-controlled flow conditions annular flow (also frequently called trickle flow) may be possible. Bubbly flow is a useful concept for packed beds.

In a number of publications, Jensen and coworkers investigated gas–liquid reactions with a heterogeneous catalyst in silicon-based microreactors. Two concepts were presented: one in which standard porous catalysts were loaded in a microreactor (either a single microchannel or 10-microchannel array) that had an inlet distribution manifold and a micromachined filter integrated in the microchannels in order to immobilize the catalyst particles [75], and the other a fully micromachined packed-bed reactor made out of silicon [76]. For the latter, a perfectly ordered array of pillars of 50 μm in diameter was etched into the reactor channels. These pillars were porosified and loaded with catalyst using conventional impregnation. Especially in the latter reactor, because of the large specific surface area, the heat losses were so large that even for exothermic reactions, heating and not cooling was required to keep the temperature constant. Mass transfer rates were characterized using the heterogeneous hydrogenation of cyclohexene to cyclohexane over Pt/Al_2O_3 catalyst beads in the packed-bed reactor and over Pt impregnated on oxidized porous silicon in the pillar microreactor. Both reactors had similar mass transfer coefficients $K_L a$ on the order of 10 per second, which was more than 100 times larger than the values reported for laboratory-scale reactors.

The findings on several other examples of packed-bed gas–liquid microreactors have been published, mainly for the purpose of catalyst testing. One example worth mentioning is the H-cube® "flow hydrogenator," which employs the electrolytic generation of hydrogen gas to be used in hydrogenation reactions [77, 78].

As can be seen in Fig. 20.1, at relatively high gas content of the gas–liquid mixture, and relatively high fluid velocities, annular flow develops, at least in the small microchannels discussed in this chapter, where capillary forces keep the liquid along the wall. Annular flow is one type of what is categorized as "stratified flow," the opposite of "dispersed flow" (e.g., bubbly flow). For most liquid–gas (but also for most liquid–liquid) combinations, gravity starts to play a role in channels larger than ~ 1 mm. Under similar conditions where annular flow is observed in smaller channels, in these larger channels the two fluids become separated, with the lighter fluid (here, the gas) running on top of the denser fluid. This, of course, only holds for horizontally aligned channels; in vertical channels, the situation is somewhat different and annular flow may also be stable in the larger channels.

Annular flow can be considered a desirable and possibly ideal situation for well-defined mass transport in a reacting system, provided that the liquid film thickness is constant or at least well-controlled all over the microchannel length. Not so many examples of microreactors exist in which this type of flow has been used, but one interesting example is the work of Kitamori and coworkers on a microchannel reactor chip for hydrogenation over a Pd catalyst [79]. These authors have used very narrow microchannels in a glass plate with a specific interfacial area of 10,000 to 50,000 m^2/m^3, in contrast to the ~ 100 m^2/m^3 for conventional reactors. To achieve a microreactor that schematically looks like the one in Fig. 20.7(a), they developed a procedure for the immobilization of Pd catalyst particles microencapsulated in a copolymer [see Fig. 20.7(b)], which after the immobilization on the microchannel wall, using amine linkers, is annealed at 150°C so that the copolymer cross-links. The successful hydrogenation of several substances, with in some cases quite high yields, was conducted with this microreactor.

An important process in fine chemicals industry is the fluorination of organic compounds. The direct use of elemental fluorine is a difficult process, for reasons of safety, but for economic reasons it would be highly desirable, and could be efficient and environmentally friendly if the HF by-product was recycled by electrolysis [80].

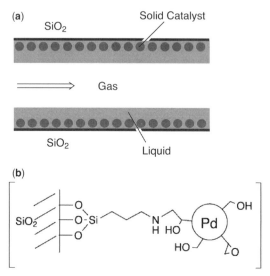

FIGURE 20.7   (a) Annular flow microchannel reactor with Pd particles immobilized at the wall. The gray area is liquid; the gas flows through the center of the channel. (b) Immobilized Pd encapsulated in a copolymer, before cross-linking (from [79]).

Therefore, several examples can be found in the literature of microreactors developed for this process. For example, Chambers et al., in a series of papers (see the references listed in [80]), have developed a modular stainless steel microreactor system (Fig. 20.8) through which gas and liquid reagents flow in annular fashion (called "pipe flow" by the authors) down the 0.5-mm-wide channels where the reaction occurs. The fluorination of diethyl malonate and Meldrum's acid with reasonable conversion was demonstrated with this reactor system, using 10% v/v F_2 in N_2 as the fluorination gas.

For the direct fluorination of aromatic compounds, a two-channel microfabricated reactor constructed of silicon and Pyrex and coated with nickel/silicon oxide thin films was developed by de Mas et al. [2, 81]. This reactor was operated mainly under annular flow conditions, although flow diagrams were constructed for Taylor flow as well. It was estimated by the authors that a reactor system consisting of 200 of these microchannels (with a total reactor volume of only 0.25 mL) would produce up to 0.4 g/h of the fluorinated product.

Jähnisch et al. have developed a falling-film microreactor for the direct fluorination of aromatic compounds [82]. This reactor looks quite similar to the one shown in Fig. 20.8, but the flow pattern is somewhat different in that the liquid flows as a thin film along one surface of the reactor, while the gas is flown along it either in cocurrent or countercurrent fashion. The falling film in the microreactor had a stable thickness of 100 μm. The most critical section of the reactor is the inlet part, which should ensure equal distribution of the liquid phase to parallel streams. The microreactor has a structured heat exchanger copper plate inserted into a

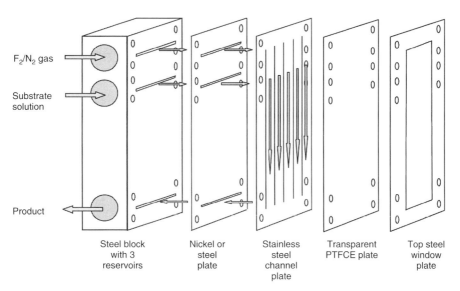

FIGURE 20.8   Schematic representation of modular microreactor device, suitable for industrial application (from [80]).

cavity beneath the falling-film plate for temperature control, and is covered by a thick glass plate that leaves open the option for photochemical reactions.

A comparison of the performance of the reactor of de Mas at al. with the microbubble column and the falling-film microreactor discussed above was given in [62]. These microreactors were benchmarked against a laboratory-scale bubble column, yielding the results listed in Table 20.1.

20.4.2   Liquid–Liquid

As was discussed in a previous paragraph, due to the interfacial tension in combination with shear forces imposed by the flow, two liquids entering a microchannel from a T-junction, in general, will break up into droplets with a radius that depends

TABLE 20.1   Benchmarking of Direct Fluorination Microreactors (from [62])

	Microbubble Microreactor [82]	Falling-Flim Microreactor [82]	Lab-Scale Bubble Column [82]	Silicon Microchannel Reactor [2]
Temperature (°C)	−15	−16	−17	Room temperature
Fluorine ratio (equiv.)	0.54	2.0	1.0	2.5
Conversion (%)	26	76	34	58
Yield (%)	11	28	22	14
Selectivity (%)	42	37	22	24

on the capillary number. Only at low flows may the two liquids remain laminated, but always the risk is present that the interface becomes unstable or starts moving. In fact, it has been observed by this author that, probably due to small defects at the microchannel walls, liquids start to turn in a corkscrew fashion, as was intentionally done in the so-called herringbone mixers of Stroock et al. [83].

In order to keep the two flow streams running in parallel in one plane, so that they may be separated at the end of the channel with a simple T-junction, special precautions have to be taken. The problem is exactly the same as faced in keeping apart the two liquids for extraction, as was encountered in Chapter 3. Inserting a membrane between the two phases significantly limits mass transport and therefore is not preferable.

Kitamori and coworkers [84] developed a surface modification procedure that was applied to a microchannel actually composed of two parrallel channels, one deep and one shallow, which connected to one another on one side. By capillary forces, it was possible to coat only the shallow part with a hydrophobic coating, with this part guiding the gas flow in parallel with the liquid in the deeper part (or an apolar organic phase in parallel with an aqueous phase in a liquid–liquid two-phase case). The performance of the hydrophobic–hydrophilic separation was tested by measuring the leakage pressure for water flow from the hydrophilic microchannel to the hydrophobic one. This pressure was found to vary from 7.7 to 1.1 kPa when the microchannel depth changed from 8.6 to 39 μm, which agreed very well with the pressures predicted by the Young–Laplace equation.

With the use of such microreactors diazocoupling reactions, phase transfer catalysis (see also Section 20.3.5 above) was performed in an ethyl acetate–water combination [85]. The same concept was used for phase transfer catalysis in a microfluidic microchannel network on a glass chip in which a 2×2 matrix of amines and acid chlorides was used, to demonstrate the use of glass chips for combinatorial synthesis [86]. Furthermore, the single-channel phase transfer catalysis chip of [85] was replicated 10 times and stacked in a "pile-up" microreactor, which offered for the amineacid chloride reaction mentioned above production of 1.7 g of crude product (90% pure) per hour [87]. The latter demonstrates that this concept may have a value for the production of fine chemicals.

Finally, a two-phase system of the type discussed above, with parallel streams of either cyclohexane and water or 1-hexanol and formamide, was used to prepare titania nanoparticles with a size less than 10 nm [88]. The reaction carried out was the hydrolysis of titanium alkoxide, which is a very fast reaction and therefore suitable for a microchannel microreactor. Reaction is assumed to take place in the very narrow diffusion region between the two immiscible layers, which is on the order of a few nm.

20.5 CONCLUDING REMARKS

In this chapter, the benefits of a number of two-phase flow patterns (annular, two parallel streams, and Taylor) in microchannel microreactors have been discussed. It was shown that the benefits mainly arise from the large interfacial areas that can be achieved between the two phases, ensuring enhanced mass transport, and the excellent control on residence time distribution that may be obtained, which is of importance for the selectivity of chemical reactions. Special types of reactions discussed were phase transfer catalysis, which is a class of reactions that typically benefit from large interfacial area, and the preparation of monodisperse particles, crystals or droplets, which typically benefit from narrow residence time distribution.

BIBLIOGRAPHY

1. Kandlikar, S. G., Garimella, S. G., Li, D., Colin, S., and King, M. R. (2006). *Heat Transfer and Fluid Flow in Minichannels and Microchannels*. Amsterdam: Elsevier.

2. de Mas, N., Günther, A., Schmidt, M. A., and Jensen, K. F. (2003). Microfabricated multiphase reactors for the selective direct fluorination of aromatics. *Ind. Eng. Chem. Res.*, 42: 698–710.

3. Triplett, K. A., Ghiaasiaan, S. M., Abdel-Khalik, S. I., and Sadowski, D. L. (1999). Gas-liquid two-phase flow in microchannels. Part I: Two-phase flow patterns. *Int. J. Multiphase Flow*, 25: 377–394.

4. Kreutzer, M. T., Kapteijn, F., Moulijn, J. A., and Heiszwolf, J. J. (2005). Multiphase monolith reactors: Chemical reaction engineering of segmented flow in microchannels. *Chem. Eng. Sci.*, 60: 5895–5916.

5. Jensen, M. K., Peles, Y., Borca-Tasciuc, T., and Kandlikar, S. G. (2006). Multiphase flow, evaporation, and condensation at the microscale. In *Micro Process Engineering, Vol. 5: Fundamentals, Devices, Fabrication and Applications*, N. Kockmann, ed., Chapter 4. Weinheim: Wiley-VCH.

6. Löb, P., Pennemann, H., and Hessel, V. (2004). G/L-dispersion in interdigital micromixers with different mixing chamber geometries. *Chem. Eng. J.*, 101: 75–85.

7. Bensakhria, A., Peysson, Y., and Antonini, G. (2004). Experimental study of the pipeline lubrication for heavy oil transport. *Oil & Gas Sci. Technol.—Rev. IFP*, 59: 523–533.

8. Wegmann, A., and Rudolf von Rohr, P. (2006). Two phase liquid–liquid flows in pipes of small diameters. *Int. J. Multiphase Flow*, 32: 1017–1028.

9. Hisamoto, H., Saito, T., Tokeshi, M., Hibara, A., and Kitamori, T. (2001). Fast and high conversion phase-transfer synthesis exploiting the liquid–liquid interface formed in a microchannel chip. *Chem. Commun.*, 2662–2663.

10. Squires, T. M., and Quake, S. R. (2005). Microfluidics: Fluid physics at the nanoliter scale. *Rev. Mod. Phys.*, 77: 977–1026.

11. Berčič, G., and Pintar, A. (1997). The role of gas bubbles and liquid slug lengths on mass transport in the Taylor flow through capillaries. *Chem. Eng. Sci.*, 52: 3709–3719.

12. Van Baten, J. M., and Krishna, R. (2004). CFD simulations of mass transfer from Taylor bubbles rising in circular capillaries. *Chem. Eng. Sci.*, 59: 2535–2545.

13. Kreutzer, M. T., Heiszwolf, J. J., Kapteijn, F., and Moulijn, J. A. (2003). Pressure drop of Taylor flow in capillaries: Impact of slug length. In *Proceedings of 1st International Conference on Microchannels and Minichannels*, pp.153–159. New York: ASME.

14. Thulasidas, T. C., Abraham, M. A., and Cerro, R. L. (1999). Dispersion during bubble-train flow in capillaries. *Chem. Eng. Sci.*, 54: 61–76.

15. Salman, W., Angeli, P., and Gavriilidis, A. (2005). Sample pulse broadening in Taylor flow microchannels for screening applications. *Chem. Eng. Technol.*, 28: 509–514.

16. Edvinsson Albers, R., Nyström, M., Siverström, M., Sellin, A., Dellve, A.-C., Andersson, U., et al. (2001). Development of a monolith-based process for H_2O_2 production: From idea to large-scale implementation. *Cat. Today*, 69: 247–252.

17. Kawakami, K., Kawasaki, K., Shiraishi, F., and Kusunoki, K. (1989). Performance of a honeycomb monolith bioreactor in a gas-liquid-solid three-phase system. *Ind. Eng. Chem. Res.*, 28: 394–400.

18. Liu, W., Addiego, W. P., Sorensen, C. M., and Boger, T. (2002). Monolith reactor for the dehydrogenation of ethylbenzene to styrene. *Ind. Eng. Chem. Res.*, 41: 3131–3138.

19. de Deugd, R. M., Kapteijn, F., and Moulijn, J. A. (2003). Using monolithic catalysts for highly selective Fischer-Tropsch synthesis. *Cat. Today*, 79: 495–501.

20. Lapkin, A. A., Bozkaya, B., and Plucinski, P. K. (2006). Selective oxidation of 1-butene by molecular oxygen in a porous membrane Taylor flow reactor. *Ind. Eng. Chem. Res.*, 45: 2220–2228.

21. Hessel, V., Ehrfeld, W., Herweck, T., Haverkamp, V., Löwe, H., Schiewe, J., et al. (2000). Gas/liquid microreactors: Hydrodynamics and mass transfer. In *Proceedings of 4th International Conference on Microreaction Technology, IMRET4*, AIChE Topical Conference Proceedings, Atlanta, GA, March 5–9, 2000, pp. 174–186.

22. Hessel, V., Ehrfeld, W., Golbig, K., Haverkamp, V., Löwe, H., Storz, M., et al. (1999). Gas/liquid microreactors for direct fluorination of aromatic compounds using elemental fluorine. In *Proceedings of 3rd International Conference on Microreaction Technology, IMRET3*, Frankfurt, Germany, April 18–21, 1999, pp. 526–540.

23. Murray, C. B., Kagan, C. R., and Bawendi, M. G. (2000). Synthesis and characterization of monodisperse nanocrystals and close-packed nanocrystal assemblies. *Ann. Rev. Mater. Sci.*, 30: 545–610.

24. Khan, S. A., Günther, A., Schmidt, M. A., and Jensen, K. F. (2004). Microfluidic synthesis of colloidal silica. *Langmuir*, 20: 8604–8611.

25. Yen, B. K. H., Günther, A., Schmidt, M. A., Jensen, K. F., and Bawendi, M. G. (2005). A microfabricated gas–liquid segmented flow reactor for high-temperature synthesis: The case of CdSe quantum dots. *Angew. Chem. Int. Ed.*, 44: 5447–5451.

26. Yen, B. K. H., Stott, N. E., Jensen, K. F., and Bawendi, M. G. (2003). A continuous-flow microcapillary reactor for the preparation of a size series of CdSe nanocrystals. *Adv. Mat.*, 15: 1858–1862.

27. Günther, A., Khan, S. A., Thalmann, M., Trachsel, F., and Jensen, K. F. (2004). Transport and reaction in microscale segmented gas-liquid flow. *Lab Chip*, 4: 278–286.

28. Chmela, E., Tijssen, R., Blom, M. T., Gardeniers, J. G. E., and van den Berg, A. (2002). A chip system for size separation of macromolecules and particles by hydrodynamic chromatography. *Anal. Chem.*, 74: 3470–3475.

29. Fabel, S., Niessner, R., and Weller, M. G. (2005). Effect-directed analysis by high-performance liquid chromatography with gas-segmented enzyme inhibition. *J. Chrom.*, A1099: 103–110.

30. De Lathouder, K. M., Bakker, J. J. W., Kreutzer, M. T., Kapteijn, F., Moulijn, J. A., and Wallin, S. A. (2004). Structured reactors for enzyme immobilization: Advantages of tuning the wall morphology. *Chem. Eng. Sci.*, 59: 5027–5033.

31. Chan, E. M., Mathies, R. A., and Alivisatos, A. P. (2003). Size-controlled growth of CdSe nanocrystals in microfluidic reactors. *Nanoletters*, 3: 199–201.

32. Chan, E. M., Alivisatos, A. P., and Mathies, R. A. (2005). High-temperature microfluidic synthesis of CdSe nanocrystals in nanoliter droplets. *J. Amer. Chem. Soc.*, 127: 13854–13861.

33. Song, H., Tice, J. D., and Ismagilov, R. F. (2003). A microfluidic system for controlling reaction networks in time. *Angew. Chem. Int. Ed.*, 42: 767–772.

34. Song, H., and Ismagilov, R. F. (2003). Millisecond kinetics on a microfluidic chip using nanoliters of reagents. *J. Am. Chem. Soc.*, 125: 14613–14619.

35. Günther, A., Jhunjhunwala, M., Thalmann, M., Schmidt, M. A., and Jensen, K. F. (2005). Micromixing of miscible liquids in segmented gas-liquid flow. *Langmuir*, 21: 1547–1555.

36. Kashid, M. N., Gerlach, I., Goetz, S., Franzke, J., Acker, J. F., Platte, F., et al. (2005). Internal circulation within the liquid slugs of a liquid–liquid slug-flow capillary micro-reactor. *Ind. Eng. Chem. Res.*, 44: 5003–5010.

37. Stone, Z. B., Stone, H. A. (2005). Imaging and quantifying mixing in a model droplet micromixer. *Phys. Fluids*, 17: 063103-1–063103-11.

38. Song, H., Bringer, M. R., Tice, J. D., Gerdts, C. J., and Ismagilov, R. F. (2003). Experimental test of scaling of mixing by chaotic advection in droplets moving through microfluidic channels. *Appl. Phys. Lett.*, 83: 4664–4666.

39. Tice, J. D., Lyon, A. D., and Ismagilov, R. F. (2004). Effects of viscosity on droplet formation and mixing in microfluidic channels. *Anal. Chim. Acta*, 507: 73–77.

40. George, A., and Wilson, W. W. (1994). Predicting protein crystallization from a dilute solution property. *Acta Crystallogr.*, D50: 361–365.

41. Yadav, M. K., Gerdts, C. J., Sanishvili, R., Smith, W. W., Roach, L. S., Ismagilov, R. F., et al. (2005). In situ data collection and structure refinement from microcapillary protein crystallization. *J. Appl. Cryst.*, 38: 900–905.

42. Li, L., Mustafi, D., Fu, Q., Tereshko, V., Chen, D. L., Tice, J. D., et al. (2006). Nanoliter microfluidic hybrid method for simultaneous screening and optimization validated with crystallization of membrane proteins. *PNAS*, 103: 19243–19248.

43. Song, H., Chen, D. L., and Ismagilov, R. F. (2006). Reactions in droplets in microfluidic channels. *Angew. Chem. Int. Ed.*, 45: 7336–7356.

44. Chen, D. L., Gerdts, C. J., and Ismagilov, R. F. (2005). Using microfluidics to observe the effect of mixing on nucleation of protein crystals. *J. Am. Chem. Soc.*, 127: 9672–9673.

45. Zheng, B., Gerdts, C. J., and Ismagilov, R. F. (2005). Using nanoliter plugs in microfluidics to facilitate and understand protein crystallization. *Curr. Opin. Struct. Biol.*, 15: 548–555.

46. Voloshin, Y., Lawal, A., and Panikov, N. S. (2005). Continuous plug-flow bioreactor: Experimental testing with *Pseudomonas putida* culture grown on benzoate. *Biotechnol. Bioeng.*, 91: 254–259.

47. Günther, A., and Jensen, K. F. (2006). Multiphase microfluidics: From flow characteristics to chemical and materials synthesis. *Lab Chip*, 6: 1487–1503.

48. Önal, Y., Lucas, M., and Claus, P. (2005). Application of a capillary microreactor for selective hydrogenation of α, β-unsaturated aldehydes in aqueous multiphase catalysis. *Chem. Eng. Technol.*, 28: 972–978.

49. Joscelyne, S. M., and Trägårdh, G. (2000). Membrane emulsification—A literature review. *J. Membrane Sci.*, 169: 107–117.

50. de Jong, J., Lammertink, R. G. H., and Wessling, M. (2006). Membranes and microfluidics: A review. *Lab Chip*, 6: 1125–1139.

51. van Rijn, and C. J. M. (2004). *Nano and Micro Engineered Membrane Technology.* Membrane Science and Technology Series, Vol. 10, Amsterdam: Elsevier.

52. Sugiura, S., Nakajima, M., Iwamoto, S., and Seki, M. (2001). Interfacial tension driven monodispersed droplet formation from microfabricated channel array. *Langmuir*, 17: 5562–5566.

53. Ikkai, F., Iwamoto, S., Adachi, E., and Nakajima, M. (2005). New method of producing mono-sized polymer gel particles using microchannel emulsification and UV irradiation. *Colloid. Polym. Sci.*, 283: 1149–1153.

54. Takasi Nisisako, T., Torii, T., and Higuchi, T. (2004). Novel microreactors for functional polymer beads. *Chem. Eng. J.*, 101: 23–29.

55. Nie, Z., Xu, S., Seo, M., Lewis, P. C., and Kumacheva, E. (2005). Polymer particles with various shapes and morphologies produced in continuous microfluidic reactors. *J. Amer. Chem. Soc.*, 127: 8058–8063.

56. Seo, M., Nie, Z., Xu, S., Mok, M., Lewis, P. C., Graham, R., et al. (2005). Continuous microfluidic reactors for polymer particles. *Langmuir*, 21: 11614–11622.

57. Dendukuri, D., Tsoi, K., Hatton, T. A., and Doyle, P. S. (2005). Controlled synthesis of nonspherical microparticles using microfluidics. *Langmuir*, 21: 2113–2116.

58. Jeong, W., Kim, J., Kim, S., Lee, S., Mensing, G., and Beebe, D. J. (2004). Hydrodynamic microfabrication via "on the fly" photopolymerization of microscale fibers and tubes. *Lab Chip*, 4: 576–580.

59. Okushima, S., Nisisako, T., Torii, T., and Higuchi, T. (2004). Controlled production of monodisperse double emulsions by two-step droplet breakup in microfluidic devices. *Langmuir*, 20: 9905–9908.

60. Haverkamp, V., Ehrfeld, W., Gebauer, K., Hessel, V., Löwe, H., Richter, R., et al. (1999). The potential of micromixers for contacting of disperse liquid phases. *Fresenius J. Anal. Chem.*, 364: 617–624.

61. Silva, L., Tonkovich, L., Lerou, J., and Daymo, E. (2006). Superior emulsion formation in microchannel architecture. Paper presented at Achema 2006, 28th International Exhibition-Congress on Chemical Engineering, Environmental Protection and Biotechnology, Frankfurt, Germany, May 15–19, 2006. Abstract available at http://www.achema.de.

62. Pennemann, H., Watts, P., Haswell, S. J., Hessel, V., and Löwe, H. (2004). Benchmarking of microreactor applications. *Org. Proc. Res. Develop.*, 8: 422–439.

63. Dummann, G., Quittmann, U., Gröschel, L., Agar, D. W., Wörz, O., and Morgenschweis, K. (2003). The capillary-microreactor: A new reactor concept for the intensification of heat and mass transfer in liquid–liquid reactions. *Catal. Today*, 79–80: 433–439.

64. Hessel, V., Hardt, S., and Löwe, H. (2004). *Chemical Micro Process Engineering. Fundamentals, Modelling and Reactions.* Weinheim, Germany: Wiley-VCH.

65. Starks, C. M. (1971). Phase-transfer catalysis. I. Heterogeneous reactions involving anion transfer by quaternary ammonium and phosphonium salts. *J. Am. Chem. Soc.*, 93: 195–199.

66. Makosza, M. (2000). Phase-transfer catalysis. A general green methodology in organic synthesis. *Pure Appl. Chem.*, 72: 1399–1403.

67. Ueno, M., Hisamoto, H., Kitamori, T., and Kobayashi, S. (2003). Phase-transfer alkylation reactions using microreactors. *Chem. Commun.*, 936–937.

68. Okamoto, H. (2006). Effect of alternating pumping of two reactants into a microchannel on a phase transfer reaction. *Chem. Eng. Technol.*, 29: 504–506.

69. Mu, J.-X., Yin, X.-F., and Wang, Y.-G. (2005). The Claisen-Schmidt reaction carried out in microfluidic chips. *Synlett*, 3163–3165.

70. Adams, D. J., Dyson, P. J., and Taverner, S. J. (2003). *Chemistry in Alternative Reaction Media*. Hoboken, NJ: John Wiley & Sons.

71. Wasserscheid, P., and Welton, T. (2002). *Ionic Liquids in Synthesis*. Wiley-VCH.

72. Fukuyama, T., Shinmen, M., Nishitani, S., Sato, M., and Ryu, I. (2002). A copper-free Sonogashira coupling reaction in ionic liquids and its application to a microflow system for efficient catalyst recycling. *Org. Lett.*, 4: 1691–1694.

73. Liu, S., Fukuyama, T., Sato, M., and Ryu, I. (2004). Continuous microflow synthesis of butyl cinnamate by a Mizoroki-Heck reaction using a low-viscosity ionic liquid as the recycling reaction medium. *Org. Proc. Res. Dev.*, 8: 477–481.

74. Rahman, M. T., Fukuyama, T., Kamata, N., Sato, M., and Ryu, I. (2006). Low pressure Pd-catalyzed carbonylation in an ionic liquid using a multiphase microflow system. *Chem. Commun.*, 2236–2238.

75. Losey, M. W., Schmidt, M. A., and Jensen, K. F. (2001). Microfabricated multiphase packed-bed reactors: Characterization of mass transfer and reactions. *Ind. Eng. Chem. Res.*, 40: 2555–2562.

76. Losey, M. W., Jackman, R. J., Firebaugh, S. L., Schmidt, M. A., and Jensen, K. F. (2002). Design and fabrication of microfluidic devices for multiphase mixing reaction. *J. Microelectromech. Syst.*, 11: 709–717.

77. Baxendale, I. R., Deeley, J., Griffiths-Jones, C. M., Ley, S. V., Saaby, S., and Tranmer, G. K. (2006). A flow process for the multi-step synthesis of the alkaloid natural product oxomaritidine: A new paradigm for molecular assembly. *Chem. Commun.*, 2566–2568.

78. Saaby, S., Rahbek Knudsen, K., Ladlowb, M., and Ley, S. V. (2005). The use of a continuous flow-reactor employing a mixed hydrogen–liquid flow stream for the efficient reduction of imines to amines. *Chem. Commun.*, 2909–2911.

79. Kobayashi, J., Mori, Y., Okamoto, K., Akiyama, R., Ueno, M., Kitamori, T., et al. (2004). A microfluidic device for conducting gas-liquid-solid hydrogenation reactions, Science, 304: 1305–1308.

80. Chambers, R. D., Fox, M. A., Holling, D., Nakano, T., and Sandford, G. (2005). Versatile gas/liquid microreactors for industry. *Chem. Eng. Technol.*, 28: 344–352.

81. de Mas, N., Günther, A., Schmidt, M. A., and Jensen, K. F. (2003). Scalable microfabricated multiphase reactors for direct fluorination reactions In *Digest of Technical Papers from IEEE Solid-State Sensor and Actuator Conference Transducers '03*, Boston, MA Piscataway, NJ: IEEE, June 8–12, 2003, pp. 655–658.

82. Jähnisch, K., Baerns, M., Hessel, V., Ehrfeld, W., Haverkamp, W., Löwe, H., et al. (2000). Direct fluorination of toluene using elemental fluorine in gas/liquid microreactors. *J. Fluorine Chem.*, 105: 117–128.

83. Stroock, A. D., Dertinger, S. K. W., Ajdari, A., Mezic, I., Stone, H. A., and Whitesides, G. M. (2002). Chaotic mixer for microchannels. *Science*, 295: 647–651.

84. Hibara, A., Iwayama, S., Matsuoka, S., Ueno, M., Kikutani, Y., Tokeshi, M., et al. (2005). Surface modification method of microchannels for gas-liquid two-phase flow in microchips. *Anal. Chem.*, 77: 943–947.

85. Hisamoto, H., Saito, T., Tokeshi, M., Hibara, A., and Kitamori, T. (2001). Fast and high conversion phase-transfer synthesis exploiting the liquid–liquid interface formed in a microchannel chip. *Chem. Commun.*, 2662–2663.

86. Kikutani, Y., Horiuchi, T., Uchiyama, K., Hisamoto, H., Tokeshi, M., and Kitamori, T. (2002). Glass microchip with three-dimensional microchannel network for 2×2 parallel synthesis. *Lab Chip*, 2: 188–192.

87. Kikutani, Y., Hibara, A., Uchiyama, K., Hisamoto, H., Tokeshi, M., and Kitamori, T. (2002). Pile-up glass microreactor. *Lab Chip*, 2: 193–196.

88. Wang, H., Nakamura, H., Uehara, M., Miyazaki, M., and Maeda, H. (2002). Preparation of titania particles utilizing the insoluble phase interface in a microchannel reactor. *Chem. Commun.*, 1462–1463.

INDEX

Microchemical Engineering in Practice. Edited by Thomas R. Dietrich
Copyright © 2009 John Wiley & Sons, Inc.